Human Anatomy Laboratory Workbook

SIXTH EDITION

Lucia J. Tranel, Prof. Emeritus

Benjamin C. Jellen, Ph.D.

With Contributions From

Alicia T. Pate, Ph.D.

Sarah Schlereth, M.S.

Anthony Wilmes, M.S.

Sarah C. Schriewer, B.S.

Kendall Hunt
publishing company

Kendall Hunt
publishing company

www.kendallhunt.com
Send all inquiries to:
4050 Westmark Drive
Dubuque, IA 52004-1840

PAK ISBN: 978-1-7924-6890-2
Text alone ISBN: 978-1-7924-6891-9

Published in the United States of America

Contents

Acknowledgments

There are always many people who conribute in valuable ways during the development of such a project. This edition represents an evolution of development with consideration to what helps a student learn.

We are indebted to our colleagues and professional mentors, who provide for us invaluable guidance and insight into the art of teaching.

We appreciate our families for being supportive.

We thank all our students who provide us excellent insight into what helps them learn, and we owe special thanks to the members of Human Anatomy Class of Spring, 2005, who helped design the format of this laboratory workbook.

We acknowledge Kendall Hunt for their encouragement and expertise.

Owner's Manual

The *Human Anatomy Laboratory Workbook* is designed to encourage you to interact with the exercises as much as possible. You are enhancing your cognitive awareness of the material every time you draw, label, color, dissect, and record an observation. Here are some important directions.

1. Always read the assigned Exercise **before** coming to lab. Use all resources to complete as much of the labeling as possible.

2. Each Exercise is written in an outline format. The major topics are lettered, and under each lettered topic there are numbered **OUTCOMES**. The Outcomes include expectations, assessment modalities, and study hints. The Outcomes include material for which you will be held accountable.

3. Notice that a **TOOLBOX** is included with each Outcome. The Toolbox is designed to advise you of the resources you need to complete the Outcome. Each Toolbox has one or more of the following icons:

Find the image in your **TEXTBOOK** that relates or corresponds to the title in the Toolbox. The textbook's labeled image will appear to be similar to the unlabeled image in your Workbook. Use the textbook's image as a guide for locating the structures listed on the page of your Workbook. *Write in the labels* on your unlabeled image. Searching for the labeled image in the textbook and then writing that label in your lab book page will help you begin to learn the anatomical structures. You may even want to color your Workbook image. Be creative! Much of this work is completed **prior** to coming to lab in order to prepare you for the exercise.

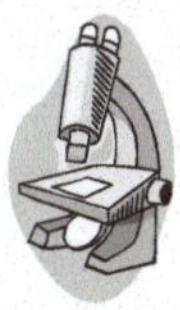

The Microscope icon will inform you to use your **MICROSCOPE** for help when examining microscope slides and other laboratory materials.

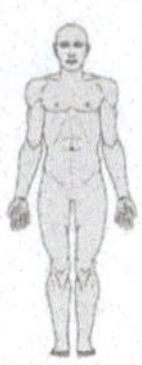

This icon will be the symbol for an **ANATOMICAL MODEL** that you will use to locate the structures on your list.

The scissors icon alerts you to **DISSECTION** activities.

EXERCISE 1

Basic and Medical Terminology

A. *Introduction*

Precision of language is extremely important in the study of anatomy. Words such as under, above, behind, down, or up become very ambiguous under certain circumstances. We have all heard those horror stories about a person going into the hospital to have a left kidney removed and getting the right one removed instead! Exercise 1 will familiarize you with anatomical terminology in seven basic areas:

1. **Anatomical regions and landmarks**
2. **Anatomical directions**
3. **Anatomical planes and sections**
4. **Body cavities**
5. **Serous membranes**
6. **Mucous membranes**
7. **Abdominopelvic quadrants and regions**

Try to apply the words in context as you learn the terminology. Think of examples of how to use the words. It is important to understand at the start that all terminology assumes that the body is in **anatomical position. In anatomical position, the body is standing erect, the face is forward (anteriorly), the arms are at the sides, the palms face forward (anteriorly), the legs are together, and the feet face forward.** Put yourself in anatomical position. Notice that it is not exactly a relaxed position. **Supine** refers to a person lying down, facing upward, while in anatomical position. **Prone** refers to a person lying down, facing downward, while in anatomical position. The concept of **left** and **right** is another point to remember. Left and right always refer to the **subject's left and right, not yours, unless you are the subject, of course.**

B. *Outcomes for Exercise 1*

1. Define and describe what is meant by "anatomical position."
2. Identify and locate common anatomical regions and landmarks on diagrams and models.
3. Memorize the definitions of regional and directional terms. If provided with a definition, identify the term. Apply the terms and provide examples of each.
4. Memorize the definitions of planes and sections. If provided with a definition, identify the term. Apply the terms and provide examples of each.
5. Describe and locate the body cavities. Provide examples of organs within each body cavity.
6. Describe and locate serous membranes. What are their functions? Name the serous membranes.
7. Describe and locate mucous membranes. What are their functions? Provide some examples of mucous membranes.
8. How are serous and mucous membranes similar? How are they different?
9. Describe and locate the abdominopelvic quadrants and regions. Determine what major organs lie deep to those regions.
10. Identify and apply medical terminology, including root words, prefixes, and suffixes.

C. *Learning Basic Terminology*

OUTCOME 1: Define and describe what is meant by "anatomical position."

- ***Read the introduction of Exercise 1 to find the definition of anatomical position. List the characteristics of this position.***
  -
  -
  -
  -
  -
  -

- ***Define supine***

- ***Define prone***

OUTCOME 2: Identify and locate common anatomical regions and landmarks on diagrams and models.

TOOLBOX

Anatomical Landmarks

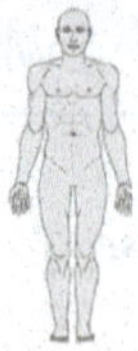

Mr. Muscle

ANTERIOR VIEW		
COMMON NAME	**ANATOMICAL NOUN**	**ANATOMICAL ADJECTIVE**
head	cephalon	cephalic
skull	cranium	cranial
face	facies	facial
forehead	frons	frontal
eye	oculus	ocular, orbital
ear	auris	otic
nose	nasus	nasal
cheek	bucca	buccal
mouth	oris	oral
chin	mentis	mental
neck	cervix	cervical
chest	thoracis/thorax	pectoral, thoracic
armpit	axilla	axillary
breast	mamma	mammary
upper arm	brachium	brachial
front of elbow	antecubitus	antecubital
forearm	antebrachium	antebrachial
wrist	carpus	carpal
hand	manus	manual
thumb	pollex	—
palm	palma	palmar
fingers	digits, phalanges	digital, phalangeal
abdomen	abdomen	abdominal
navel	umbilicus	umbilical
pelvis	pelvis	pelvic
groin	inguen	inguinal
pubic	pubis	pubic
thigh	femur	femoral
kneecap	patella	patellar
shin	crus	crural
ankle	tarsus	tarsal
foot	pes	pedal
top of foot	dorsum	—
toes	digits, phalanges	digital, phalangeal
big toe	hallux	—

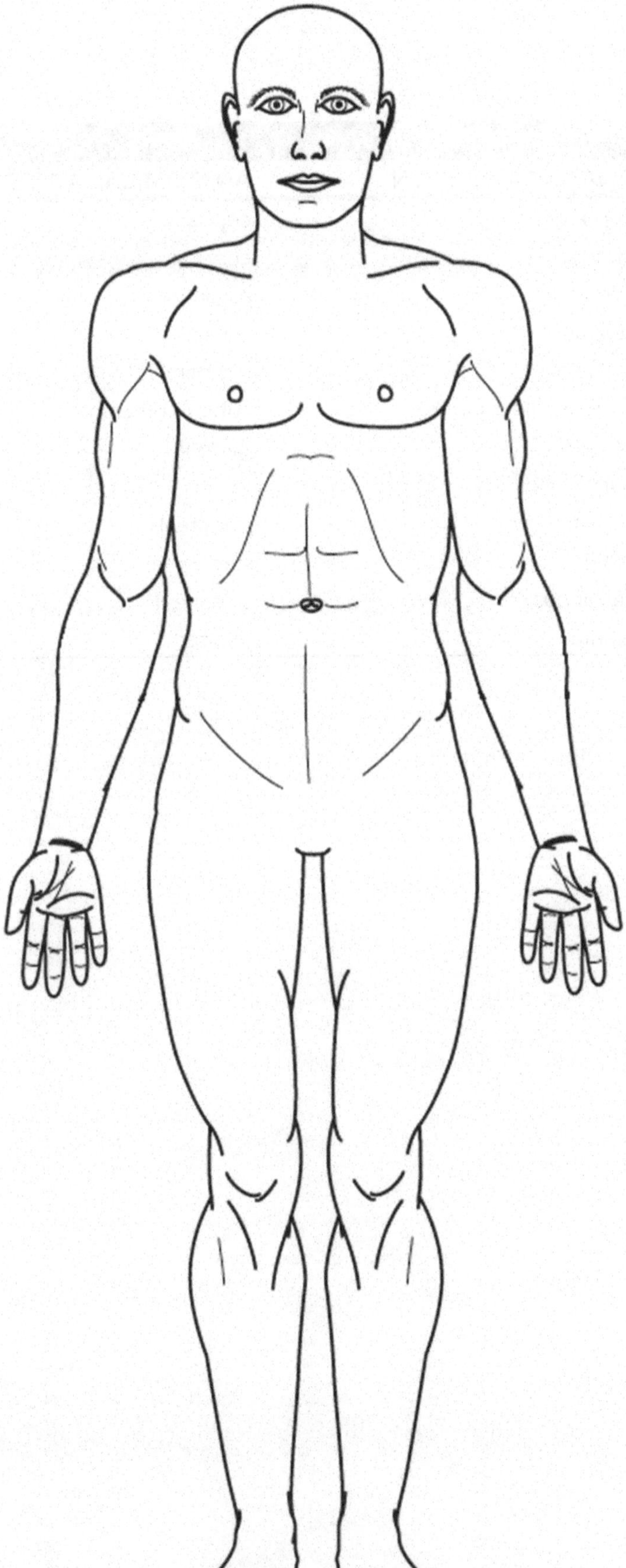

Anterior or Ventral View

POSTERIOR VIEW		
COMMON NAME	**ANATOMICAL NOUN**	**ANATOMICAL ADJECTIVE**
head	cephalon	cephalic
neck	cervix	cervical
shoulder	acromion	acromial
back	dorsum	dorsal
loin	lumbus	lumbar
back of elbow	cubitus, olecranon	cubital, olecranal
buttock	gluteus	gluteal
back of thigh	ham	—
back of knee	popliteus	popliteal
calf	sura	sural
heel	calcaneus	calcaneal
sole	planta	plantar

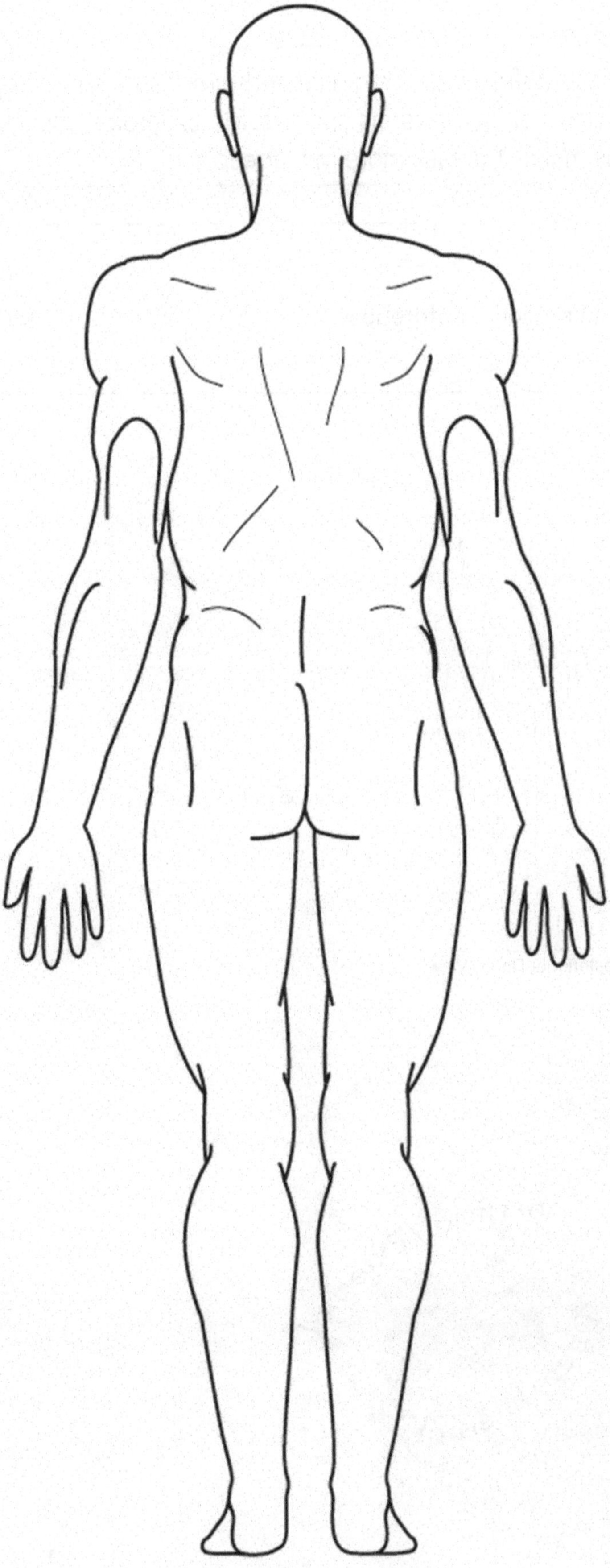

Posterior or Dorsal View

OUTCOME 3: Memorize the definitions of regional and directional terms. If provided with a definition, identify the term. Apply the terms and provide examples of each. Learning the terms in pairs of antonyms makes the task more manageable.

TOOLBOX

Directional References

Mr. Muscle

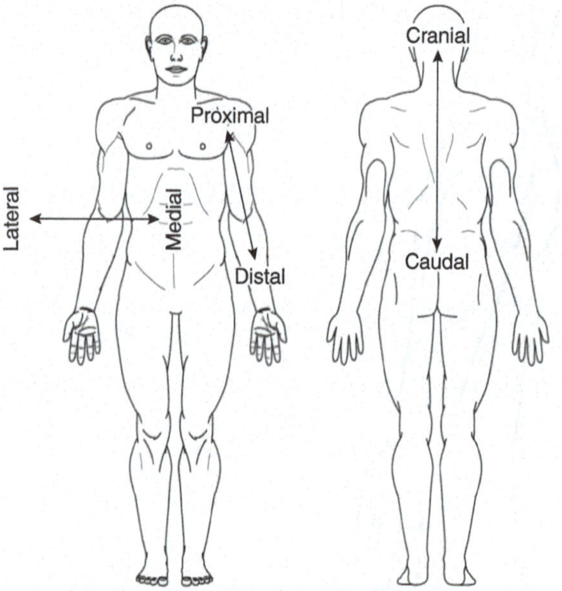

Anterior or Ventral View　　Posterior or Dorsal View

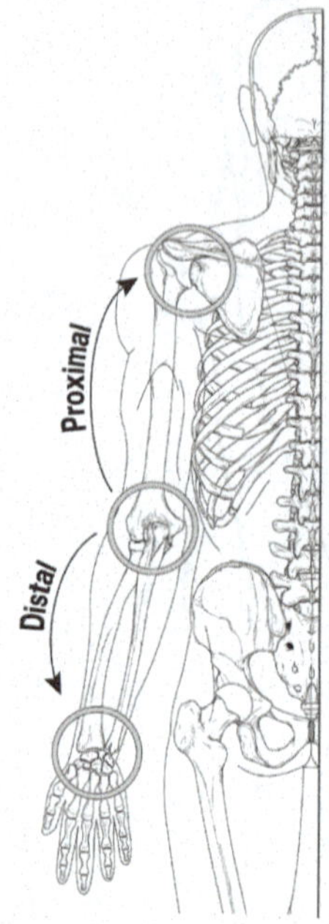

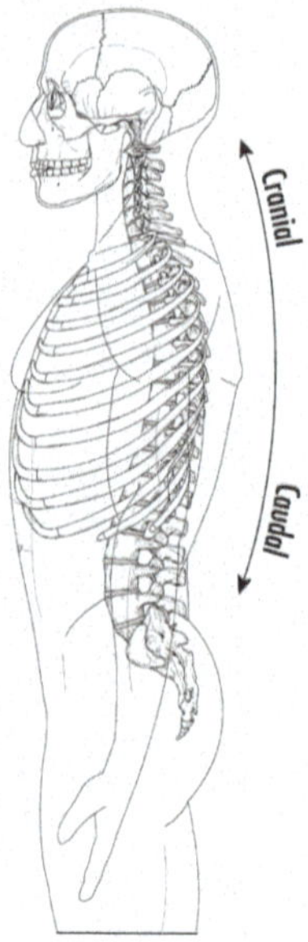

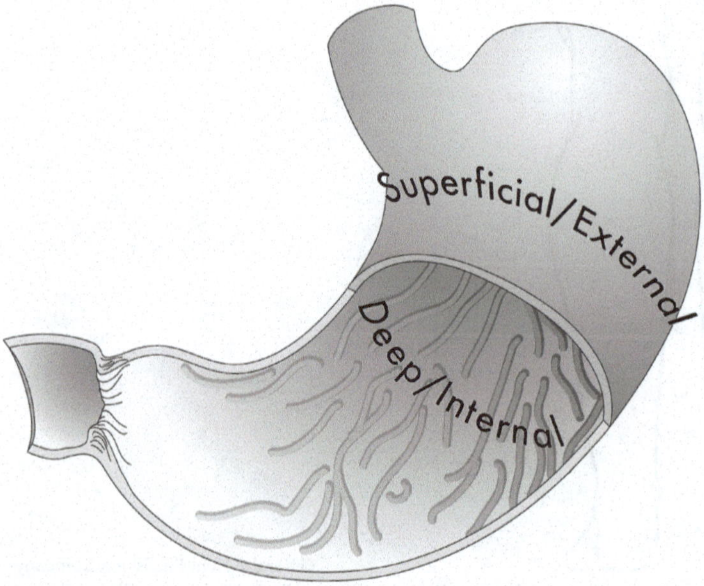

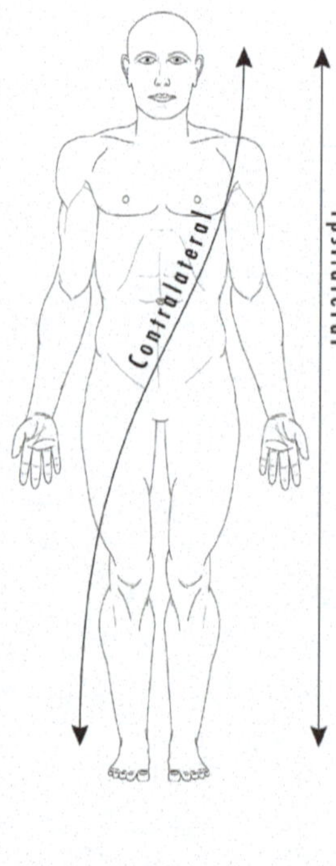

Illustration by Jamey Garbett. © 2003 Mark Nielsen.

TERM	DEFINITION	PROVIDE SOME EXAMPLE
Anterior	before, the region of the body that advances first	
Posterior	behind	
Dorsal	back	
Ventral	belly side	
Cranial	toward the head	
Caudal	toward the tail	
Superior	above, at a higher level	
Inferior	below, at a lower level	
Medial	toward the midline or longitudinal axis	
Lateral	away from the midline, toward the side	
Proximal	toward an attached base	
Distal	away from an attached base	
Superficial	at, near, or relatively close to the surface	
Deep	toward the interior, further from the surface	
Ipsilateral	on the same side	
Contralateral	on opposite sides	

OUTCOME 4: Memorize the definitions of planes and sections. If provided with a definition, identify the term. Apply the terms and provide examples of each.

TOOLBOX

Planes of Sections

TERM	DEFINITION	EXAMPLE
Transverse or Cross	perpendicular to the long axis	
Longitudinal and Sagittal	parallel to the long axis and dividing into left and right portions	
Longitudinal and Midsagittal	parallel to the long axis and dividing into exactly left and right halves	
Longitudinal and Frontal or Coronal	parallel to long axis and dividing into front and back portions	
Serial	a series of sections at small intervals in one sectional plane	
Oblique	at a diagonal angle to the long axis	

■ ***Label the illustrations to indicate the plane or section.***

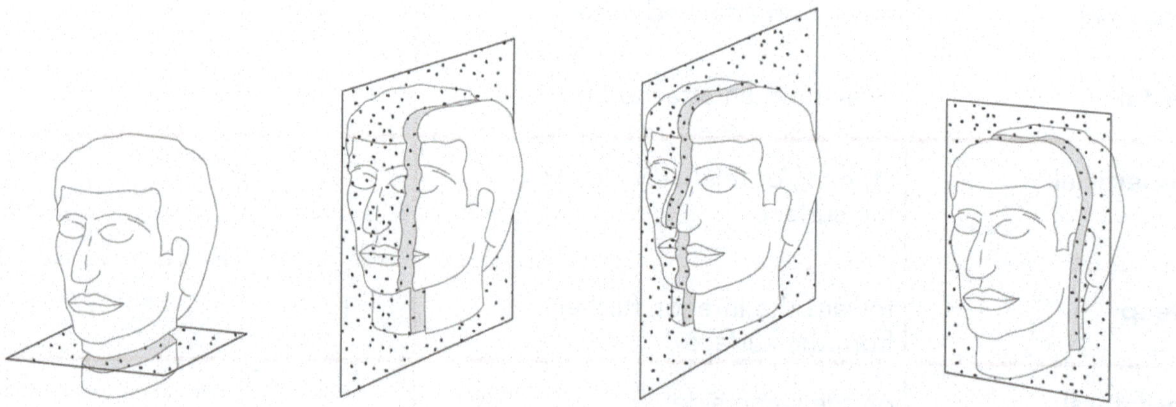

______________ ______________ ______________ ______________

■ ***Sagittal, Midsagittal, and Frontal Sections are parallel to the long axis. How are they different?***

Visualizing Planes of Sections

■ ***Imagine cutting a banana with sectional planes A, B, and C, as shown. Draw the resulting cut surface; that is, draw what you would see if you looked at the surface that the knife passed through. On the line to the right, label the drawings to indicated the plane of section.***

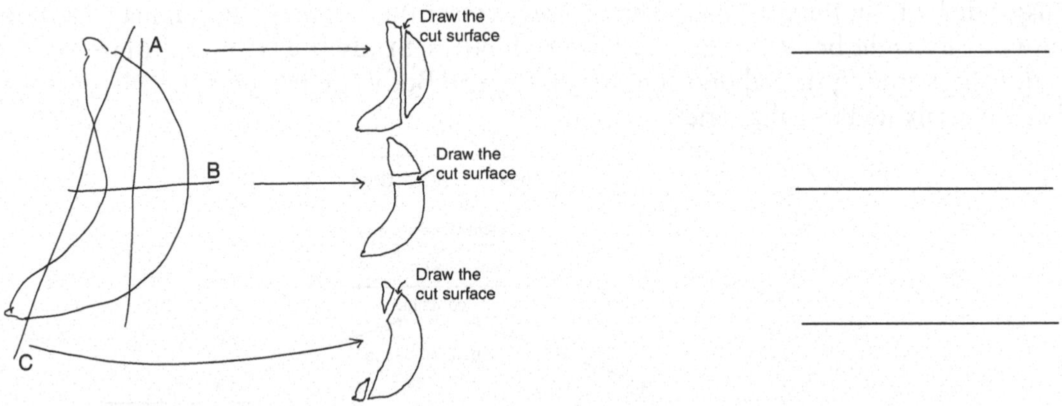

■ ***Repeat the exercise with the sectional cuts through a hollow structure. Imagine slicing through a stomach utilizing the various sectional planes. In each case, draw the resulting cut surface: i.e., draw what you see if you looked at the surface that the knife went through.***

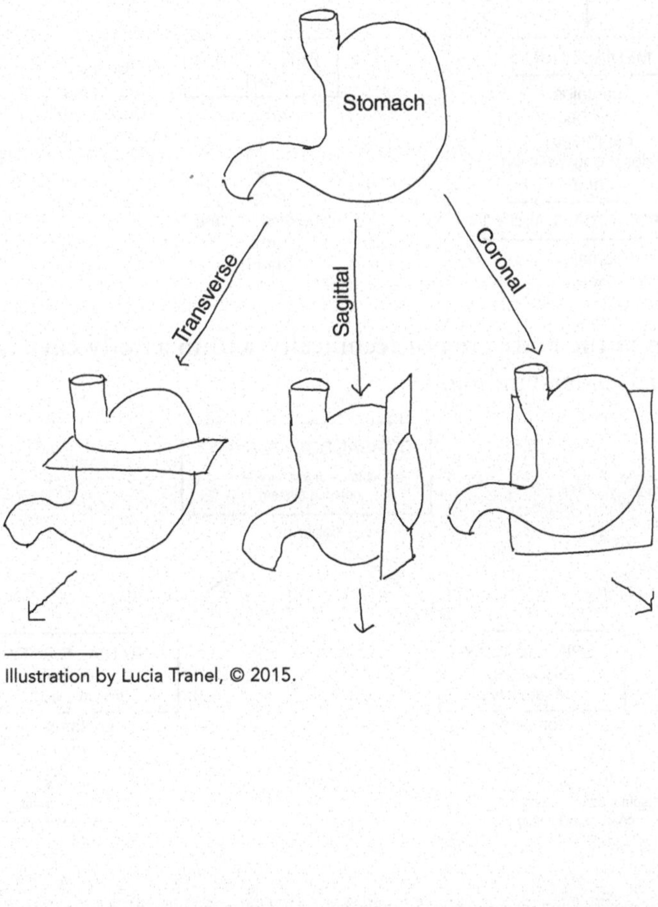

Illustration by Lucia Tranel, © 2015.

OUTCOME 5: Describe and locate the body cavities. Provide examples of organs within the body cavities.

TOOLBOX

Body Cavities

The body is not solid, but it contains hollow spaces known as **cavities.** The organs, or **viscera,** hang suspended within these cavities. The cavities protect the viscera from trauma and allow the viscera to move easily as the body changes positions and moves. **Study the "Concept Map." It illustrates the relationships among the subdivisions of the Ventral Body Cavity.** Note which organs might be located in each cavity and its subcavities.

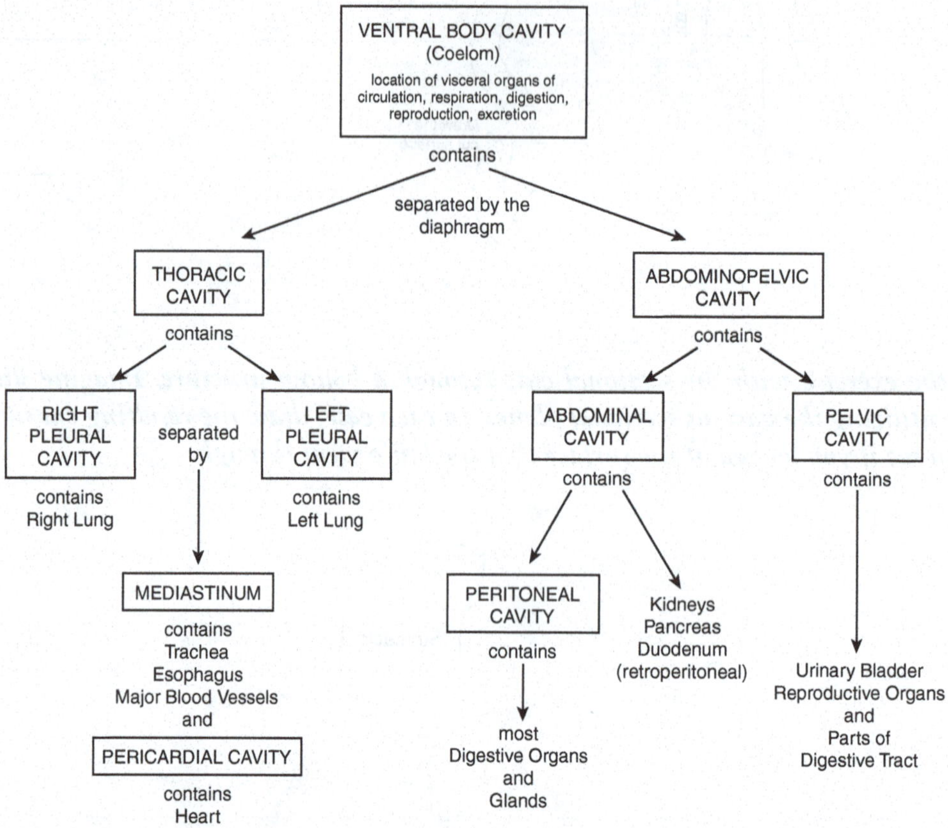

The organs in the dorsum of the body are not technically within a body cavity, but are often described as such to aid in conceptual understanding.

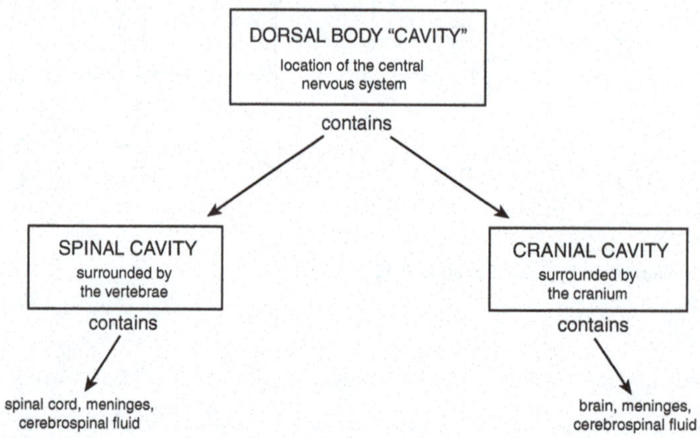

OUTCOME 6: Describe and locate serous membranes. Name the serous membranes.

TOOLBOX

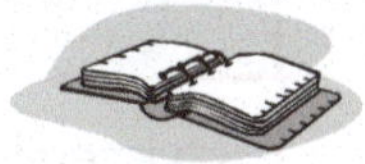

Body Cavities

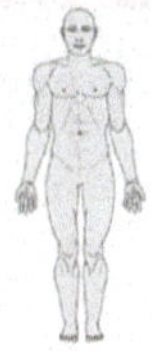

Mr. Flip Chart

Sheets of tissues known as **serous membranes** cover the walls of the cavities and the organs within the body cavities. The definition of a **serous membrane** is a **continuous sheet of simple squamous epithelial cells with their underlying loose connective tissue.** The serous membranes are located on the **walls of the pleural, pericardial, and peritoneal cavities. They are also on the surface of the viscera within each of these cavities.** A **parietal** membrane adheres to the *wall of the cavity.* A **visceral** membrane adheres to the *surface of the organ.* The **cavity** is the space between the two membranes. The primary fuction of these membranes is to reduce friction as organs rub against each other and the cavity walls. Examine the diagram of the fist moving into the balloon. It demonstrates the concept of parietal and visceral membranes.

Application example:
Imagine the skin on your hand as the visceral membrane. Then, put on a glove, which represents the parietal membrane.

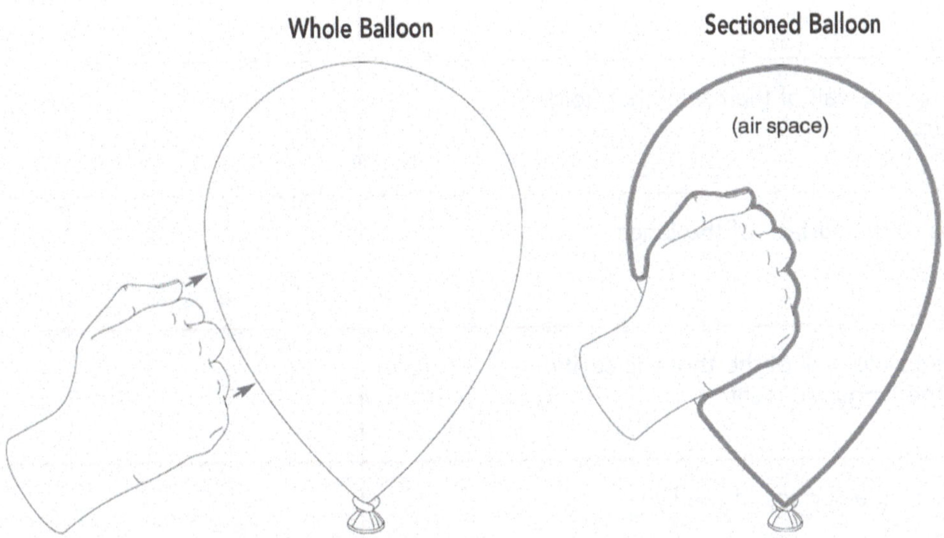

Illustration by Jamey Garbett. © 2003 Mark Nielsen.

Carefully read the description of serous membranes. On the diagram of the *Sectioned Ballon*:

- ***Label the membrane that represents the visceral membrane.***
- ***Label the membrane that represents the parietal membrane.***
- ***Label the space that represents the cavity.***

This table summarizes the six different **Serous Membranes:**

LOCATION	PLEURAL CAVITY	PERICARDIAL CAVITY	PERITONEAL CAVITY
Adhering to the Wall of the Cavity	parietal pleura	parietal pericardium	parietal peritoneum
Adhering to the Organ's Surface	visceral pleura	visceral pericardium	visceral peritoneum

Remember, when identifying a serous membrane, you must have two words:

1. **One word that designates whether the membrane is on the *wall* of the cavity or adhering to the surface of the *organ.***
2. **A second word that designates in *which* of the three cavities the membrane is located.**

■ ***Use the table to practice identifying the names of some serous membranes:***

THE MEMBRANE THAT . . .	IS CALLED . . .
adheres to the surface of the heart	
makes up the wall of the cavity that holds the heart	
adheres to the surface of the lungs	
adheres to the wall of the thoracic cavity where the lungs are located	
adheres to the surface of the liver	
adheres to the surface of the stomach	
adheres to the wall within the abdominal cavity	

OUTCOME 7: Describe and locate mucous membranes. Name some mucous membranes.

A **Mucous Membrane** is also a sheet of epithelial cells, but it differs from a serous membrane in that

1. **The type of epithelium may not consist of simple squamous epithelium; and**
2. **these membranes line spaces and cavities that are continuous with the outside of the body.**

A mucous membrane is any membrane reachable through a natural opening of the body. Sometimes mucous membranes have special cells and glands that produce mucus to keep them lubricated, but this is not always the case. Mucous membranes are involved with secretion, absorption, and lubrication. They also act as a barrier.

■ ***List some examples of mucous membranes. Be as detailed as possible in your descriptions.***

1.

2.

3.

4.

OUTCOME 8: How are serous and mucous membranes similar? How are they different?

■ ***Summarize what you have learned by filling in the table.***

	SIMILARITIES	DIFFERENCES
SEROUS		
MUCOUS		

OUTCOME 9: Describe and locate the abdominopelvic quadrants and regions. Determine what major organs lie deep to those regions.

TOOLBOX

Abdominopelvic Quadrants and Regions

The abdominopelvic surface is quite large, so clinicians and anatomists find it helpful to use terms that indicate specific locations on the abdominopelvic surface. It is possible to locate organs deep to the abdominal and pelvic regions more precisely by using anatomical landmarks on the surface. The abdominopelvic surface is divided into **Quadrants** or into **nine Regions.**

- *Draw in the lines.*
- *Label the quadrants and regions on each of the diagrams. Make sure to designate left or right.*
- *Make note of which organs lie deep to each quadrant and region.*

1. Right Upper Quadrant
2. Left Upper Quadrant
3. Right Lower Quadrant
4. Left Lower Quadrant

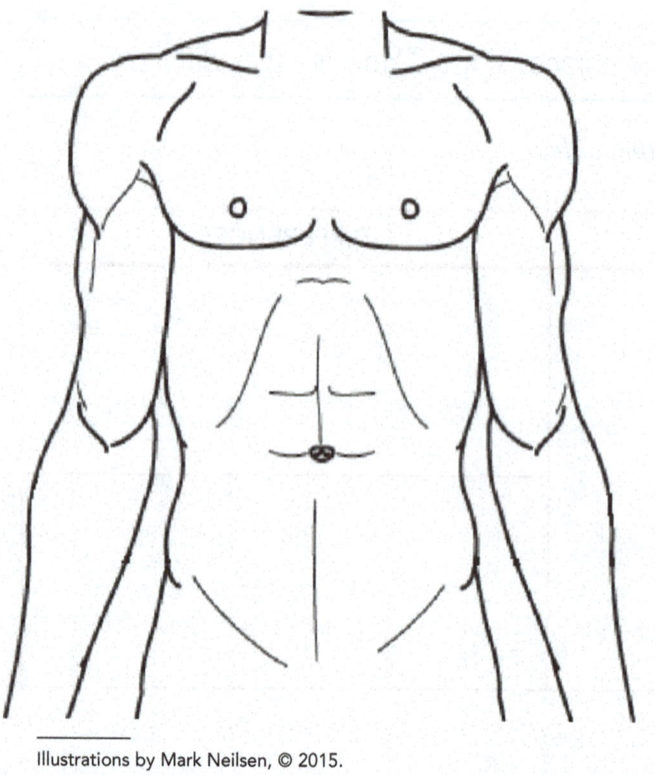

1. Right Hypochondriac Region
2. Epigastric Region
3. Left Hypochondriac Region
4. Right Lumbar Region
5. Umbilical Region
6. Left Lumbar Region
7. Right Inguinal Region
8. Hypogastric Region
9. Left Inguinal Region

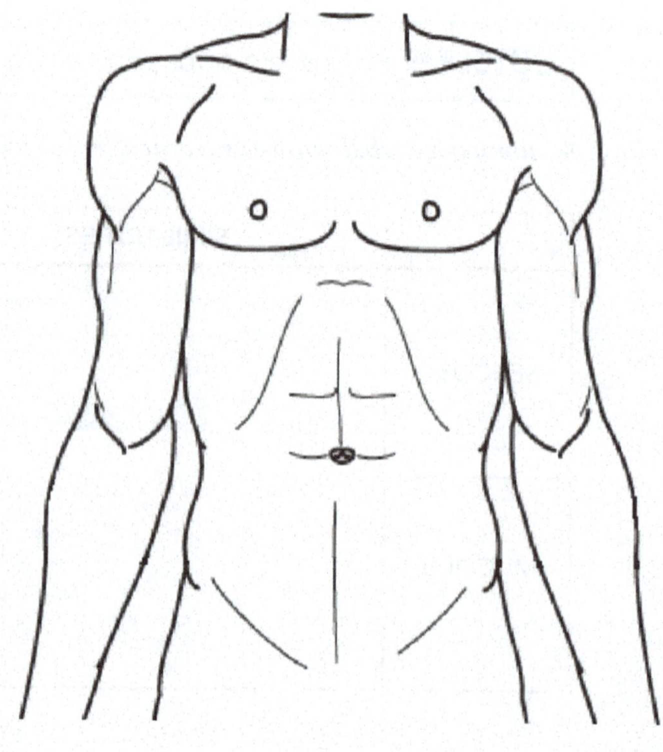

OUTCOME 10: Identify and apply medical terminology, including root words, prefixes, and suffixes.

Medical Terminology

The specialized set of vocabulary utilized throughout the medical field serves as a universal communication method. This terminology can be broken down into Greek and Latin root words, prefixes, and suffixes. Understanding the components of the complex terms that appear throughout the study of anatomy will provide a substantial foundation for learning.

The **word root** is the basic core of any word. The word may consist of a single word root alone, or the word root may have its meaning modified or altered slightly by the addition of a suffix, a prefix, one or more additional word roots, or some combination of these elements.

A **prefix** is a preposition, an adjective, or an adverb placed before words or roots to show various relationships. They are never used independently. Most prefixes are used widely in everyday language and are not limited to medical or scientific terminology; however, their use may indicate a greater degree of precision.

A **suffix** is an element added to a word root to modify the meaning or form a new word. The suffix usually indicates the part of speech (noun, verb, adjective, adverb) of the completed word.

After one has learned the commonly used roots, suffixes, and prefixes, the student is then able to deduce the meanings of new medical words by analyzing the composition of the term and the context in which the term is used.

TERM	DEFINITION	EXAMPLE
Cyto/cyte	Cell	Cytology: science of the formation, structure, and function of cells Osteocyte: bone cell
Hist/histo	Tissue	Histology: study of tissues
Ecto/exo/extra	Without, outside of	Ectoderm: outer layer of cells in a developing embryo Exocrine: external secretion of gland Extracellular: outside of the cell
Endo	Within, inner, inside	Endocarditis: inflammation of the inner lining of the heart
Enter/entero	Intestine	Enteric: pertaining to the small intestine Parenteral: any medication route other than the alimentary canal
Epi	Upon, over, on	Epigastric: pertaining to the region over the stomach
Peri	Around, near	Periosteum: thick fibrous membrane covering the surface of a bone Pericellular: around the cell
Path/pathy/patho	Disease	Neuropathy: disease of the nerve system Pathogen: microorganism capable of producing disease

TERM	DEFINITION	EXAMPLE
Gastro	Stomach	Gastroenteritis: inflammation of stomach and intestinal tract
Cardi/cardio	Heart	Cardiac: pertaining to the heart Cardiology: study of the physiology and pathology of the heart
Hepato/hepat	Liver	Hepatotoxic: referring to an agent that damages the liver Hepatology: study of the liver
Reno/reni/ren	Kidney	Renal: pertaining to or of the kidney
Osteo	Bone	Osteogenesis: formation of bone
Homo	Like, similar, same	Homogeneous: of uniform structure or composition throughout
Hetero	Other, different	Heterocellular: composed of different kinds of cells
Syn	With, together	Syndrome: the aggregate of signs and symptoms associated with any disease
Anti	Against, opposing	Anticonvulsant: agent that prevents or reduces the severity of convulsions
Macro	Large	Macrocyte: abnormally large erythrocyte
Micro	Small	Microcyte: abnormally small red blood cell
Inter	Between	Intercellular: between or among cells Intercostal: between the ribs
Intra	Within	Intracellular: within a cell or cells Intra-alveolar: inside of an alveolus
Glyco	Sweet, sugars	Glycolysis: energy-yielding conversion of glucose to 2 pyruvic acid molecules in cells Glycolipid: lipid containing sugar molecule found in the myelin sheath of nerves
Hydro	Water	Hydrolysis: chemical breakdown of a molecule involving water
Hyper	Above normal, beyond, excessive	Hypertrophy: increase in the size of a tissue or organ due to growth and an increase in cellular components
Hypo	Below normal, under, deficient, beneath	Hypotonic: describes a solution having a lower osmotic pressure than another solution Hypoxia: an oxygen deficiency

TERM	DEFINITION	EXAMPLE
Poly	Many	Polyphagia: excessive eating, in multiple sittings
Lysis	Process of breaking down	Hemolysis: destruction of red blood cells with liberation of hemoglobin in the surrounding medium
Lytic	Pertaining to that which causes lysis	Hemolytic: relating to that which is destructive of red blood cells Thrombolytic: describing an agent that breaks up blood clots
Som/soma/somato	Body	Somatic: relating to the trunk, the wall of the body cavity, or to the body in general Lysosome: membrane-brane cytoplasmic organelle which contains hydrolytic enzymes
Erythro	Red	Erythrocyte: red blood cell
Leuk/leu	White	Leukocyte: white blood cell
Heme/hemo/hemato	Blood	Hematology: study of blood and blood-forming tissues and the disorders associated with them
Plasia	Process of cellular multiplication, formation	Hyperplasia: increase production and growth of normal cells resulting in an abnormal increase in size of the tissue or organ
Trophy	Nourishment, development, growth	Hypertrophy: increase in the size of a tissue or organ due to growth and an increase in cellular components
Blast	Immature precursor cell, pertaining to cells that build/form	Fibroblast: an elongated cell in connective tissue that is capable of forming collagen fibers Osteoblast: cell that is capable of building bone, precursor to osteocytes
Clast	Pertaining to cells that break	Osteoclast: cell that is capable of breaking done bone, releasing calcium into the bloodstream
Genesis	Production, formation, development	Osteogenesis: formation of bond
Penia	Condition of lack, deficiency	Granulocytopenia: condition in which the number of granulocytes in the blood is lower than normal
Philia/phil	To love, have affinity for	Hydrophilic: to have affinity for water, water loving
Phobia/phobe	Fear, have aversion to	Hydrophobic: to have aversion to water, water fearing
Rrhage/rrhagia	Condition of excessive or abnormal discharge	Hemorrhage: excessive bleeding

(*continued*)

TERM	DEFINITION	EXAMPLE
Derma/dermato	Skin	Dermatology: medical specialty concerned with diagnosis and treatment of skin disorders
Lemma	Sheath or envelope	Neurolemma: the sheath enclosing the axon of a nerve fiber
Sarco	Denotes muscular substance; flesh	Sarcolemma: plasma membrane sheath enveloping a muscle fiber
Ectomy	Excision, surgical removal of a segment of entire part of an organ	Gastrectomy: surgical removal of the stomach
Chondro	Cartilage	Chrondrocyte: cartilage cell, found in the lacunae of the matrix of cartilage
Necro	Death, dead	Necrosis: pathologic death of one or more cells, usually in a localized area
Onco	Tumor, bulk, mass	Oncology: branch of medication dealing with tumors
Pre	Before, in front of	Presynaptic: situated before or proximal to a synapse
Post	Behind, after	Postsynaptic: situated beyond or distal to a synapse
Brady	Slow	Bradycardia: abnormally slow heartbeat
Tachy	Rapid, quick	Tachycardia: abnormally rapid heartbeat
Chrono	Time	Chronic: relating to health-related state or condition persisting over a long period of time
Athero	Deposit of soft, pasty material	Atherosclerosis: a type of arteriosclerosis characterized by irregularly distributed lipid deposits that could eventually lead to restriction of blood flow
Arthro	Joint, jointed	Arthritis: inflammation of the joints
Vaso	Blood vessel	Vasoconstriction: tightening or narrowing of blood vessel
Itis	Indicates a specific condition of inflammation	Gastritis: inflammation of the stomach

EXERCISE 2

Histology

A. *Introduction*

Cells of living organisms exist in organized **functional** and **structural** groups called **tissues.** That is, a tissue is a group of cells working together to perform a common function. **Histology** is the study of these tissues. The tissues do not exist in isolation. They are made up of cells and they associate with one another to form **organs** and **organs systems.** There are four basic tissues or **histological categories** in the body: **epithelial, connective, muscle,** and **nervous tissue.** Each of these tissues consists of a collection of cells that sit in a background of an extracellular substance known as a **matrix.** It may seem odd, given the variable structure and function of the organs in the body, that only four tissue types compose the organs. Think about how different a liver looks from a heart, or the skin looks compared to a bone; yet, the same four histological categories make up all these organs! The key to the unique look and function of these organs is the types of cells that make up the tissues, the manner in which the tissues combine, and their association with the extracellular matrix.

B. *Outcomes for Exercise 2*

1. Define TISSUE. Identify where tissues fit in the structural hierarchy of living things. Memorize the definitions of each level of organization.

2. List the histological categories. Describe the general functional and structural characteristics of each. Name at least one location where each is found within the body.

3. How are epithelial tissues classified? Examine different types of epithelial tissue under the microscope. Using the tables provided and your textbook, create a labeled drawing noting any unique characteristics of what you see. Provide a specific location within the body and specific function for each type.

4. How are connective tissues classified? Examine different types of connective tissue under the microscope. Using the tables provided and your textbook, create a labeled drawing noting any unique characteristics of what you see. Provide a specific location within the body and specific function for each type.

5. Label the details of an osteon. Locate each detail on the microscope.

6. How are muscle tissues classified? Examine different types of muscle tissue under the microscope. Using the tables provided and your textbook, create a labeled drawing noting any unique characteristics of what you see. Provide a specific location within the body and specific function for each type.

7. Identify three functional types of neurons and describe their locations and functions. Determine if the neutron is structurally classified as pseudounipolar or multipolar. For each structural type, identify its cell parts and the function of each of those parts. Make sure you understand the difference between functional classification and structural classification.

C. *The Hierarchial Structure of Living Things*

OUTCOME 1: Define Tissue. Identify where tissues fit in the structural hierarchy of living things. Memorize the definitions of each level of organization. Explain what is meant by structural hierarchy.

- ***Provide the word for each definition of the organization of living things, and give a specific example.***

STRUCTURAL LEVEL OF ORGANIZATION	DEFINITION	EXAMPLE
	The smallest living unit in the body. Made of internal organized molecules known as organelles.	
	An aggregate of cells and their matrix combined in such a way as to perform a unique function.	
	An aggregate of two or more tissues combined in such a way as to perform a unique function.	
	An aggregate of two or more organs combined in such a way as to perform a unique function.	
	An individual composed of interacting organs systems which maintain homeostasis within a changing environment.	

- ***What levels of structural organization are more complex than tissues?***

- ***What levels of structural organization are less complex than tissues?***

Structural organization is simply a way of categorizing or classifying levels of complexity. More important than knowing the levels of complexity is understanding HIERARCHY.

Hierarchical organization tells us something about how each level of complexity relates to the other levels. Each less complex level combines to form the level that is more complex, and each more complex level can be dissected apart to create the levels that are less complex. Thus, hierarchical organization assumes an **interdependency** and **interrelationship** among the levels.

Complete the following exercise.

- ***The girls in the photograph represent the*** ***organism level*** ***of structural organization. Label each of the eight drawings to the left and right of the photograph to indicate which level of organization the drawings represent.***

- ***In each group, next to your labels, number the level of complexity from 1– 4, with 1 being the least complex and 4 being the most complex***

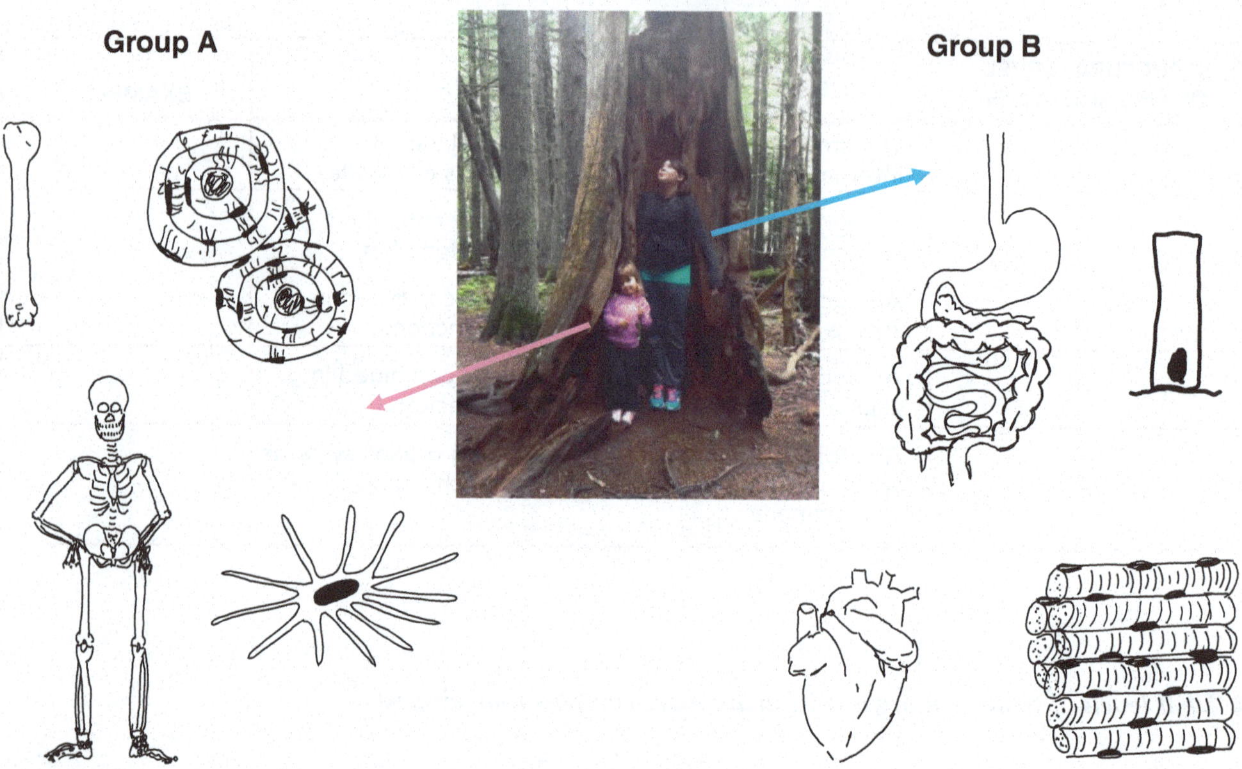

Illustration by Lucia Tranel, © 2015.

- ***Which one of the groupings represents a hierarchy of structural organization?***

- ***Explain your answer. How do you know that the group you chose represents hierarchy? How do you know the other group does not?***

D. *An Overview of the Four Histological Categories*

OUTCOME 2: List the histological categories. Describe the general functional and structural characteristics of each. Name some at least one location where each is found within the body.

TOOLBOX

An Orientation to the Tissues of the Body

Epithelial Tissue

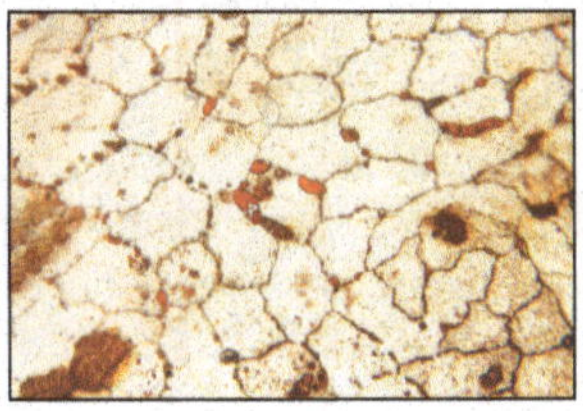

Epithelial tissues derive from all three embryonic germ layers and **consist of closely aggregated cells with very little intercellular matrix.** Epithelium generally functions to **cover surfaces and to act as a barrier.** An extracellular **basal lamina** always attaches the cells to the subjacent connective tissue. The **lamina propria** is the name of the connective tissue beneath the epithelial layer.

Connective Tissue

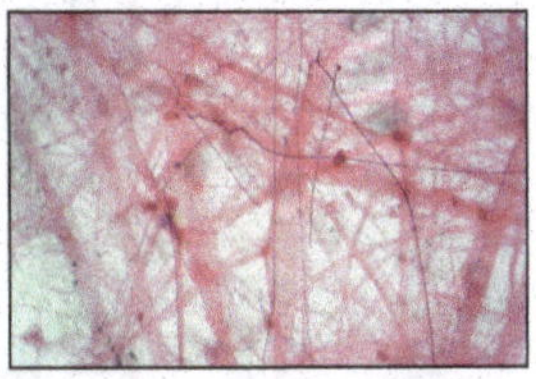

There are many different types of connective tissues, but they generally function to **provide structural support and framework, to protect against trauma, to bind layers of tissues together, and to protect against invading pathogens.** Connective tissues derive from mesodermal tissues and contain three structural components: **cells, matrix,** and **fibers.** Connective tissue cells tend to be spread apart, unlike epithelium. The cells may appear fixed or mobile. They sit within an **acellular background** or **extracellular matrix.**

Muscle Tissue

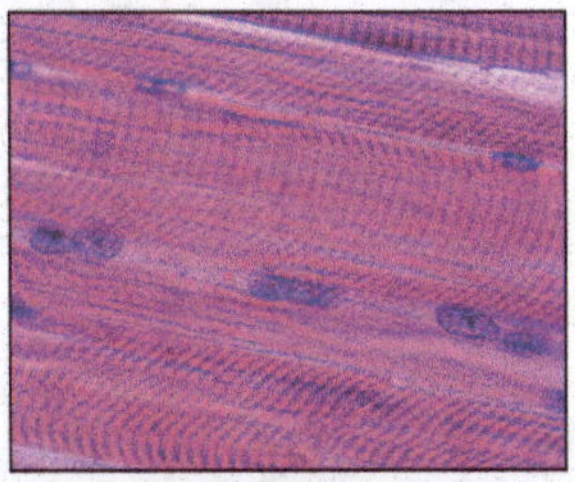

Muscle tissue is the only tissue in the body that can **provide movement by contracting.** It derives from the mesodermal germ layer and consists of highly specialized cells that demonstrate **four unique characteristics: irritability, contractility, extensibility, and elasticity.** The elongated cells are frequently referred to as **fibers.**

Nervous Tissue

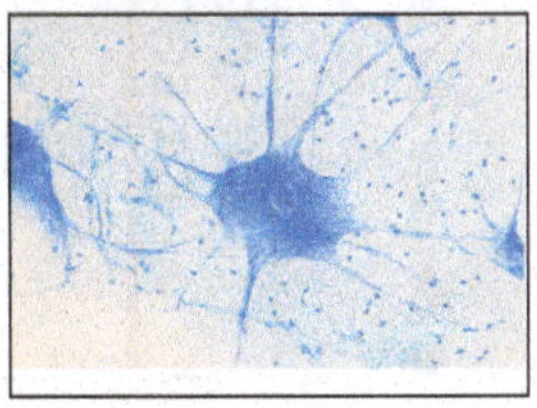

Nervous tissue consists of two populations of cells: nerve cells **(neurons)** and glial cells **(neuroglia)**. Neuroglia provide structural support, immune protection, and nutritional support for the neurons. Neurons are the cells capable of **receiving, processing, and conducting electrical impulses throughout the body.**

■ *Review the summary of the 4 principal types of tissues:*

HISTOLOGICAL CATEGORY	GENERAL FUNCTION	CELLS	EXTRACELLULAR MATRIX	GENERAL LOCATION
Epithilial Tissue	Acts as a barrier by covering or lining free or exposed surfaces. Absorption. Glandular secretion. Sensation.	Polyhedral shaped. Collective mass or thin sheet, usually called a membrane.	Very small amount.	Lines cavities and hollow organs. Covers surfaces. Avascular. Often makes up the **parenchyma** of an organ.
Connective Tissue	Binds layers together. Supports. Pads. Protects against trauma and pathogens.	Varied in shape and function. May appear fixed or wandering.	Abundant amount. Matrix can vary in consistency, from liquid to semi-liquid to solid. The matrix often includes fibers.	Between layers of other tissues. Often makes up the **stroma** of an organ.
Muscle Tissue	Contracts to enable movement.	Elongated. Cells are often referred to as fibers.	Moderate amount.	Within the walls of hollow organs. Attached to bones. Within the wall of the heart.
Nervous Tissue	Can generate and conduct an electrical impulse or action potential.	Two categories: Neuroglia vary in structure, function, and location. Neurons perform the general function of NervousTissue and are composed of dendrites, cell bodies, and axons.	None.	Within the brain and spinal cord. Attached from sensory receptors in PNS to the CNS (sensory or afferent), or from the CNS to effector organs (motor or efferent) in the PNS.

E. *Microscopy Examination of Tissues*

During this laboratory exercise, you will examine each of the four histological categories, and you will look at different types of cells within each of those categories.

Getting Ready

The microscope slides you will examine are in the box on the lab table. You also will find lens paper for cleaning the slides and a list of the slides in numbered order. **Please take care to return the slides in the correct order when you have finished the laboratory exercise.**

What to Do

Goals and Tools:

1. Read the description and locate each tissue on the microscope slide 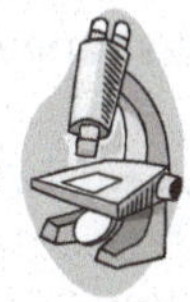 .
2. For each tissue, draw and label what you see.
3. You will be asked to identify each tissue and its details on an assessment.

   Use your textbook's  histology chapter as a reference.
4. Make sure you are *specific* when you must provide a location and a function for a particular tissue. ***Organs consist of multiple tissues; you must indicate where the tissue is in the organ.*** The textbook will provide the information that you need.

If what you see does not look like the pictures in the textbook or in this manual, you are probably looking at the wrong thing. Remember that many of the slides are sections through whole organs. By definition, an organ is made up of two or more tissues. Always start with your microscope at low magnification and move the slide around to get a general "lay of the land." IN GENERAL:

Epithelial Tissue — **is on a free surface, so it will always have a "white space" adjacent to it.**

Connective Tissue — **will be deep to a surface. It usually contains cells, fibers, and an extracellular matrix.**

Muscle Tissue — **will appear "stringy."**

Nervous Tissue — **will be individual cells, as described.**

F. *Epithelial Tissue*

OUTCOME 3: How are epithelial tissues classified? Examine different types of epithelial tissue under the microscope. Using the tables provided and your textbook, create a labeled drawing noting any unique characteristics of what you see. Provide a specific location within the body and specific function for each type.

The **basis of classification** for most epithelial tissues consists of **two features:**

1. whether the cells are in a single layer or in multiple layers and
2. the general shape of the cells.

1. Number of Layers:

simple (one cell layer)

stratified (multiple layers)

2. Cell Shape:

squamous (flat)

cuboidal (cube-shaped)

columnar (tall, elongated)

The features of number of layers and cell shape can be combined to create classifications of epithelial tissues, as follows:

CELL SHAPE	NUMBER OF LAYERS	
	SIMPLE	STRATIFIED
SQUAMOUS	simple squamous	stratified squamous
CUBOIDAL	simple cuboidal	stratified cuboidal
COLUMNAR	simple columnar	stratified columnar

CHECKLIST: Summary of Epithelial Tissues to Be Examined

■ ***Before coming to lab, use your* TEXTBOOK *to help you describe a specific location and function for each epithelial tissue you will examine.***

REMEMBER: Your named location and described function must be very specific. Simply naming an organ is *not* a specific location; you must name where in the organ. Likewise, the described function should *not* be the general function of the histological category; you must describe the specific function of the tissue at the location you named.

The first one has been completed for you.

EPITHELIAL TISSUE	SPECIFIC LOCATION	SPECIFIC FUNCTION
Simple squamous	Composing the wall of an alveolus	Allows for rapid diffusion of respiratory gases
Stratified squamous non-keratinized		
Stratified squamous keratinized		
Simple cuboidal		
Stratified cuboidal		
Simple columnar		
Ciliated pseudostratified columnar		
Transitional		

The microscopic examination of these epithelial tissues will occur in the lab.

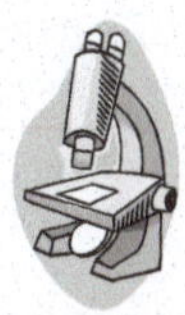

1. Simple Squamous Epithelium

squamous epithelium, amphibian w.m.

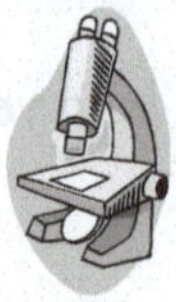

This slide gives you an idea of the appearance of a sheet of squamous epithelium from a superior aspect or surface view. The dark dot in each cell is the **nucleus.** It is surrounded by **cytoplasm.** Notice how compact the cells are; there is little visible intercellular space between the cells.

IMAGE	LABELED DRAWING AND NOTES	DETAILS TO LOCATE
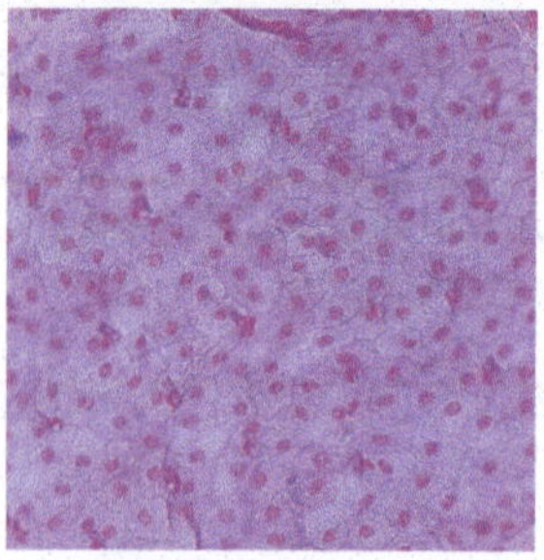		cytoplasm nucleus cell membrane

simple squamous epithelium, x.s. (kidney, Bowman's capsule)

simple squamous epithelium, x.s. (lung, alveolus)

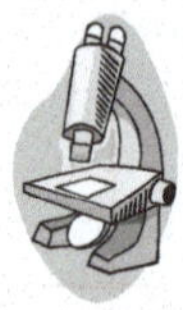

Both these slides have good examples of simple squamous epithelium. Look at the lung slide first. Using low magnification, find an area that looks like a section or a slice through a sponge. The thin-walled spaces are the air sacs of the lungs, the **alveoli.** At high magnification, you will see that the cells are very thin. The **nucleus** looks squashed. Sometimes all that's visible is the nucleus. The cells touch an open space called the **lumen.** Examine the second slide. The lining of Bowman's capsule is another good example of this tissue.

IMAGE	LABELED DRAWING AND NOTES	DETAILS TO LOCATE
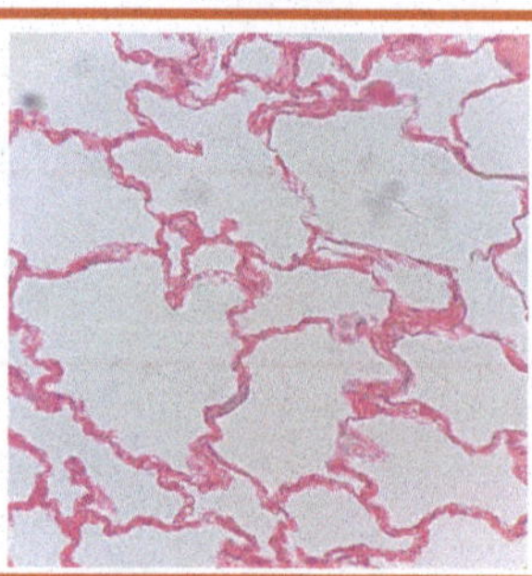		alveolus alveolar lumen cell nucleus

2. Stratified Squamous Epithelium (Non-Keratinized)

stratified squamous epithelium, x.s. esophagus

The epithelium seen on this slide consists of squamous cells, but there are many cells stacked on one another. Notice that the cells that touch the **basal lamina** and **lamina propria** are more cubed-shaped. This is the **mitotic layer.** As these cells go through mitosis and produce more cells, the older cells get pushed up toward the **lumen or exposed surface.** The mature cells become more squamous-shaped.

IMAGE	LABELED DRAWING AND NOTES	DETAILS TO LOCATE
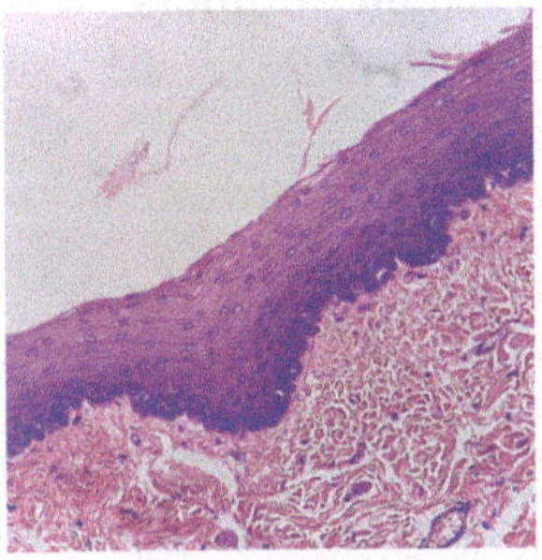		squamous cell basal lamina mitotic layer adlumenal surface

3. Stratified Squamous Epithelium (Keratinized)

Cornified skin layer

In locations of the body where there is the potential for water loss or abrasion, you will find a layer of stratified squamous epithelium that contains a protective protein called **keratin.** This keratinized stratified squamous epithelium can be found on the outermost surface of your skin. The palms of your hand and soles of your feet have an especially large amount of keratin as these areas have a high potential for abrasion.

IMAGE	LABELED DRAWING AND NOTES	DETAILS TO LOCATE
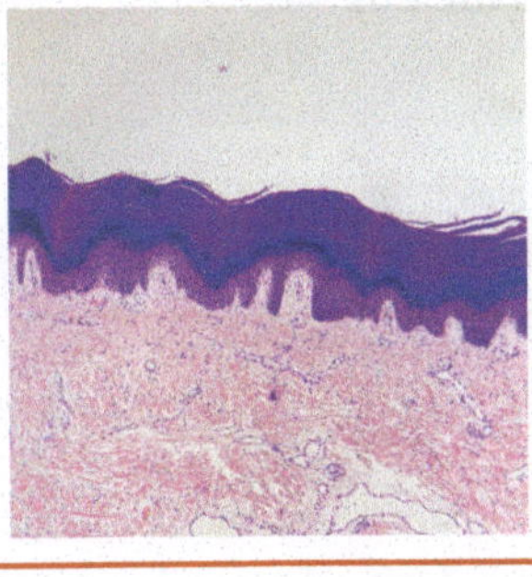		Squamous cells Nucleus Mitotic layer (stratum basale) Adlumenal surface (apical surface) Basal lamina (basal surface)

4. Simple Cuboidal Epithelium

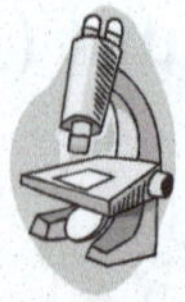

simple cuboidal epithelium, renal tubules

At high magnification, cube-shaped cells with very round **nuclei** in the centers of the cells are visible. Simple cuboidal epithelium lines the renal tubules and the ducts of exocrine glands.

IMAGE	LABELED DRAWING AND NOTES	DETAILS TO LOCATE
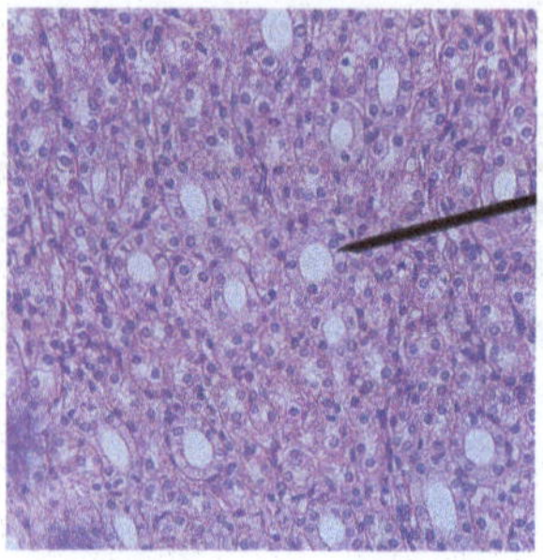		cuboidal cell basal lamina nucleus lumen of tubule nucleus

5. Stratified Cuboidal Epithelium

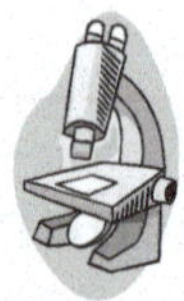

Thick and thin skin

Similar to simple cuboidal epithelium, stratified cuboidal epithelium also contains cube-shaped cells; however, because they are stratified, the cells are more than one layer thick. This type of tissue specializes in secretion and selective reabsorption and is typically found in glands. A specific example you will observe is in the lining of eccrine sweat glands that can be found in the dermis throughout the body.

IMAGE	LABELED DRAWING AND NOTES	DETAILS TO LOCATE
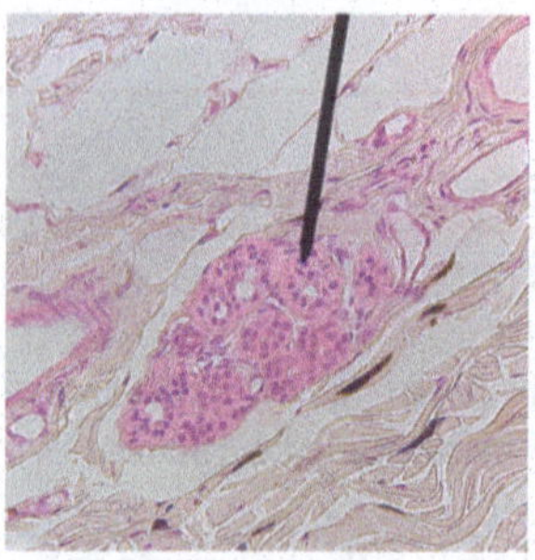		**Basal surface** **Cuboidal epithelial cell** **Apical surface** **Lumen** **Nucleus**

6. Simple Columnar Epithelium

simple columnar epithelium, goblet cells
simple columnar epithelium

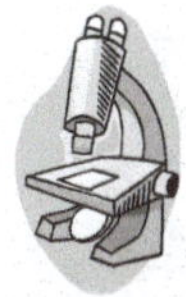

Simple columnar epithelium cells line the **lumen** of the digestive tract from the stomach to the rectum. Keep in mind as you examine the x.s. of the intestine that the lining of the intestine has many folds to help increase surface area. There are finger-like projections, or **villi,** on each fold. The simple columnar epithelium lines the villi and touches the lumen. Notice that the nucleus of each cell is located near the **basal lamina.** You might observe an occasional cell that appears to have a clear space or balloon within its cytoplasm. These cells, called **goblet cells,** are adapted especially to produce and secrete **mucus.** The second slide of the intestine is just like the previous slide, except that it is stained to enhance the view of the goblet-shaped vacuoles (orange) within the goblet cells.

IMAGE	LABELED DRAWING AND NOTES	DETAILS TO LOCATE
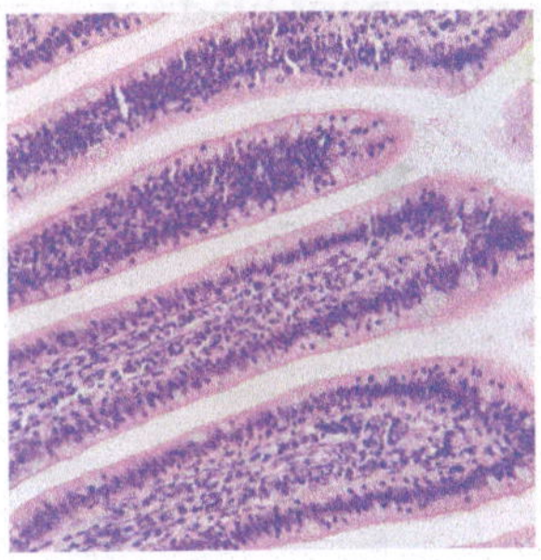		columnar cell basal lamina nucleus adlumenal surface goblet cell villus

7. Ciliated Pseudostratified Columnar Epithelium

ciliated pseudostratified columnar epithelium, x.s. trachea

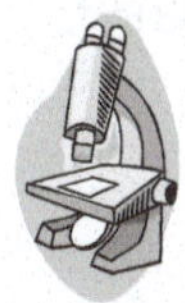

This tissue is so named because the **nuclei** of its cells are in different levels. This gives the tissue the appearance that it consists of many cell layers but in actuality consists of only one cell layer. Every cell touches the **basal lamina,** though not all cells will touch the **lumen.** This particular tissue appears ciliated as well as pseudostratified. Examine the **adlumenal** surface of these cells; locate the **cilia.** How do the cilia function? What advantage would exist to having ciliated epithelium line the respiratory passages?

IMAGE	LABELED DRAWING AND NOTES	DETAILS TO LOCATE
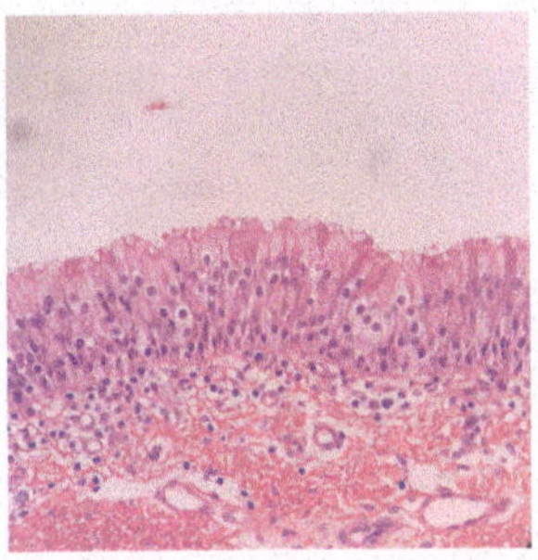		c.p.c.e. cell basal lamina cilia adlumenal surface

---

8. Transitional Epithelium

Mammal Urinary Bladder Collapsed c.s.

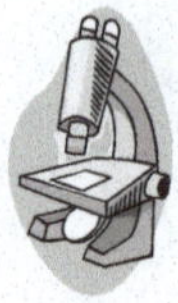

Transitional epithelium is a stratified epithelium that gets its name because of the ability of its cells to transition in appearance depending on whether or not the tissue is stretched or relaxed. This specific tissue type can be found lining the organs of the urinary system including the urinary bladder (the slide you will observe). The cells near the basal lamina typically appear cuboidal or columnar in appearance, while the more superficial cells touching the lumen can "transition" from cuboidal, when the tissue is relaxed, to squamous, when the tissue is stretched. Can you think of a reason why this type of tissue lines the urinary bladder?

IMAGE	LABELED DRAWING AND NOTES	DETAILS TO LOCATE
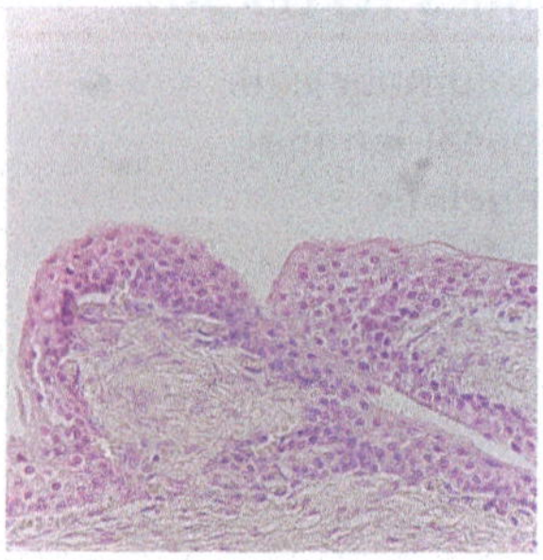		Dome-shaped epithelial cells Lumen of urinary bladder Nucleus

Illustrations Contributed by Tony Wilmes.

G. *Connective Tissue*

OUTCOME 4: How are connective tissues classified? Examine different types of connective tissue under the microscope. Using the tables provided and your textbook, create a labeled drawing noting any unique characteristics of what you see. Provide a specific location within the body and specific function for each type.

The matrix of connective tissue varies in its structure, ranging from liquid (**blood plasma**), to semi-solid and flexible (**cartilage**), or dense and rigid (**bone**). The matrix not only provides support but is the **extracellular environment through which nutrients and metabolic materials diffuse.** There is often a collection of protein fibers (**collagen, elastic, reticular**) within the matrix. The fibers provide extra support, elasticity, and strength. The variety of connective tissues results from the variety of cell types, fiber types, and matrix composition and how those three components are put together.

CHECKLIST: Summary of Connective Tissues to Be Examined

- *Before coming to lab, use your* TEXTBOOK *to help you describe a specific location and function for each connective tissue you will examine.*

CONNECTIVE TISSUE	SPECIFIC LOCATION	SPECIFIC FUNCTION
Loose irregular tissue (areolar)		
Dense irregular connective tissue		
Dense Regular C.T.		
Adipose		
Blood		
Hyaline Cartilage		
Elastic Cartilage		
Fibrocartilage		
Bone		

The microscopic examination of these connective tissues will occur in the lab.

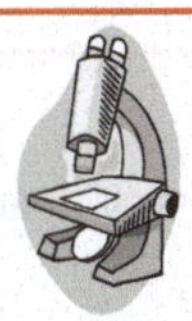

1. Loose Irregular Connective Tissue

loose irregular connective tissue (areolar)

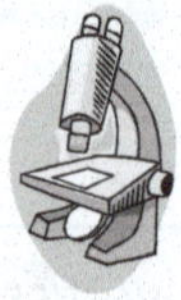

Look for the **cells, fibers,** and **matrix** as you examine this slide . This tissue gets its name because the matrix is a semi-solid colloid (like Jell-O®), and the fibers are in an irregular pattern. There are two types of fibers. The thin, blue-black squiggly fibers are **elastic fibers.** They allow the tissue to return to its original shape after being stretched. The pale, pink, thick fibers are **collagen fibers.** They add strength and support to the matrix. If there is a higher density of fibers, the tissue is **dense irregular connective tissue.** If the fibers run in a parallel arrangement, the tissue is **regular connective tissue.**

IMAGE	LABELED DRAWING AND NOTES	DETAILS TO LOCATE
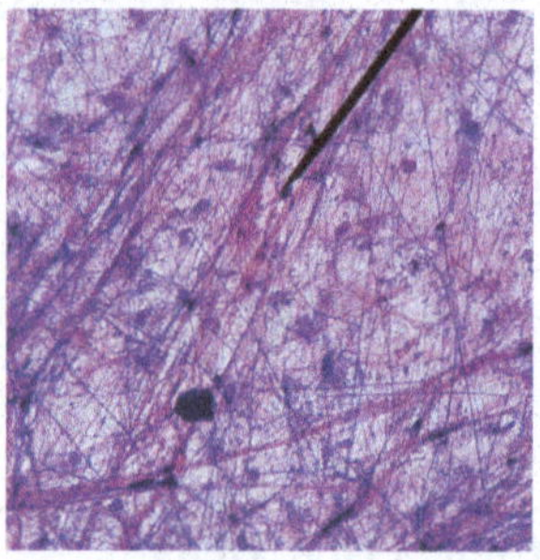		fibroblasts elastic fibers collagen fibers matrix

2. Dense Irregular Connective Tissue

Cornified Skin

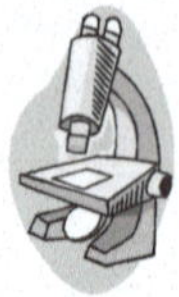

This specific tissue contains the same components as the previous tissue, collagen fibers and fibroblasts. The difference between these two tissues lies in the orientation of their fibers and cells. Dense irregular connective tissue does not have neat bundles of collagen fibers; rather, these fibers run in all directions giving them a disorganized appearance, with fibroblast cells embedded irregularly throughout the tissue. As the name implies, the fibers are more densely packed. The reticular layer of the dermis is comprised of dense irregular connective tissue.

IMAGE	LABELED DRAWING AND NOTES	DETAILS TO LOCATE
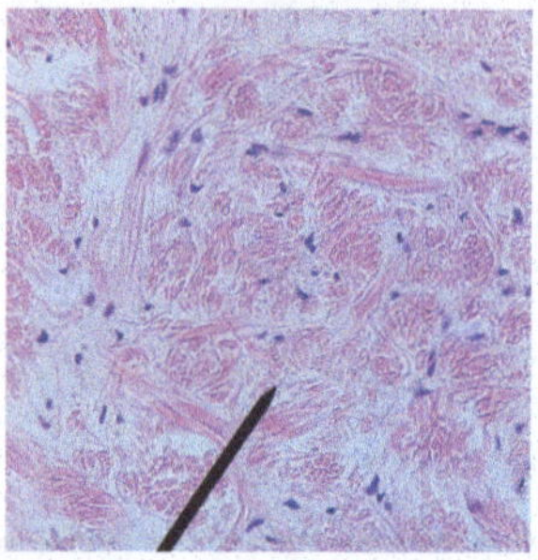		collagen fibers fibroblast nuclei

3. Dense Regular Connective Tissue

Dense Regular Connective Tissue of Tendon

Dense regular connective tissue consists primarily of bundles of collagen fibers, all oriented in the same direction, with layers of fibroblast cells separating each bundle. This tissue type can be found in tendons (cords that attach muscle to bone) and ligaments (cords that attach bones to other bones).

IMAGE	LABELED DRAWING AND NOTES	DETAILS TO LOCATE
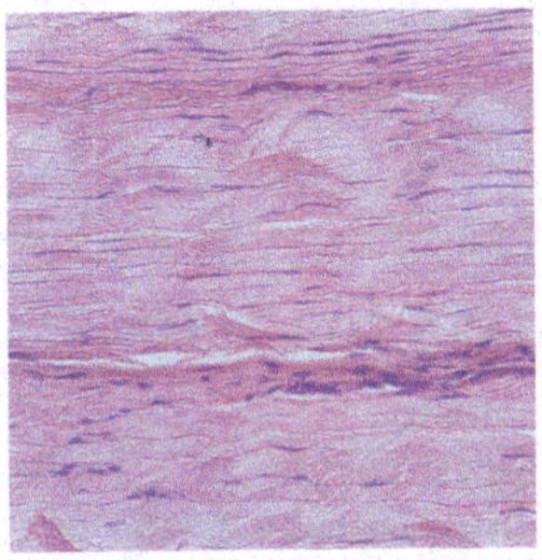		collagen fibers fibroblast nuclei

4. Adipose Connective Tissue

adipose connective tissue

Adipose tissue functions to **store lipids**. The lipids are semi-solid at body temperature; therefore, the tissue acts as an excellent **shock absorber** and often occurs around more vulnerable organs such as the eyeballs, kidneys, hearts, and joints. Adipose, which is a poor conductor of heat, acts as an insulator and exists in the subcutaneous layer. The **adipocytes** are very compact, so you will not observe matrix or fibers. Notice that the cell membranes of the individual adipocytes are noticeable, but the adipocytes appear empty. The **vacuole** that stores the lipid is colorless. It fills the cell and pushes the cytoplasm and nucleus to the periphery.

IMAGE	LABELED DRAWING AND NOTES	DETAILS TO LOCATE
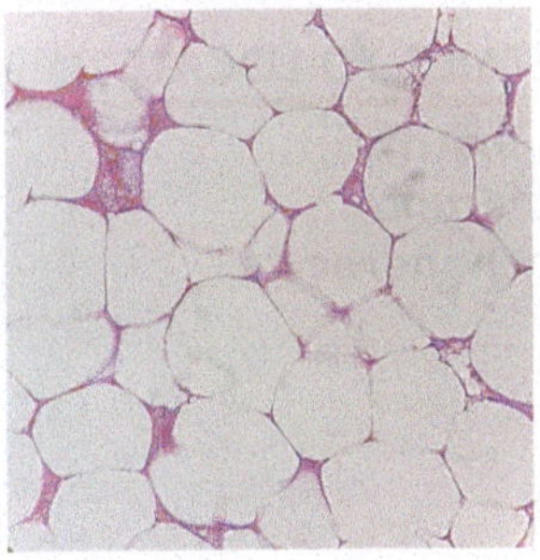		adipocyte nucleus cell membrane lipid vacuole

5. Blood

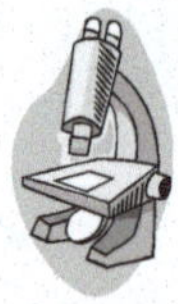

blood smear, Wright's stain

You must examine this slide under high magnification in order to see the details of the blood components. The clear background is the matrix or **plasma.** The **erythrocyte** (**red blood cell**) is the most numerous of the cells you see as you scan the slide. Notice that these cells are **enucleated.** They lose their nucleus during cellular development. As a result, the cell is "caved-in" in the center—like a jelly donut with the jelly pulled out. The resulting shape is **biconcave,** and the cell appears paler in the center, almost like a donut but the hole does not go all the way through. Erythrocytes **transport oxygen** to the tissue cells. A second type of cell that you see is a **leukocyte** (**white blood cell**). Leukocytes are **nucleated.** They function with our **immunity.** Look for **thrombocytes** (**platelets**). They are pinpoint fragments of a larger cell and are involved with the process of **blood clotting.**

IMAGE	LABELED DRAWING AND NOTES	DETAILS TO LOCATE
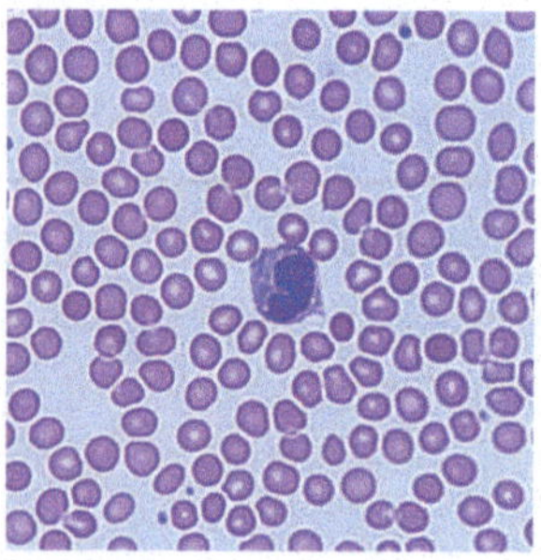		plasma erythrocyte leukocyte thrombocyte

6. Hyaline (Articular) Cartilage

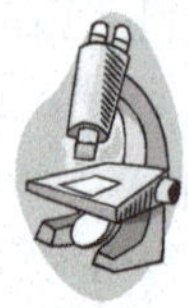

hyaline cartilage connective tissue

Hyaline cartilage is the connective tissue found at the ends of bones and within the wall of the trachea. The matrix of hyaline cartilage has no fibers and appears pale purple on the slide. Because the **matrix** is denser than the connective tissue previously observed, the **chondrocyte** (cartilage cell) must have a cavity to house it. The name of this cavity is **lacuna (pl, lacunae).**

IMAGE	LABELED DRAWING AND NOTES	DETAILS TO LOCATE
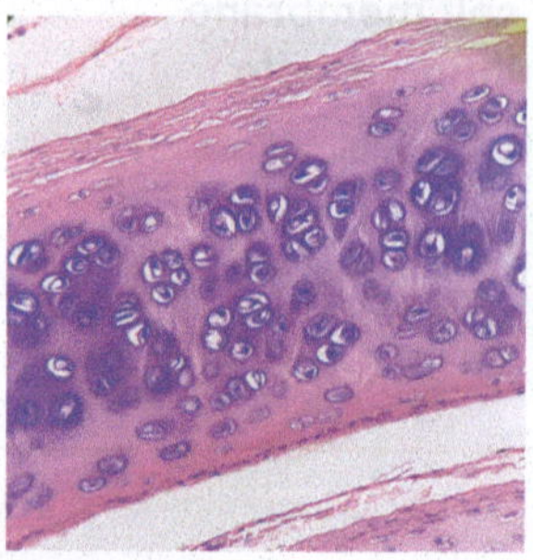		matrix lacuna chondrocyte

7. Elastic Cartilage

Elastic cartilage is a highly flexible tissue that contains a large amount of elastin fibers embedded in the extracellular matrix. Just like elastic connective tissue, these elastic fibers are easily visible as they stain dark purple. Along with the elastin fibers you will also find an abundance of lacunae containing chondrocytes throughout the matrix in a disorderly pattern similar to hyaline cartilage. Elastic cartilage can be found in the auricle of the ear, Eustachian tube, and epiglottis.

IMAGE	LABELED DRAWING AND NOTES	DETAILS TO LOCATE
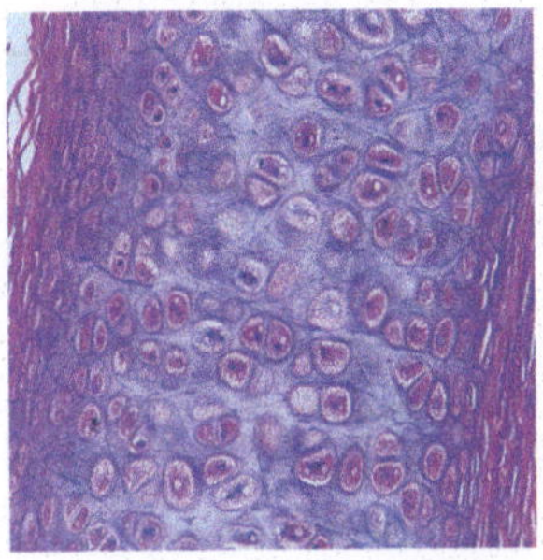		chondrocytes lacunae elastic fibers

8. Fibrocartilage (Fibrous Cartilage)

Fibrocartilage is a tough, slightly flexible tissue that contains an orderly arrangement of both collagen fibers and chondrocytes. The chondrocytes, and the lacunae they sit in, are lined up in parallel rows which allows for easy identification of this tissue compared to other cartilage tissue. The pubic symphysis and intervertebral discs are the two specific locations of this tissue. Fibrocartilage is also utilized by the body to repair hyaline cartilage that has been damaged.

IMAGE	LABELED DRAWING AND NOTES	DETAILS TO LOCATE
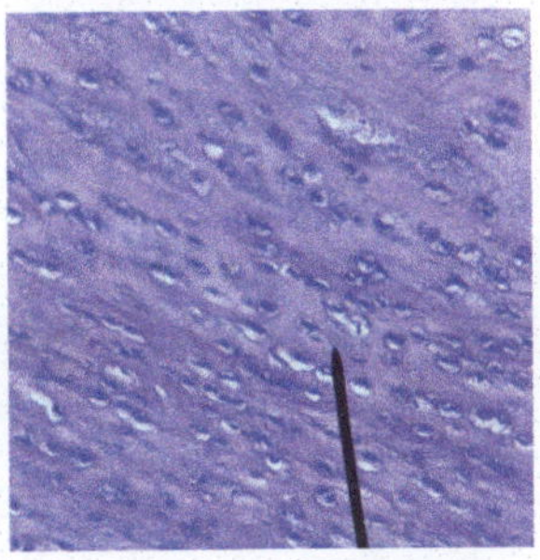		collagen fibers chondrocytes lacunae

9. Bone or Osseous Tissue

bone connective tissue, ground bone x.s.

Bone tissue is located within the bone organs of your skeletal system. It provides the strength and structure that bones (organs) need in order to provide a framework and to surround and protect many vital organs, such as the brain, spinal cord, heart, and lungs. The sample of bone on the slide comes from the **compact region** of the **diaphysis** of a **long bone.** This region appears solid when you examine the long bone, but its porous quality is evident under the microscope. You will notice as you examine the tissue on the slide that the arrangement resembles cross sections through "tree trunks." The name of each circular pattern is an **osteon** or **Haversian system.** Use the labeled drawings on the next page to help you locate all the details of an osteon.

IMAGE	LABELED DRAWING AND NOTES	DETAILS TO LOCATE
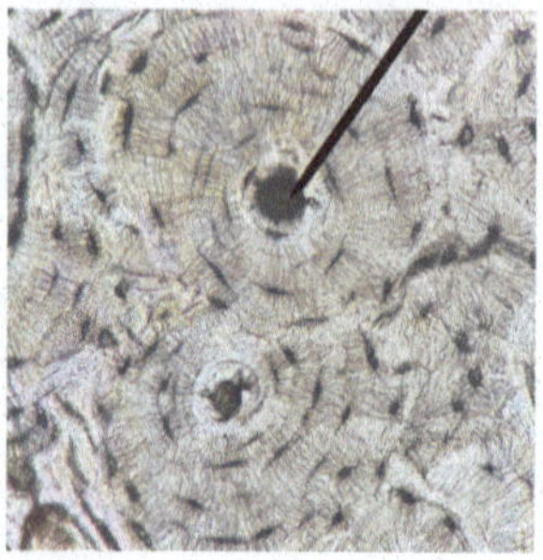		**osteon** **matrix** **concentric lamellae** **lacuna** **osteocyte** **canaliculi** **central canal**

OUTCOME 5: Label the details of an Osteon. Locate each detail on the microscope slide.

- *Before coming to lab to examine the slide of bone tissue, use your* TEXTBOOK *to complete this page. Labeling the parts of an Osteon in the images below will help you locate the detailed structures on the microscope slide.*

TOOLBOX

Anatomy and Histological Organization of Bone

bone (organ)
transverse section—cut surface view
periosteum
compact bone (tissue)
endosteum
single osteon
lamellae (concentric circles of matrix)
central canal
blood vessels within central canal
lacunae (containing osteocytes)
canaliculi ("spidery" little canals that house osteocyte extensions)
single osteocyte
osteocyte extensions

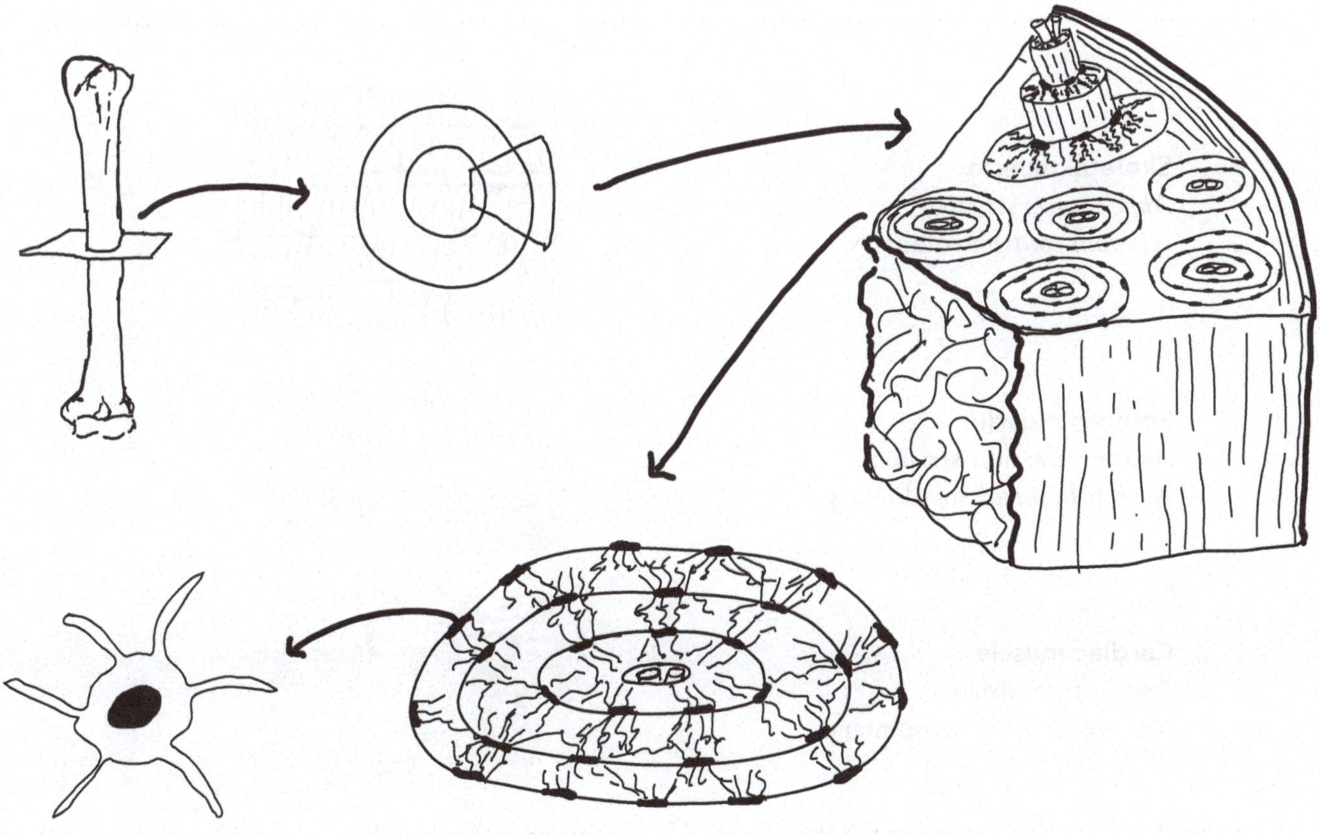

Illustration by Lucia Tranel, © 2015.

H. *Muscle Tissue*

OUTCOME 6: How are muscle tissues classified? Examine different types of muscle tissue under the microscope. Using the tables provided and your textbook, create a labeled drawing noting any unique characteristics of what you see. Provide a specific location within the body and specific function for each type.

Muscle Tissue is classified based upon its subcellular characteristics or **appearance** and its nervous **system** control. The cells, which are string-like and often called **fibers,** may appear striated or non-striated. The tissue may be controlled by the **Somatic Nervous System (voluntary)** or the **Autonomic Nervous System (involuntary).** These criteria can be combined to determine the name of the muscle tissue, as in the table below:

NERVOUS CONTROL	APPEARANCE	
	STRIATED	NON-STRIATED
VOLUNTARY (SOMATIC NERVOUS SYSTEM)	skeletal muscle tissue	NA
INVOLUNTARY (AUTONOMIC NERVOUS SYSTEM)	cardiac muscle tissue	smooth muscle tissue

Skeletal muscle
Appearance: striated
Nervous control: voluntary

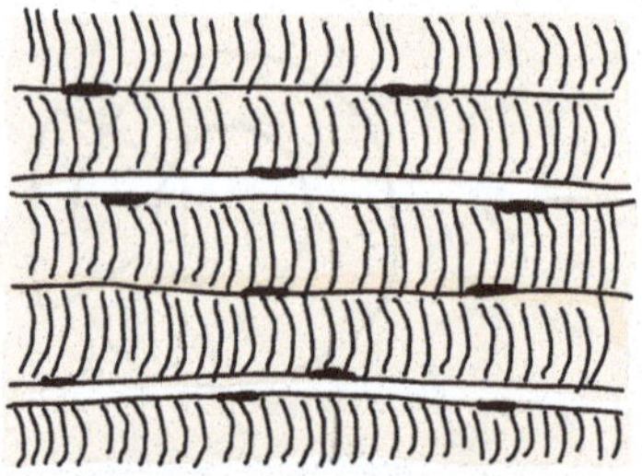

Smooth muscle
Appearance: non-striated
Nervous control: involuntary

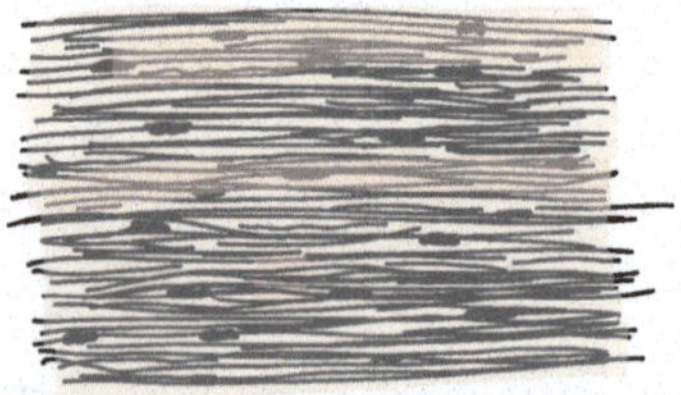

Cardiac muscle
Appearance: striated
Nervous control: involuntary

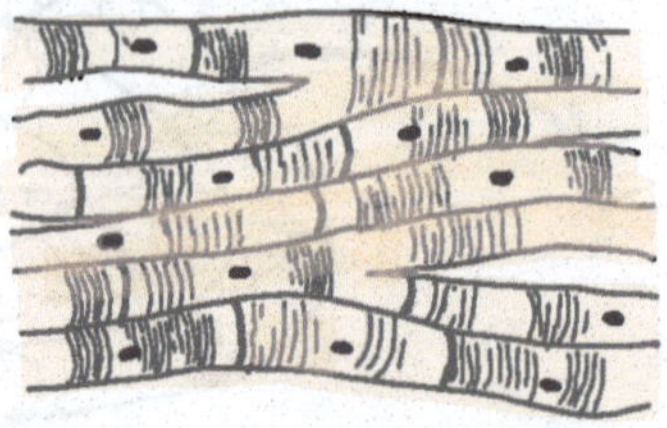

Examine the slides that represent the three types of muscle tissue. Each slide has both longitudinal and cross sections. Make sure that you focus first on the longitudinal section, so that you can observe the elongated cells known as **fibers.**

CHECKLIST: Summary of Muscle Tissues to Be Examined

■ ***Before coming to lab, use your* TEXTBOOK *to help you describe a specific location and function for each muscle tissue you will examine.***

MUSCLE TISSUE	SPECIFIC LOCATION	SPECIFIC FUNCTION
skeletal		
smooth		
cardiac		

The microscopic examination of these muscle tissues will occur in the lab.

1. Skeletal Muscle

skeletal muscle tissue, x.s. and l.s.

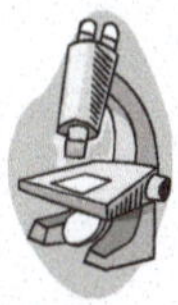

Skeletal muscle tissue is located within the muscle organs attached to the bones of our skeleton. When observed under high magnification, faint stripes or **striations** can be seen running across the length of the fiber. We have **voluntary** or conscious control over the part of the nervous system that stimulates this type of muscle. Hence, the classification for this tissue is **striated and voluntary**. Notice that each cell is **multinucleated,** and the nuclei exist in the perimeter of the cell, just under the **sarcolemma,** (cell membrane).

IMAGE	LABELED DRAWING AND NOTES	DETAILS TO LOCATE
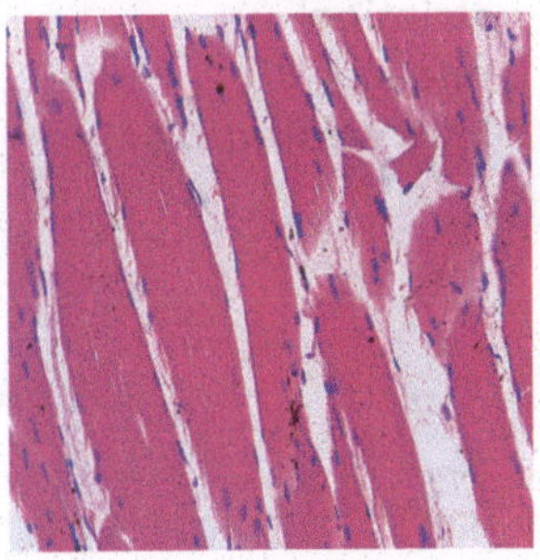		**muscle fiber** **striation** **nucleus**

Illustrations Contributed by Tony Wilmes.

2. Smooth Muscle

smooth muscle tissue

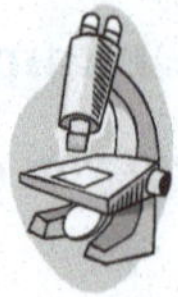

This muscle exists in the walls of hollow organs, such as in the wall of the stomach, the urinary bladder, a blood vessel, and the intestines. It has no striations, and it is controlled by the part of our nervous system that operates automatically. Hence, the classification for this muscle tissue is **nonstriated and involuntary.** Smooth muscle fibers are spindle-shaped with a single nucleus located within the center of the fiber.

IMAGE	LABELED DRAWING AND NOTES	DETAILS TO LOCATE
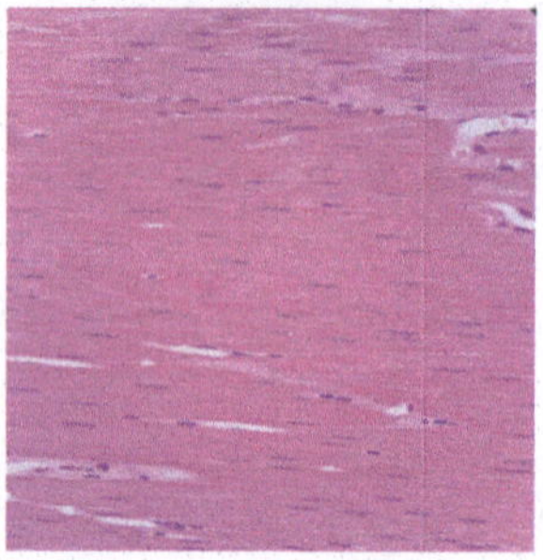		**muscle fiber** **nucleus**

3. Cardiac Muscle

cardiac muscle tissue, intercalated discs

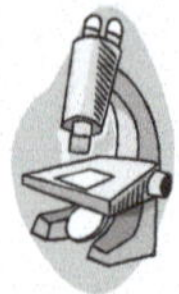

Obviously, cardiac muscle tissue is located in the heart; specifically, in the layer of the heart known as the **myocardium.** It has striations, and it is controlled by the part of our nervous system that operates automatically. Hence, the classification for this muscle tissue is **striated and involuntary.** A distinguishing characteristic of cardiac muscle is the **intercalated disc,** a special intercellular junction allowing the branching cells to be electrically connected to one another. The nucleus of a cardiac muscle fiber is centrally located.

IMAGE	LABELED DRAWING AND NOTES	DETAILS TO LOCATE
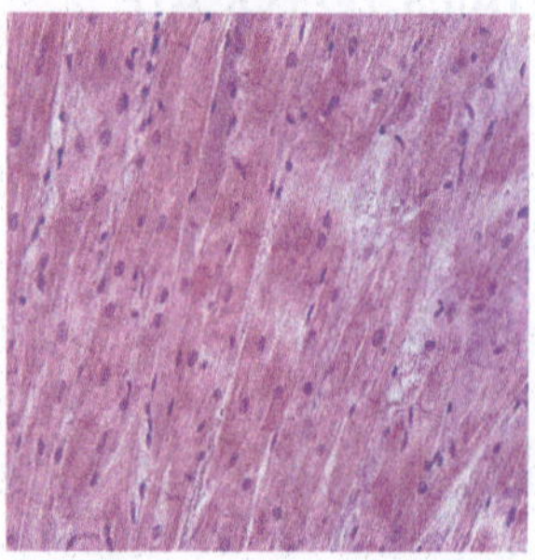		**muscle fiber** **striation** **nucleus** **intercalated disc**

I. *Neural Tissue*

OUTCOME 7: Identify three functional types of neurons and describe their locations and functions. Determine if the neuron is structurally classified as pseudounipolar or multipolar. For each structural type, identify its cell parts and the function of each of those parts. Make sure you understand the difference between functional classification and structural classification.

Neural Tissue is composed of **neurons** and **neuroglia.**

Neurons can receive, process, conduct, and generate an electrical impulse.

Neuroglia provide the environmental support for the neurons. The types of neuroglia will be examined later during the discussion of the Nervous System.

Functional Cassification of Neurons

Neurons form an integral pathway of the homeostatic mechanism of control. There are three main functional types, or **functional classifications,** of neurons; that is, what they DO:

1. **SENSORY NEURONS:** Sensory neurons **perceive** or **sense** changes, or **stimuli,** in the environment of the peripheral nervous system, and then conduct the information from **receptors** via **afferent messages** <u>into</u> the central nervous system.

2. **ASSOCIATION NEURONS:** Association neurons are located <u>within</u> the central nervous system. These neurons **integrate** that information and create **associations.** That information is passed on to the third type of neuron.

3. **MOTOR NEURONS:** Motor neurons conduct the information via **efferent messages** out of the central nervous system to an **effector organ,** usually a muscle or a gland. These effector organs will **respond** by contracting or secreting.

Structural Classification of Neurons

Neurons vary slightly in their structure; that is, what they LOOK like. Each structural type, or structural classification, is named based upon the number of extensions there are coming off of the cell body. Structurally, there are three classifications of neurons:

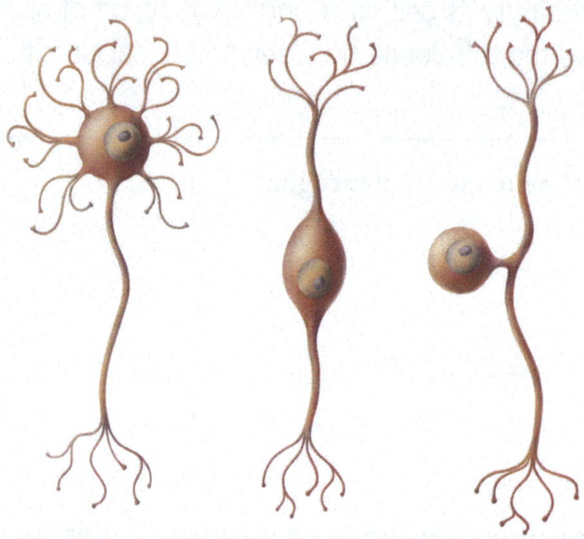

multipolar
many extensions coming off of the cell body

bipolar
Two extensions coming off of the cell body

pseudounipolar
the appearance of one extension coming off of the cell body

Regardless of what they look like, they all have the same three parts. The **dendrite** conducts the impulse from some location to the **cell body** or **perikaryon,** and the **axon** conducts the impulse away from the cell body to another location. The cell body (soma) processes the information.

In this exercise, you will examine pseudounipolar neurons and multipolar neurons. You will examine bipolar neurons when studying the retina of the eye.

- ***Label the parts of each neuron: dendrite, cell body, axon***
- ***Draw an arrow to indicate the direction the electrical impulse travels.***

pseudounipolar
(the appearance of one extension coming off of the cell body)

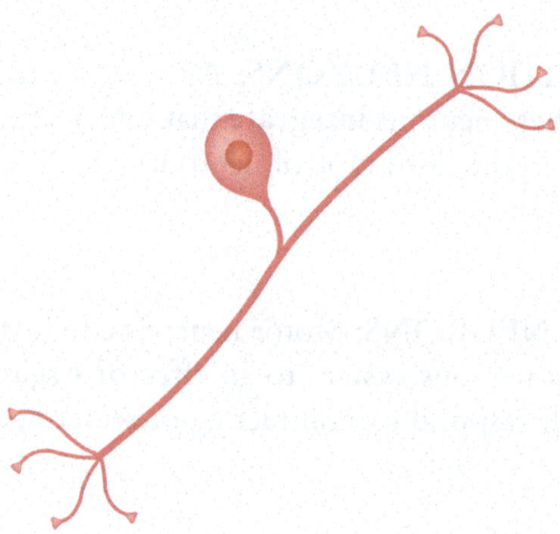

multipolar
(many extensions coming off of the cell body)

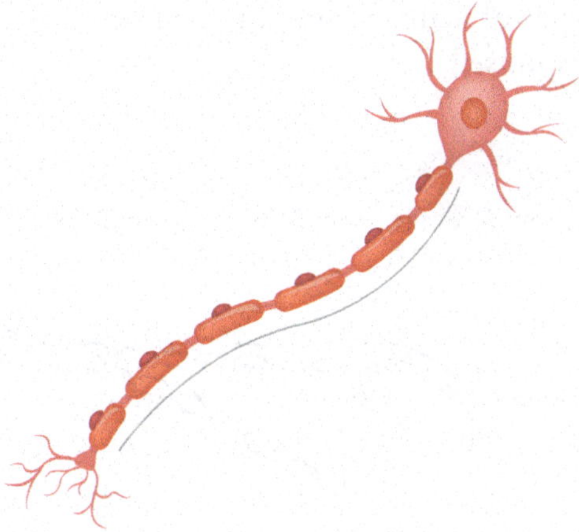

The table matches the types of neurons to their functions and structures. Notice that both Association Neurons and Motor Neurons are Multipolar.

■ ***Fill in the function of each type of neuron.***

TYPE OF NEURON	FUNCTION (What the neurons DO)	STRUCTURE (How the neurons LOOK)
Sensory neuron		pseudounipolar
Association neuron		multipolar
Motor neuron		multipolar

When the three types of neurons are put together, it creates a **reflex arc**, starting with the stimulus and ending with the response.

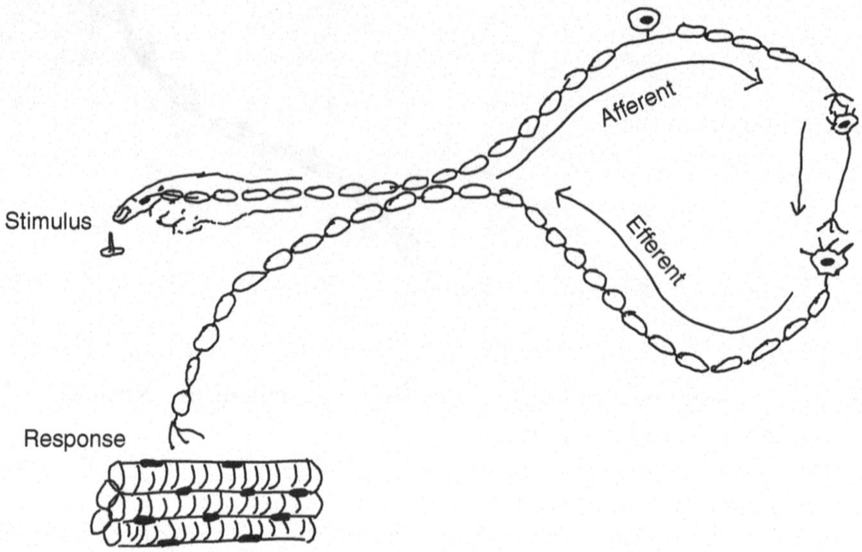

If the reflex arc is placed within the context of the spinal cord:

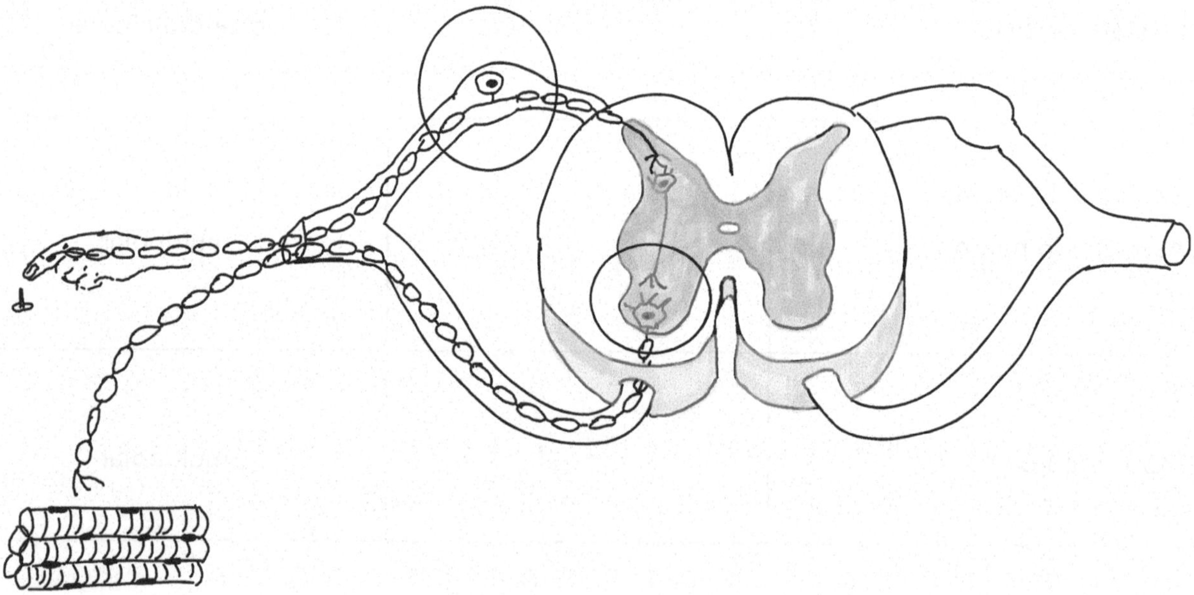

The circled areas on the diagram represent the locations that you will examine under the microscope.

Illustration by Lucia Tranel, © 2015.

You will use the same slide for examining both of these neurons. Before you put the slide on the microscope, take a moment to look at it. Hold it up against a white background for contrast. This is called "grossing the slide." It allows you to see the general anatomy of the section of spinal cord that is on the slide before putting under the microscope.

1. Sensory/Pseudounipolar Neuron (Cell Bodies)

spinal cord (c.s.), mammal 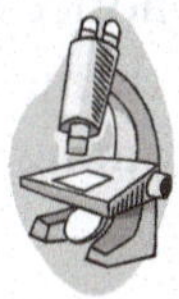

As you examine the slide on low power, find the area that corresponds to the "dorsal root ganglion." You will notice at low and medium magnifications that there are many dark, spherical bodies in this region. They are the cell bodies of the sensory neurons. The "stringy-ness" is due to dendrites or axons coming in or going out of the ganglion. Examine on high magnification as well.

IMAGE	LABELED DRAWING AND NOTES	DETAILS TO LOCATE
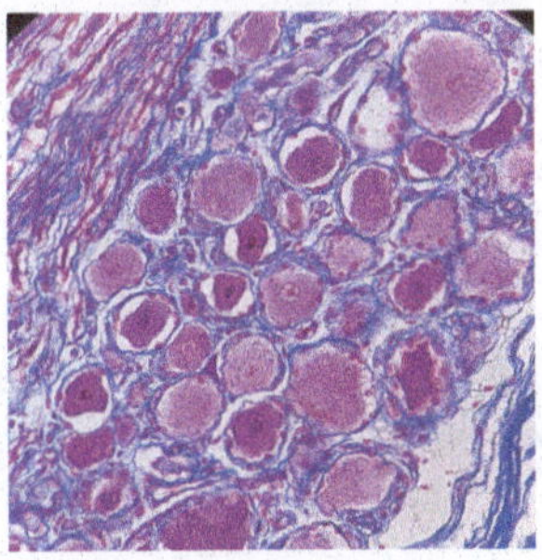		cell bodies nuclei

Illustrations Contributed by Tony Wilmes. © Kendall Hunt Publishing Company.

2. Motor/Multipolar Neuron (Cell Bodies)

spinal cord (c.s.), mammal 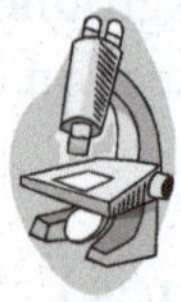

Find the area that corresponds to the "anterior gray horn." You will notice at low and medium magnifications that there are many dark bodies with fibers extending from them. They are the cell bodies of the motor neurons. The "stringy-ness" is due to dendrites or axons coming in or going out of this area. Examine them on high magnification as well.

IMAGE	LABELED DRAWING AND NOTES	DETAILS TO LOCATE
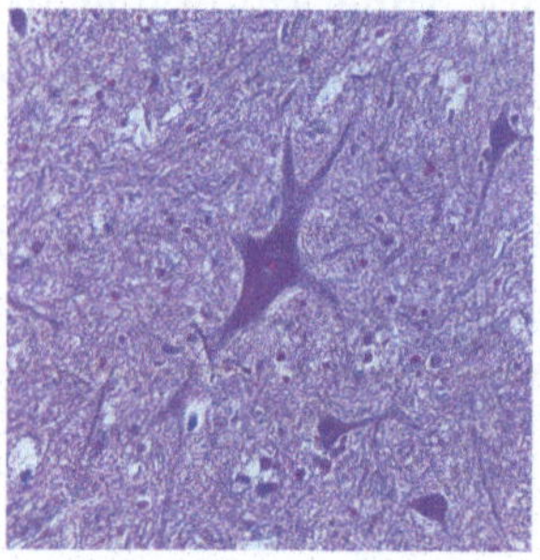		**cell bodies** **nuclei**

D. *Putting Things Away*

1. Remove slide from microscope. Place correctly in slide box.
2. Rotate nosepiece of microscope until the low power objective lens is in place.
3. Wind cord neatly.
4. Lower the stage of the microscope to its lowest level.
5. Return microscope to its designated location.

EXERCISE 3

Integumentary System

A. Introduction

We know that an organ system is made up of separate organs that work together in such a way as to perform a unique function. The Integumentary System is distinctive among other organ systems because all of its organs are located *within* its largest organ, **the skin**. Keep in mind the hierarchical organization of living things as you examine the skin and other organs. Remember that the **Integumentary System** consists of **organs** composed of **tissues**. The Integumentary System is an excellent organ system to study after having studied tissues, because within the skin one can find examples of all the histological categories discussed in Exercise 2.

Since the Integumentary System covers the surface of our body, it is in a position to be the first line of defense and provides protection in a variety of ways. Waterproofing, cushioning against trauma or abrasion, excreting wastes, controlling body temperature, detecting sensations such as touch, pressure, heat or pain, and aiding in the synthesis of vitamin D are all functions of the Integumentary System.

It might be helpful to view an organizational layout of the Integumentary System:

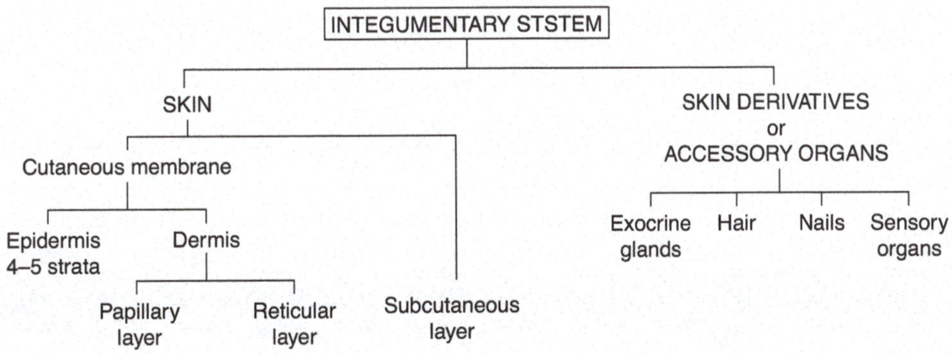

* Some authorities do not include the subcutaneous layer as part of the skin, but it is important for acting as a shock absorber, an insulator, and for attaching and stabilizing the cutaneous membrane to the deep fascia. For this reason, it is inserted in the organizational chart as one of the three major layers of the skin: epidermis, dermis, subcutaneous (sometimes referred to as "hypodermis").

B. *Outcomes for Exercise 3*

1. Identify the major layers, tissues, and structures of the skin.
2. Identify the details of the epidermis and dermis.
3. Identify three types of exocrine glands in the skin. Where are they located? What are their functions?
4. Identify two types of sensory receptors in the skin. Where are they located? What are their functions?
5. Examine THIN (AXILLARY) SKIN under the microscope. Identify the layers, tissues, and structures listed. Draw and label what you see.
6. Examine THICK (CORNIFIED / KERATINIZED SKIN) under the microscope. Identify the layers, tissues, and structures listed. Draw and label what you see.
7. Examine SKIN OF THE SCALP under the microscope. Identify the layers, tissues, and structures listed. Draw and label what you see.

C. *General Anatomy of the Skin*

OUTCOME 1: Identify the major layers, tissues, and structures of the skin.

TOOLBOX

Components of the Integumentary System

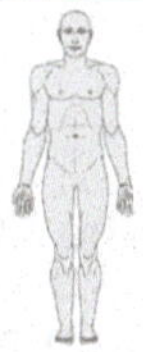

skin model

- epidermis (stratified squamous epithelium)
- dermis (irregular C.T.)
  - papillary layer (loose)
  - reticular layer (dense)
- subcutaneous or hypodermis layer (adipose tissue)
- hair shaft
- hair follicle
- sebaceous gland
- arrector pili muscle
- eccrine sweat gland
- sweat gland duct
- sweat gland pore
- apocrine sweat gland
- tactile (Meissner's) corpuscle
- lamellated (Pacinian) corpuscle
- artery
- vein
- arteriole
- venule
- capillary loop

D. *Details of the Epidermis and the Dermis*

OUTCOME 2: Identify the details of the epidermis and dermis.

TOOLBOX

The Structure and Layers of the Epidermis
Thin and Thick Skin

skin model

epidermis (stratified squamous epithelium)
- stratum corneum
- stratum lucidum
- stratum granulosum
- stratum spinosum
- stratum basale (germinativum)

dermis (irregular C.T.)
- papillary layer (loose irregular C.T.)
- dermal papilla
- capillary loop
- tactile (Meissner's) corpuscle
- reticular layer (dense irregular C.T.)

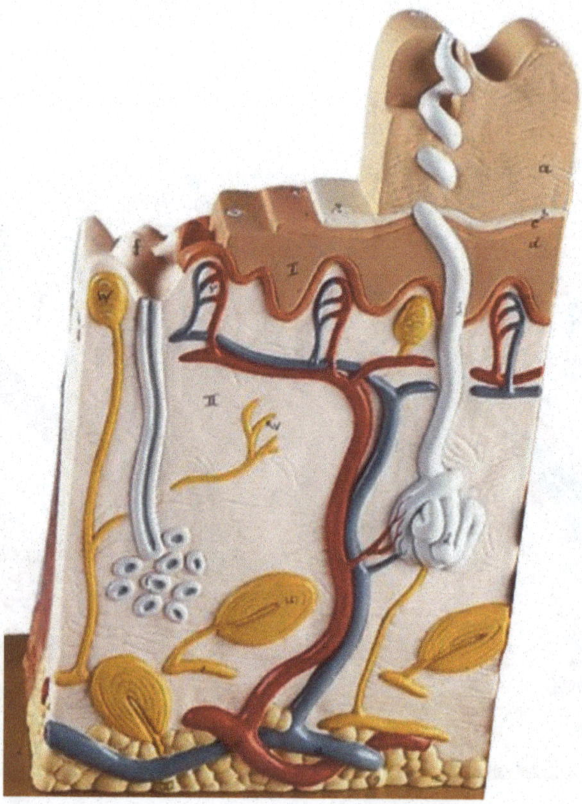

E. *Appendages of the Skin*

OUTCOME 3: Identify three types of exocrine glands in the skin. What is their method of secretion? Where are they located? What are their functions?

TOOLBOX

Classification of Exocrine Glands in the Skin

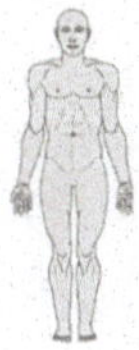

skin model

EXOCRINE GLAND	METHOD OF SECRETION	HISTOLOGICAL IMAGE	LOCATION	SPECIFIC FUNCTION
eccrine sweat gland	merocrine	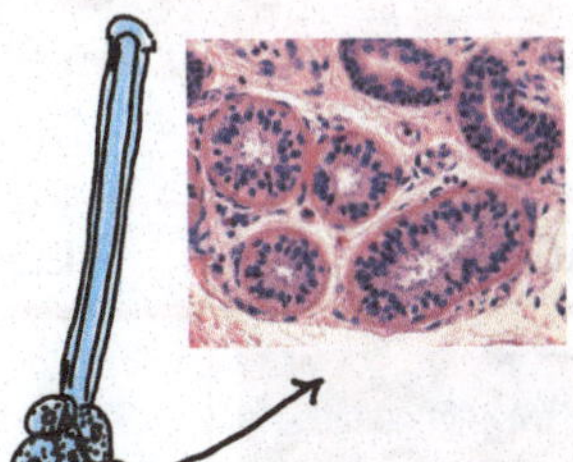 Stratified cuboidal epithelium		
apocrine sweat gland	apocrine	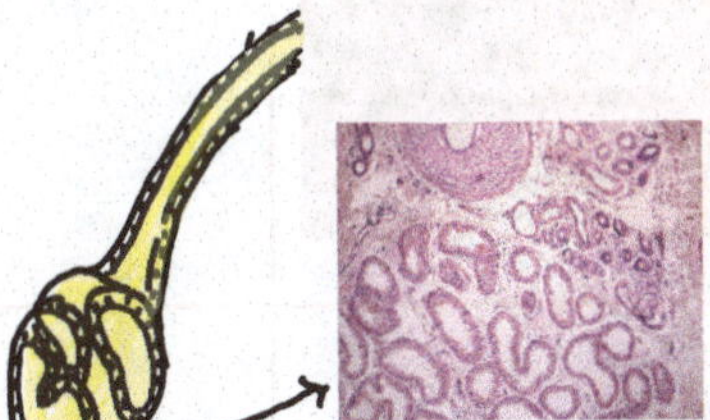 Simple cuboidal epithelium		
sebaceous gland	holocrine	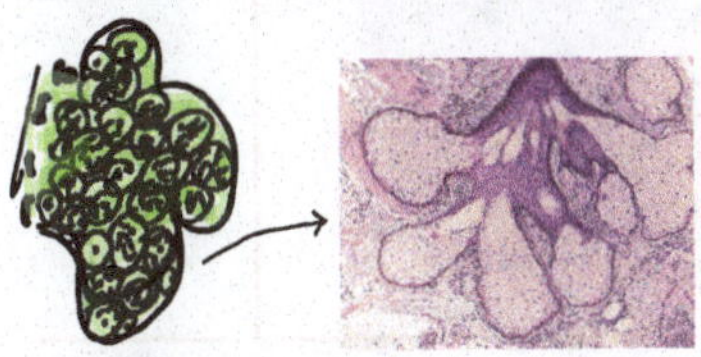 Sebaceous cells		

OUTCOME 4: Identify two types of sensory receptors in the skin. Where are they located? What are their functions?

TOOLBOX

Components of the Integumentary System

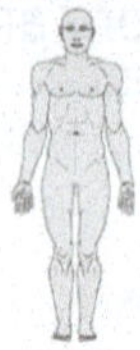

skin model

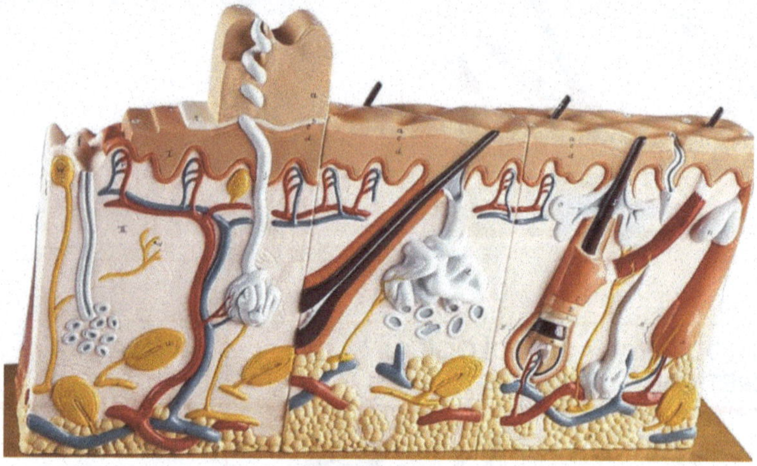

SENSORY RECEPTOR	IMAGE	LOCATION	FUNCTION
Tactile corpuscle (Meissner's)	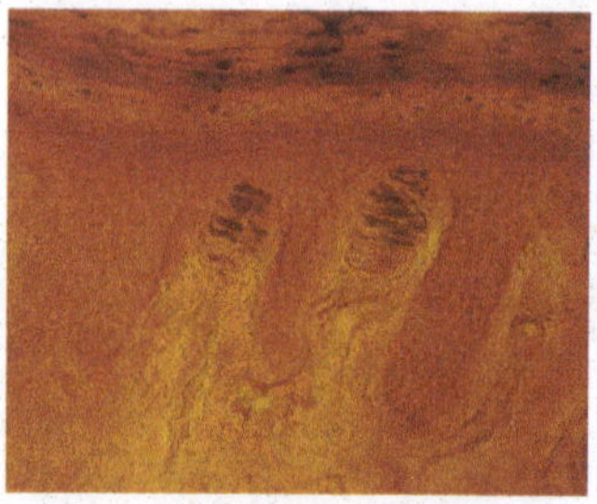		
Lamellated corpuscle (Pacinian)	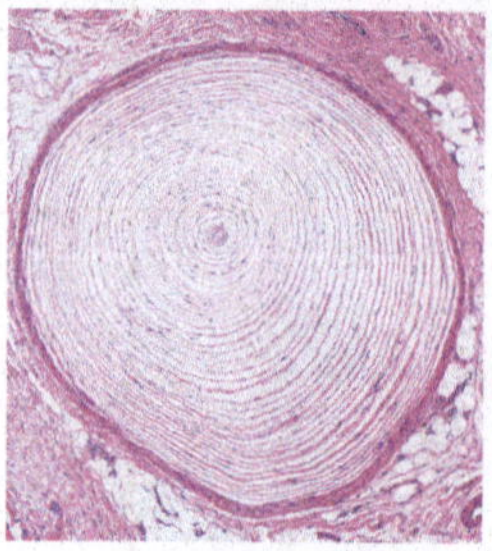		

F. *Histological Examination*

OUTCOME 5: Examine THIN (AXILLARY) SKIN under the microscope. Identify the layers, tissues, and structures listed below. Draw and label what you see.

TOOLBOX

axillary skin

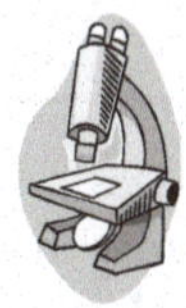

LOCATE	DRAW AND LABEL WHAT YOU SEE
epidermis (stratified squamous epi.) stratum corneum stratum granulosum stratum spinosum stratum basale	
dermis (irregular C.T.) papillary layer dermal papilla reticular layer eccrine sweat gland apocrine sweat gland	
subcutaneous layer (adipose and irregular C.T.)	

OUTCOME 6: Examine THICK (CORNIFIED / KERATINIZED) SKIN under the microscope. Identify the layers, tissues, and structures listed below. Draw and label what you see.

TOOLBOX

cornified / keratinized skin

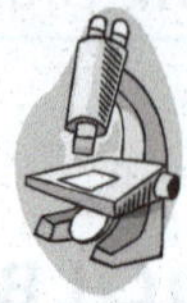

Locate the structures listed. You may have to turn the slide upside down on the microscope so that the image will appear right side up in the view.

LOCATE	DRAW AND LABEL WHAT YOU SEE
epidermis (stratified squamous epi.) stratum corneum stratum lucidum stratum granulosum stratum spinosum stratum basale	
dermis (irregular C.T.) papillary layer dermal papilla Meissner's corpuscle reticular layer Pacinian corpuscle (deep dermis)	
hypodermis (adipose and irregular C.T.)	

OUTCOME 7: Examine SKIN OF THE SCALP under the microscope. Identify the layers, tissues, and structures listed below. Draw and label what you see.

TOOLBOX

scalp

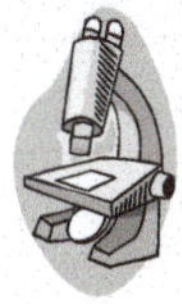

Locate the structures listed.

LOCATE	DRAW AND LABEL WHAT YOU SEE
epidermis	
dermis hair follicle sebaceous gland	
subcutaneous layer	

CHECKLIST: Summary Checklist Integument

TERM/S TRUCTURE	YOUR NOTES
Cornified (Thick) Skin	
Axillary (Thin) Skin	
Scalp (Thin) Skin	
Stratum Corneum	
Stratum Lucidum	
Stratum Granulosum	
Stratum Spinosum	
Stratum Basale	
Papillary Dermis	
Reticular Dermis	
Dermis	
Epidermis	
Hypodermis (Subcutaneous Layer)	
Dermal Papilla	
Loose Irregular Connective Tissue	
Tactile/Meissner's Corpuscle	
Lamellated/Pacinian Corpuscle	
Eccrine Sweat Gland	
Apocrine Sweat Gland	
Sebaceous Gland	
Arrector Pili Muscle	
Free Nerve Endings	
Dense Irregular Connective Tissue	
Adipose Tissue	
Stratified Squamous Epithelium	

EXERCISE 4

Skeletal System

A. Introduction

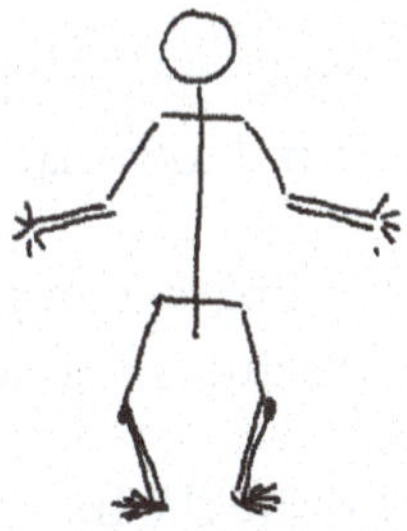

There are two divisions in the skeleton: **axial** and **appendicular.** The stick figure to the right illustrates a rough way to think about it. The lines that make up the head, neck, and spine represent the axial skeleton. The lines of the shoulders, arms, hands, hips, legs, and feet represent the appendicular skeleton.

You will examine the skeletal system in both its **articulated (jointed)** and **disarticulated (disjointed)** form. You will learn the names of all the bones comprising the articulated skeleton, and study some of the bone markings that represent locations for blood vessels, nerves, and muscle attachments. To prepare for this, you must learn some basic nomenclature, or bone marking terminology, some of which are presented in the table below. Locate a figure in your textbook that illustrates bone markings and surface features.

GENERAL DESCRIPTION	TERM	DEFINITION
"Outies"	process	any projection or structure that sticks out
	ramus	an angle extension, an arm (ram = arm)
	condyle	a smooth, rounded process that articulates with another bone
	epicondyle	pointed process or bulging knob located superior to a condyle
	styloid	thin, pointed process
	head	enlarged articular end or epiphysis
	neck	narrow connection between the epiphysis and diaphysis
	malleolus	thick, rounded process (little mallet)
"Innies"	fossa	a shallow depression
chin	foramen	a hole, usually for a blood vessel or nerve to pass thru
neck	canal	a long, narrow passage way through the bone
chest	meatus	same as canal
	sinus	a chamber or hollow space within a bone

B. Outcomes for Exercise 4

1. Identify the parts of a typical long bone and practice drawing the anatomy.
2. Examine the different types of bone under the microscope. For each type, make a labeled drawing of what you see. Make note of any unique characteristics. Provide a specific location for each type.
3. Identify the bones of the AXIAL and APPENDICULAR divisions of the articulated skeleton.
4. Identify the details of the skull, external views. Some bones are visible in multiple views, and you are responsible for knowing the structure in all views.
5. Identify the details of the skull, sectional view.
6. Locate the foramina listed in the table. Describe their locations; name the structures that pass through them.
7. Locate the paranasal sinuses and the hyoid bone.
8. Name the five different regions of the vertebral column. Locate the regions on the articulated column. How many bones in each region? What is their numerical designation?
9. Compare a single vertebra from each of the cervical, thoracic, and lumbar regions. Identify the structures that are common to each. Determine how you will distinguish one from another.
10. Locate the first two cervical vertebrae: the ATLAS (C1) and the AXIS (C2). Identify them by name and numerical designation. Describe their articulation with the occipital condyles and with each other.
11. Identify the sacrum and coccyx.
12. Identify the parts of the thoracic cage. Locate the parts of the sternum. Determine which ribs are true, false, vertebrochondral, or floating.
13. Identify the details of the pectoral girdle.
14. Identify the details of the bones of the upper appendages.
15. Identify the details of the bones of the pelvic girdle.
16. Identify the details of the bones of the lower appendages.

C. *General Anatomy of a Long Bone*

OUTCOME 1: Identify the parts of a typical long bone, and practice drawing the anatomy.

TOOLBOX

The Internal Organization in Representative Bones
Anatomy of a Representative Bone

real femur

proximal epiphysis
distal epiphysis
metaphysis
diaphysis
periosteum
medullary cavity
spongy bone
trabeculae of spongy bone
compact bone
location of red bone marrow
location of yellow bone marrow
endosteum
epiphyseal plate (young bone)/
line (mature bone)

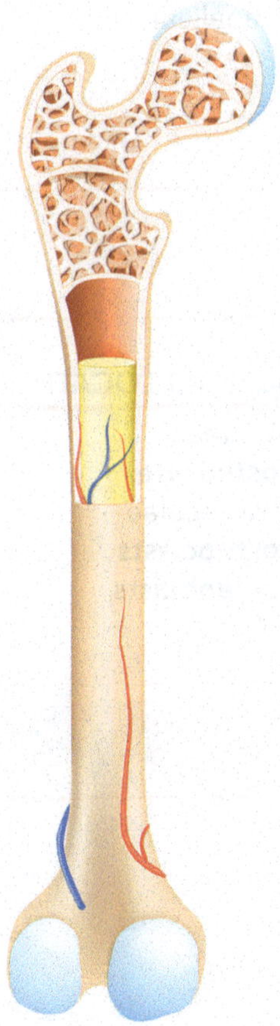

© Designua/shutterstock.com

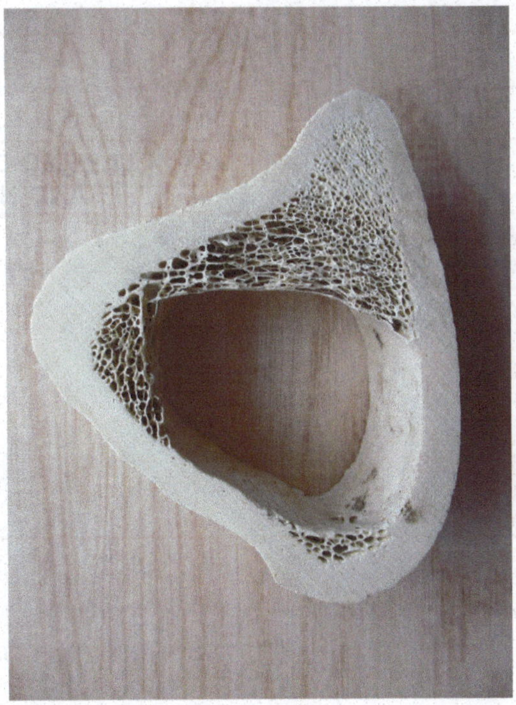

© Denise Cogliando/shutterstock.com

D. *Bone Histology*

OUTCOME 2: Examine the different types of bone under the microscope. For each type, make a labeled drawing of what you see. Make note of any unique characteristics. Provide a specific location for each type.

TOOLBOX

Histological Organization of Mature Bone

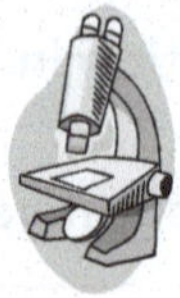

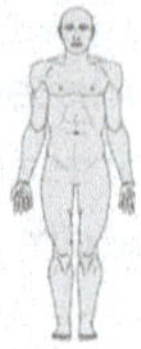

Mr. Bones

Compact Bone

IMAGE	LABELED DRAWING AND NOTES	DETAILS TO LOCATE
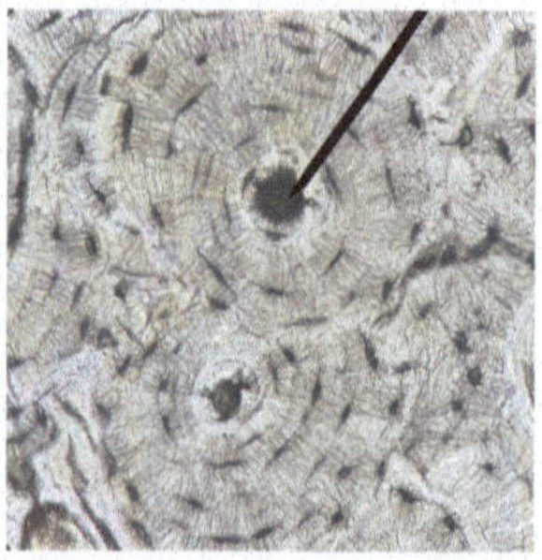		osteon central canal concentric lamellae osteocyte lacuna canaliculi

Spongy Bone

IMAGE	LABELED DRAWING AND NOTES	DETAILS TO LOCATE
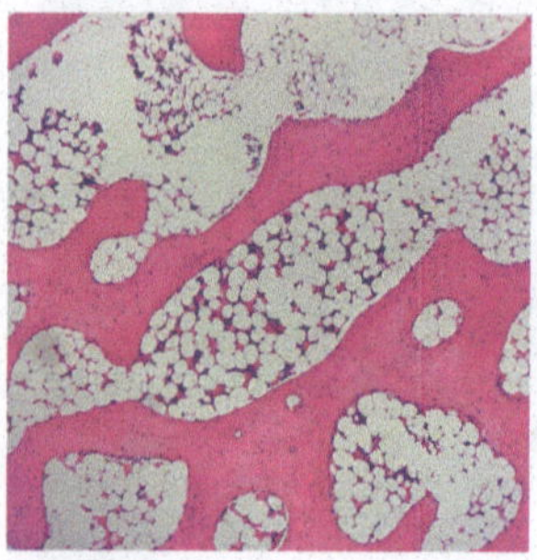		lacuna osteocyte trabeculae osteoclasts osteoblasts

E. *Divisions of the Articulated Skeleton*

OUTCOME 3: Identify the bones of the AXIAL and APPENDICULAR divisions of the articulated skeleton.

TOOLBOX

The Axial Skeleton

The Appendicular Skeleton

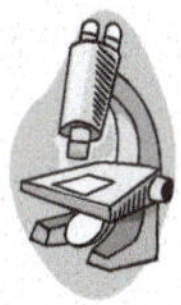

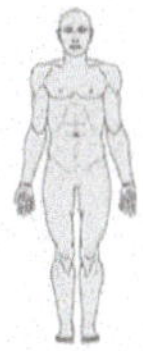

Mr. Bones

■ *Color the bones of the axial skeleton YELLOW and the bones of the appendicular skeleton GREEN.*

■ *Identify the bones listed on the diagram.*

skull

sternum

ribs

vertebrae

sacrum

coccyx

scapula

clavicle

humerus

radius

ulna

carpal bones

metacarpal bones

phalanges

os coxae
  ilium
  ischium
  pubis

femur

patella

tibia

fibula

tarsal bones

metatarsal bones

phalanges

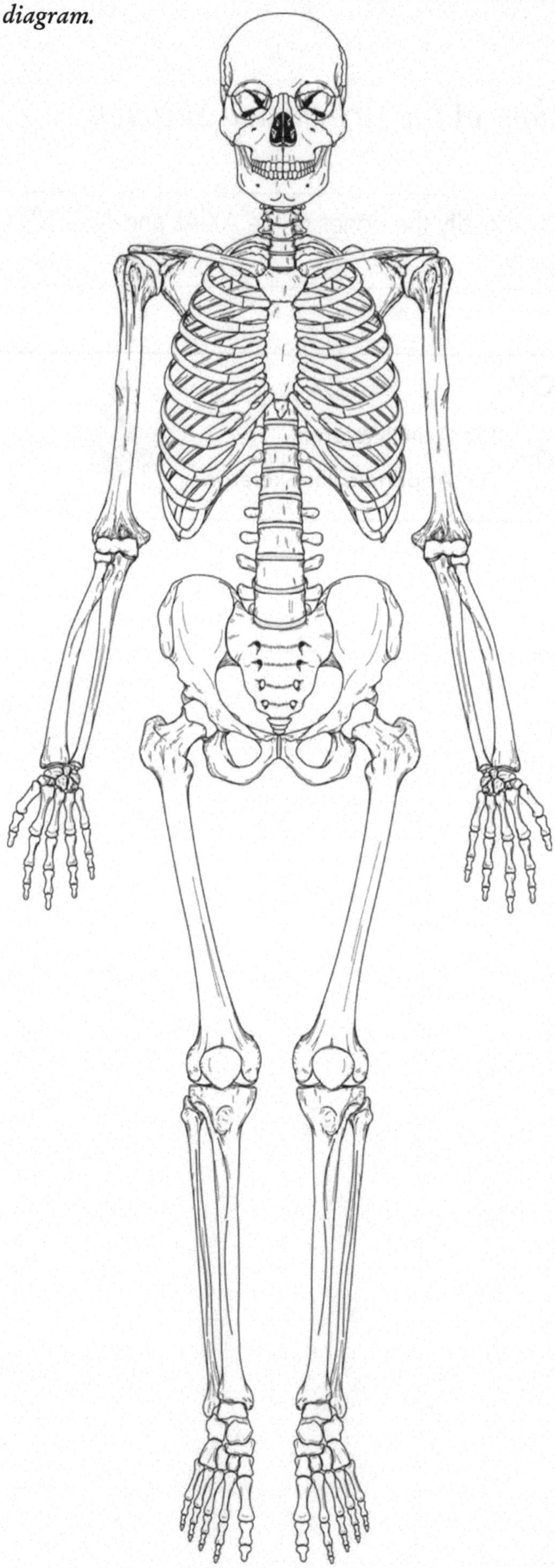

Illustration by Jamey Garbett. © 2003 Mark Nielsen.

■ *Color the bones of the axial skeleton YELLOW and the bones of the appendicular skeleton GREEN.*

■ *Identify the bones listed on the diagram.*

skull

ribs

vertebrae

sacrum

coccyx

scapula

clavicle

humerus

radius

ulna

carpal bones

metacarpal bones

phalanges

os coxae

ilium

ischium

pubis

femur

tibia

patella

tibia

fibula

tarsal bones

metatarsal bones

phalanges

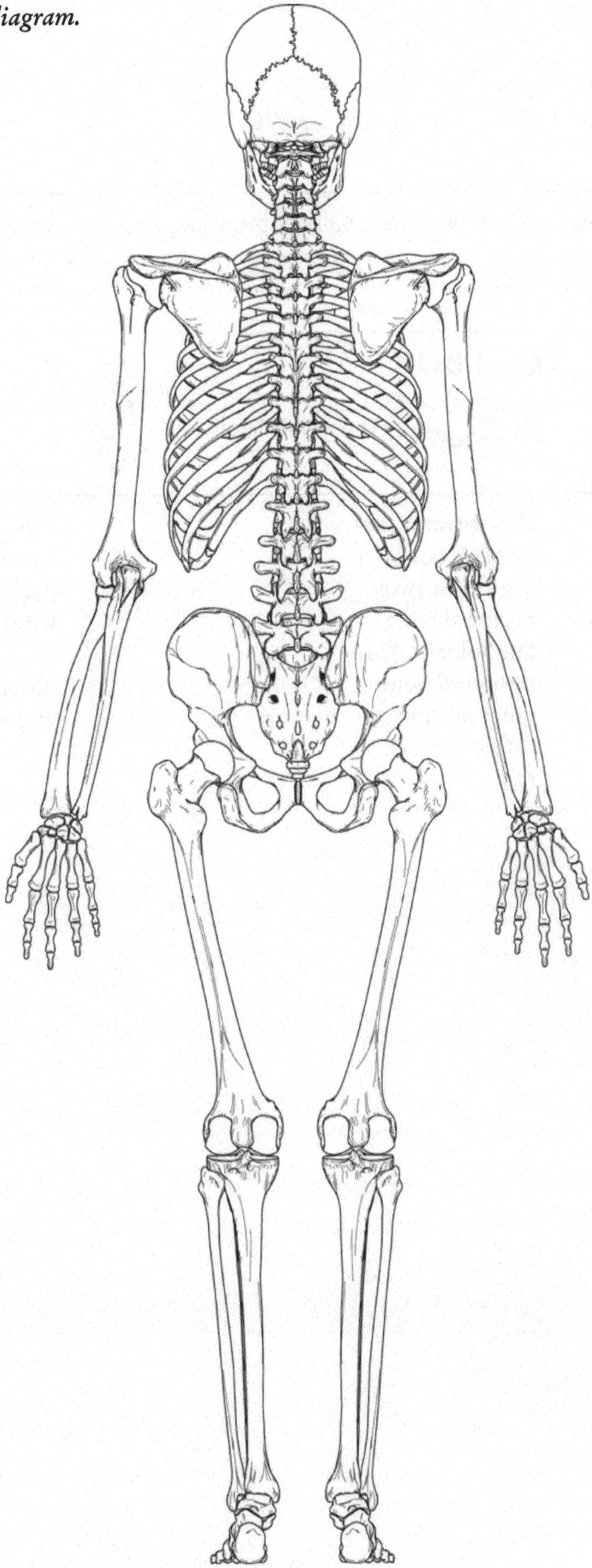

F. *Details of the Axial Skeleton*

OUTCOME 4: Identify the details of the skull, external views. Some bones are visible in multiple views, and you are responsible for knowing the structure in all views.

TOOLBOX

The Adult Skull

skull

Posterior View
parietal bones
sagittal suture
occipital bone
lamboidal suture
temporal bone
  mastoid process
  styloid process
mandible

Superior View
occipital bone
parietal bones
frontal bone
coronal/frontal suture
nasal bones
zygomatic bone

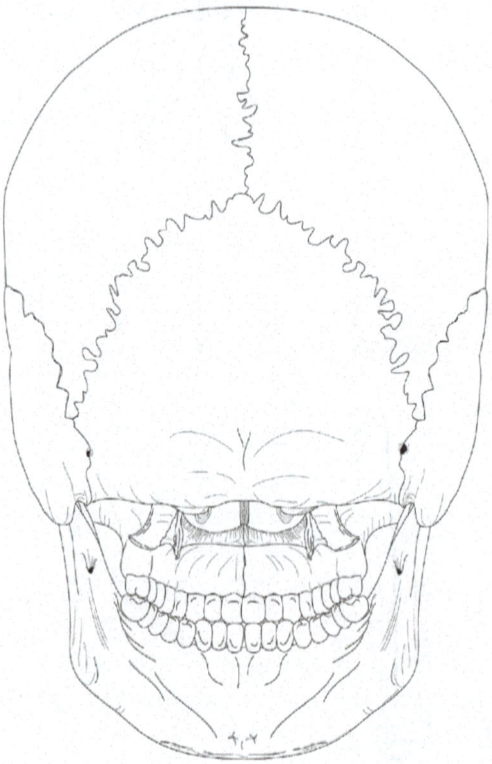

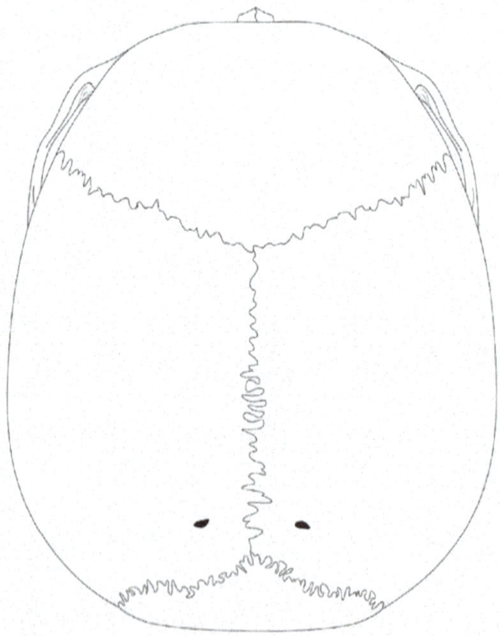

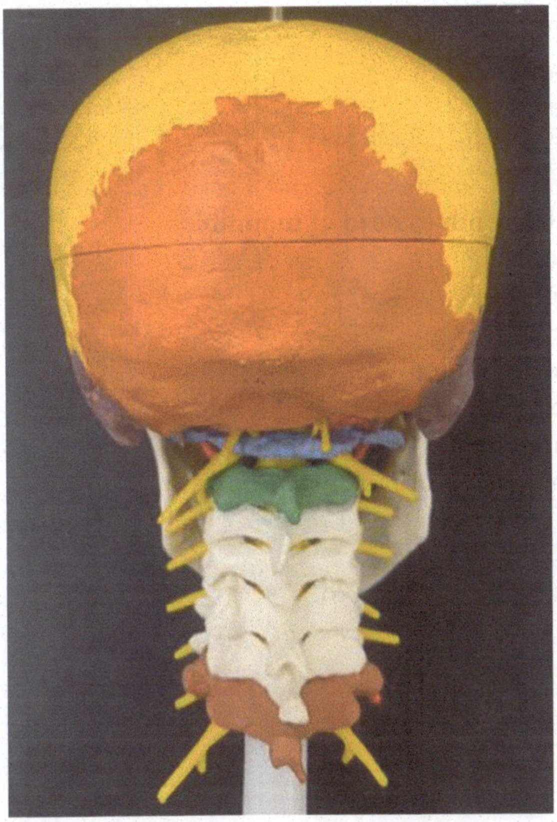

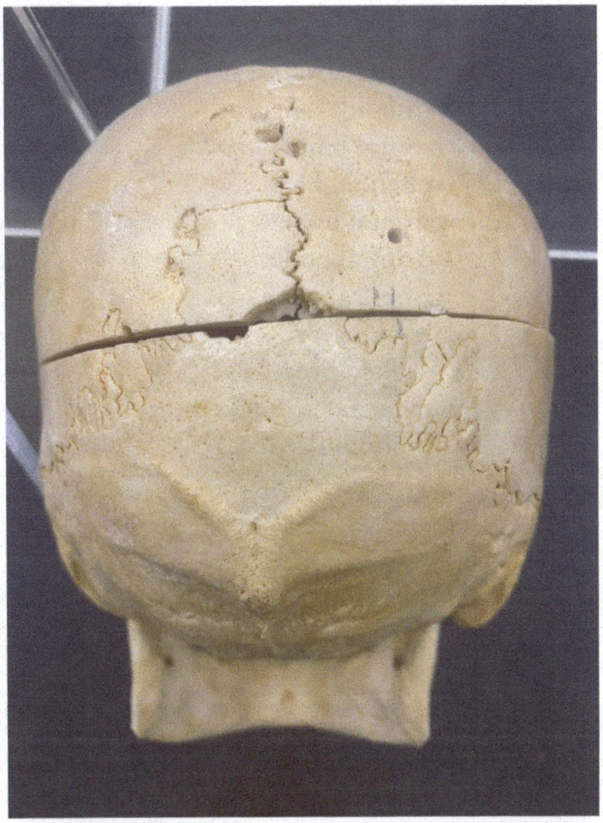

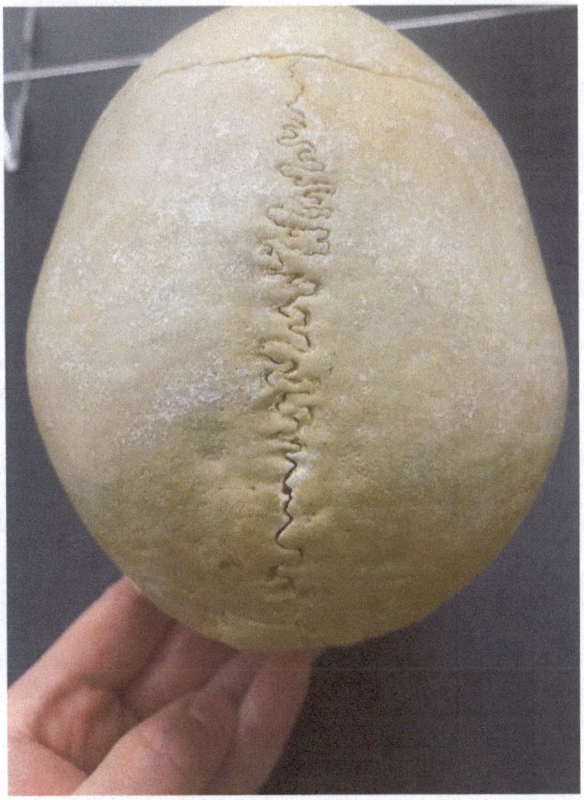

Source: Ben Jellen, 2021.

Lateral View

coronal/frontal suture
lamboidal suture
frontal bone
sphenoid bone
parietal bone
occipital bone
temporal bone
  mastoid process
  styloid process
  external acoustic meatus
  mandibular fossa
ethmoid bone
nasal bone
maxilla
mandible
  mandibular condyle/head of mandible
lacrimal bone
  lacrimal groove
zygomatic bone
temporal mandibular joint (TMJ)
zygomatic arch:
  zygomatic process of temporal bone
  temporal process of zygomatic bone

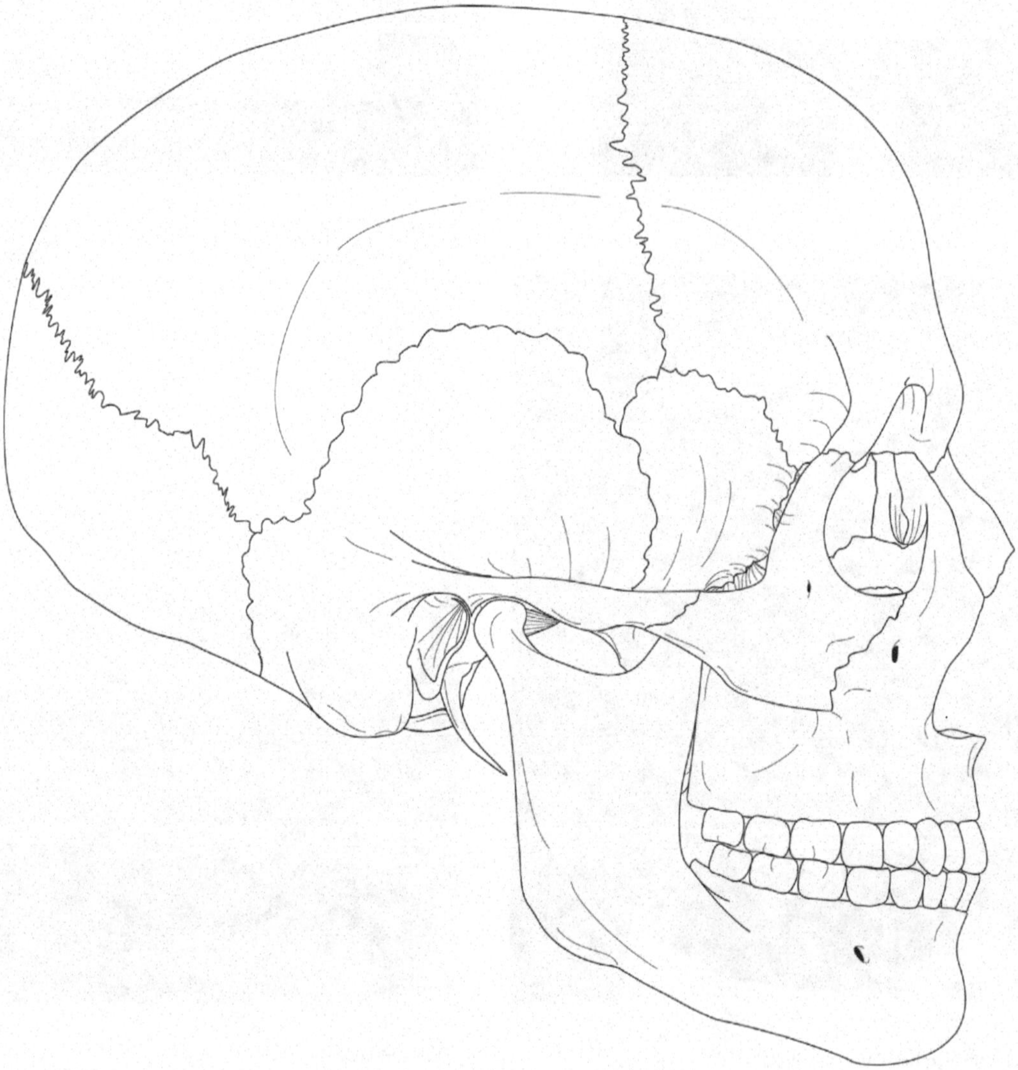

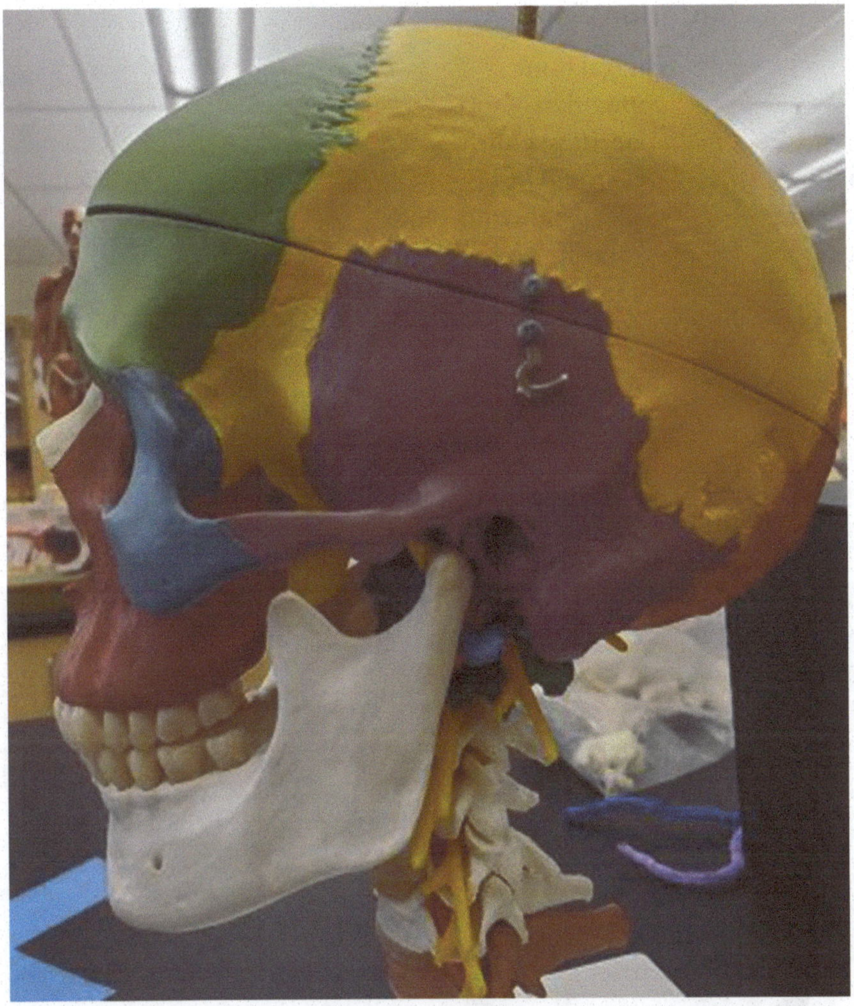

A20/2 © 3B Scientific GmbH, Germany, 2021 www.3bscientific.com

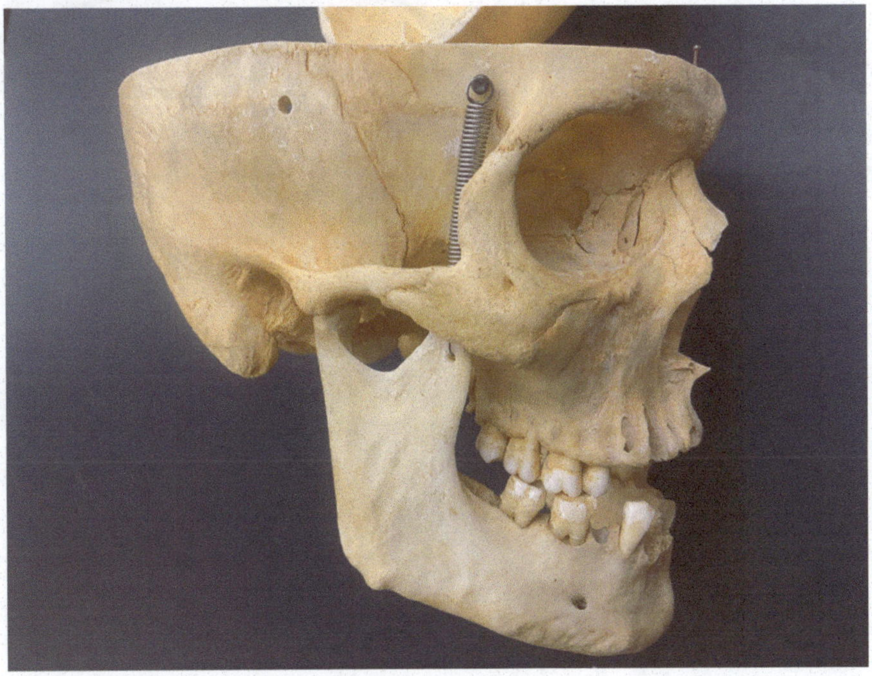

Source: Ben Jellen, 2021.

Anterior View

parietal bone
frontal bone
sphenoid bone
  inferior orbital fissure
  superior orbital fissure
  optic canal
temporal bone
zygomatic bone
nasal bone
maxilla
lacrimal bone
ethmoid bone
  perpendicular plate
palatine bone (small surface in orbit)
vomer

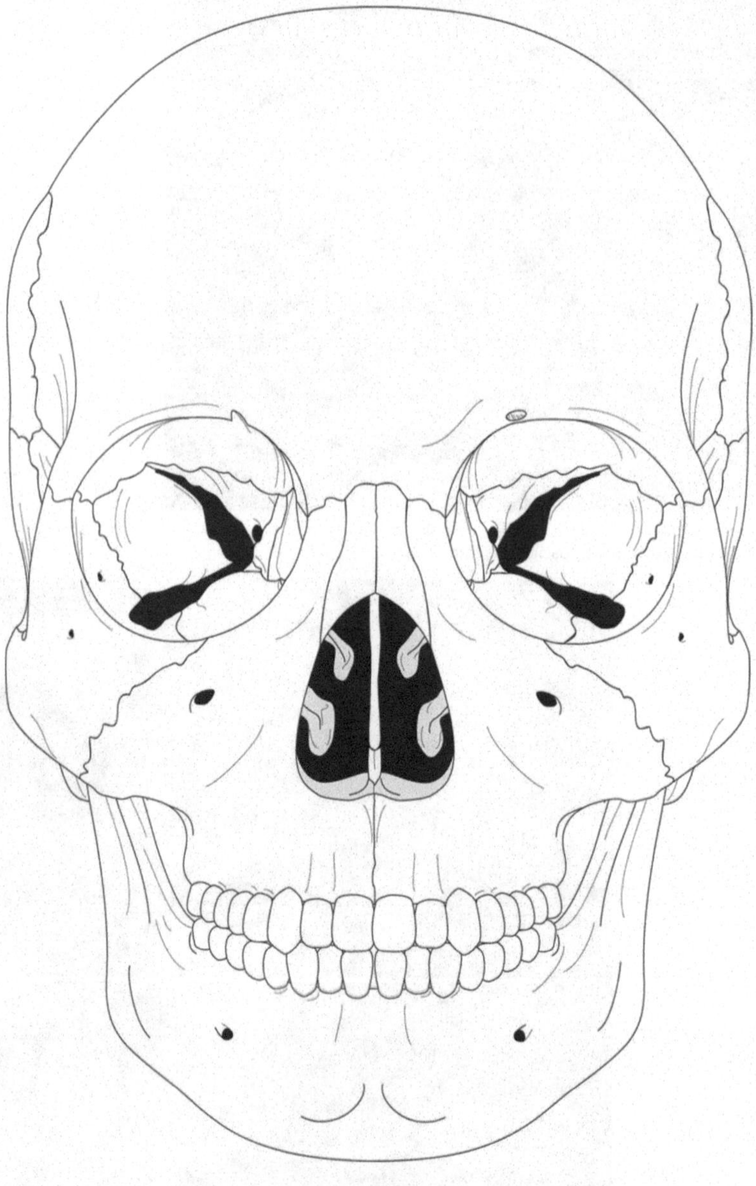

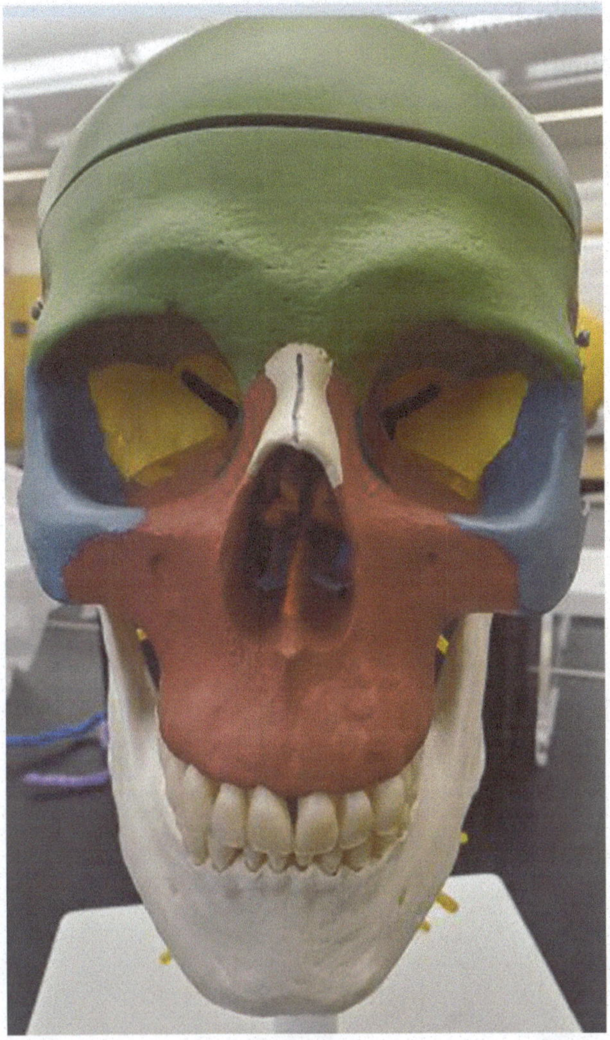

A20/2 © 3B Scientific GmbH, Germany, 2021 www.3bscientific.com

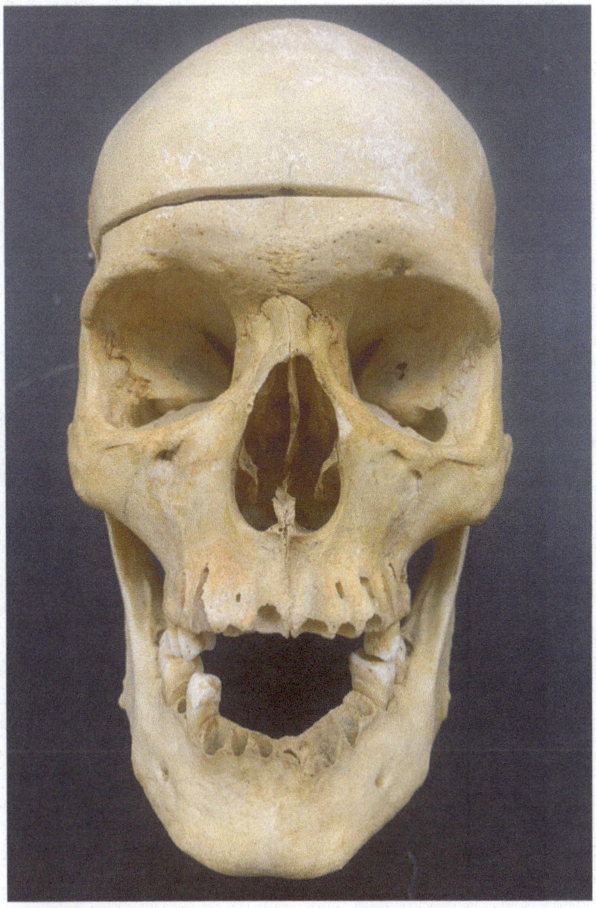

Source: Ben Jellen, 2021.

Inferior View, Mandible Removed

vomer
sphenoid
temporal bone
  styloid process
  mastoid process
  mandibular fossa
  external acoustic meatus
  carotid canal
  zygomatic process
jugular foramen
occipital bone
  foramen magnum
  occipital condyle
  hypoglossal canal (indicated by ▙====▟ red pipe cleaner)
zygomatic bone
zygomatic arch
  temporal process + zygomatic process
palatine bone
maxilla

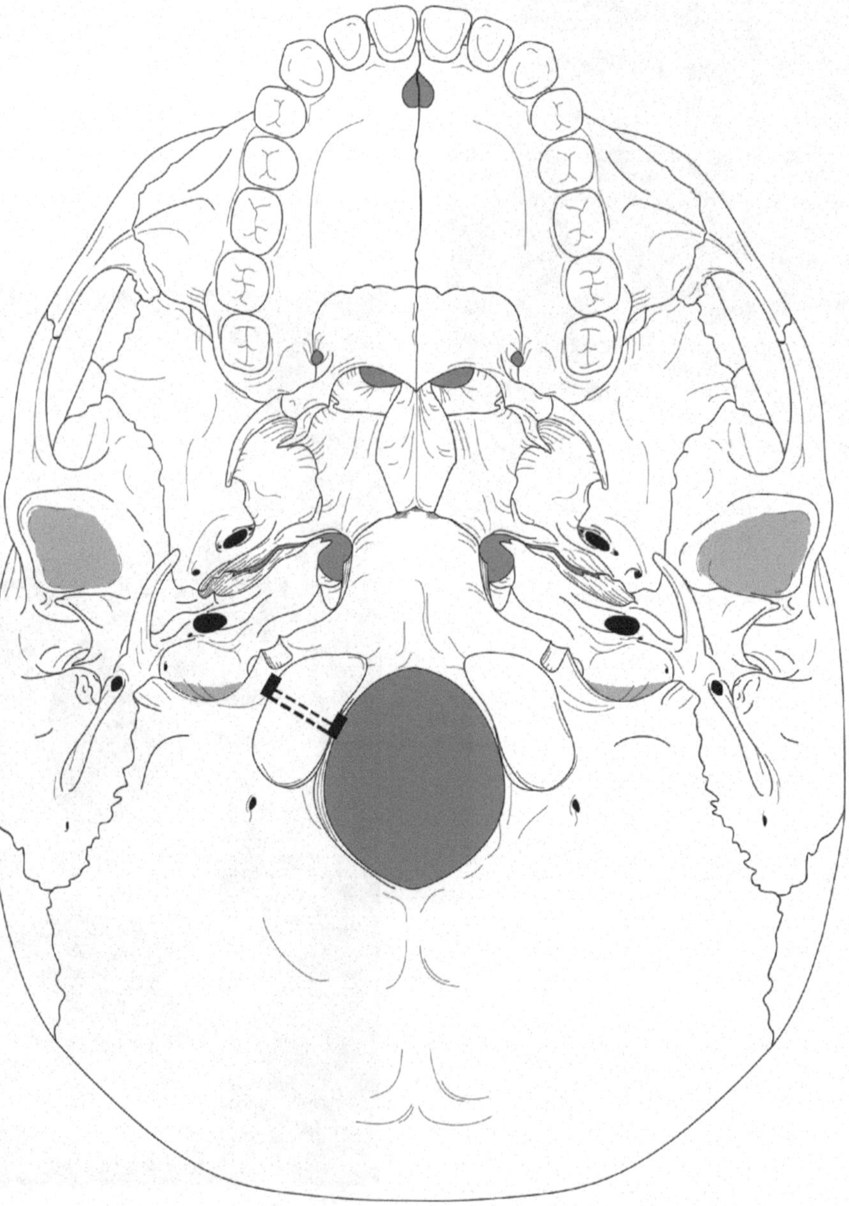

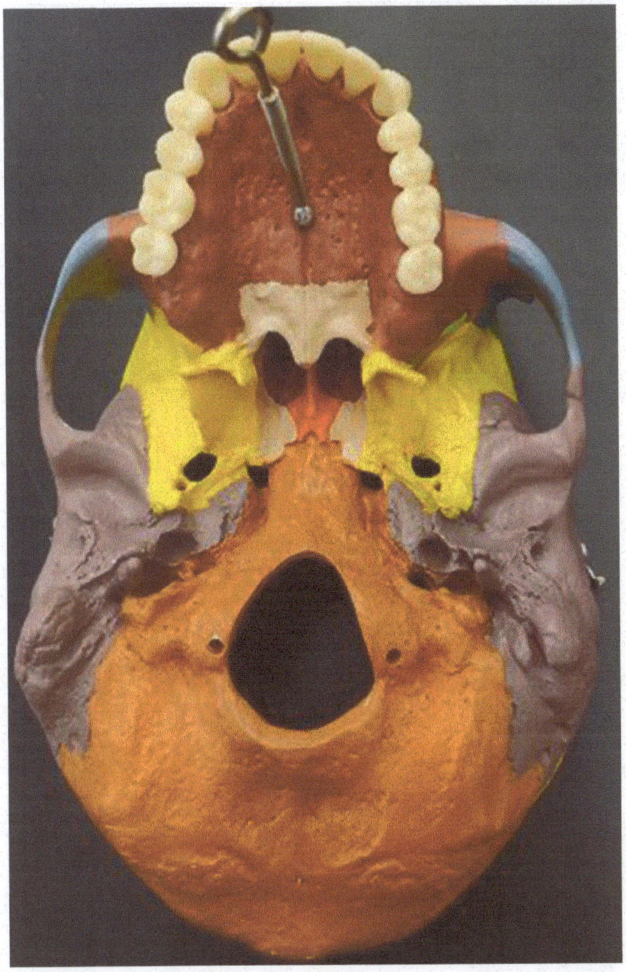

A20/2 © 3B Scientific GmbH, Germany, 2021 www.3bscientific.com

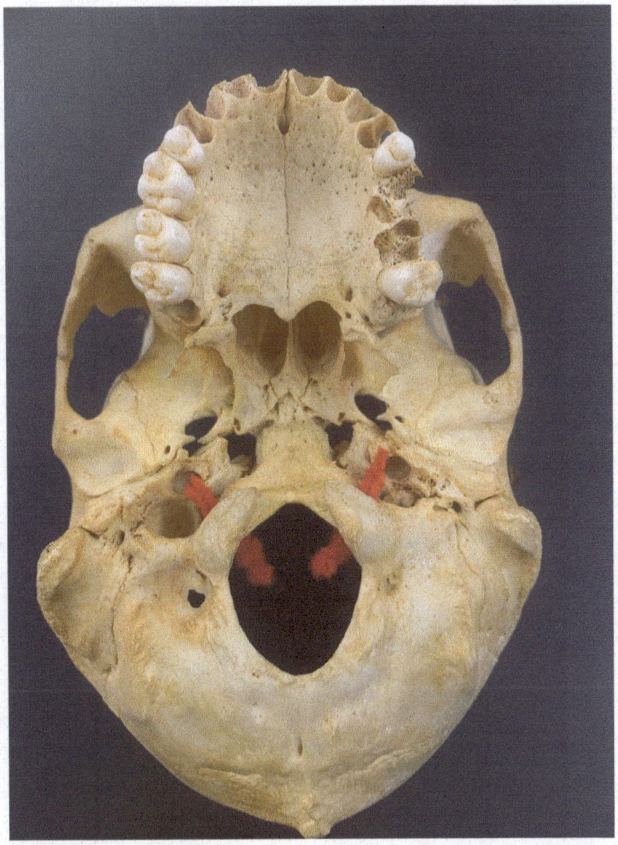

Source: Ben Jellen, 2021.

OUTCOME 5: Identify the details of the skull, sectional view.

TOOLBOX

Sectional Anatomy of the Skull

skull

frontal bone
ethmoid bone
  cribriform plate
sphenoid bone
  optic foramina/canals
  foramina ovale
  sella turcica and hypophyseal fossa

temporal bone
  carotid canal
  internal acoustic meatus
  jugular foramina
occipital bone
  foramen magnum
  hypoglossal canal
parietal bone

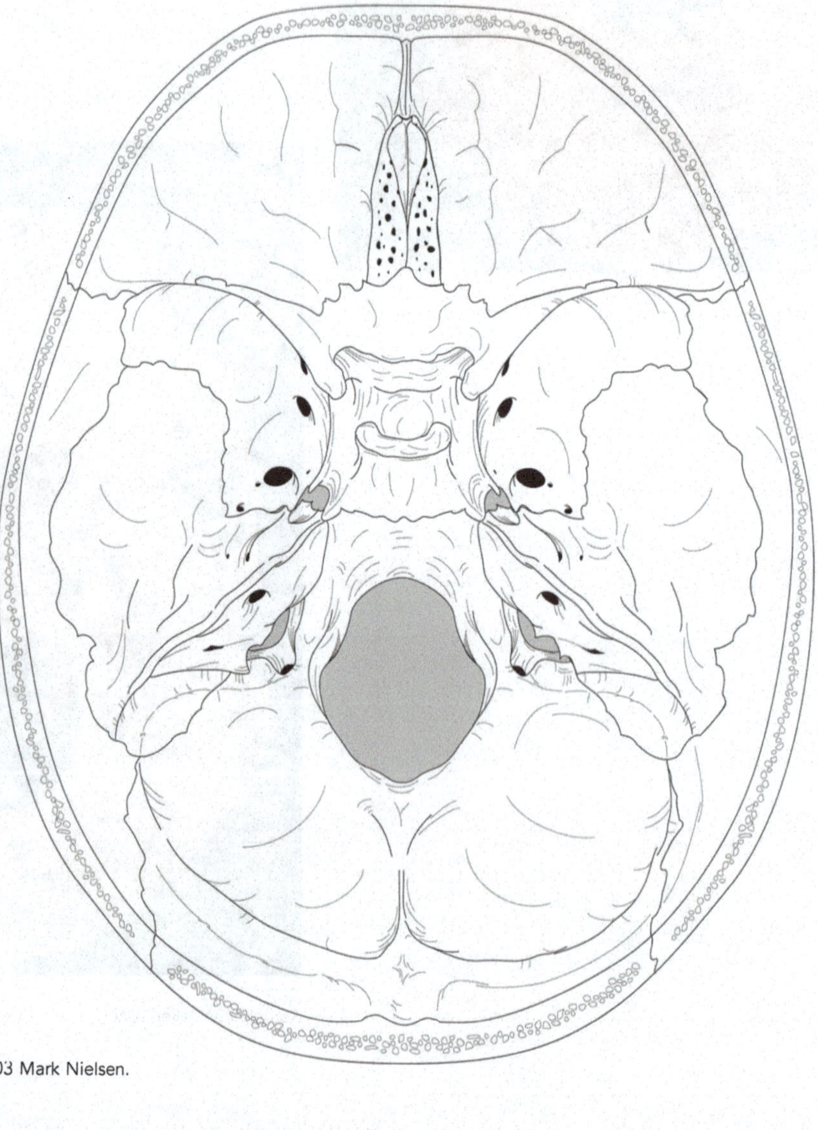

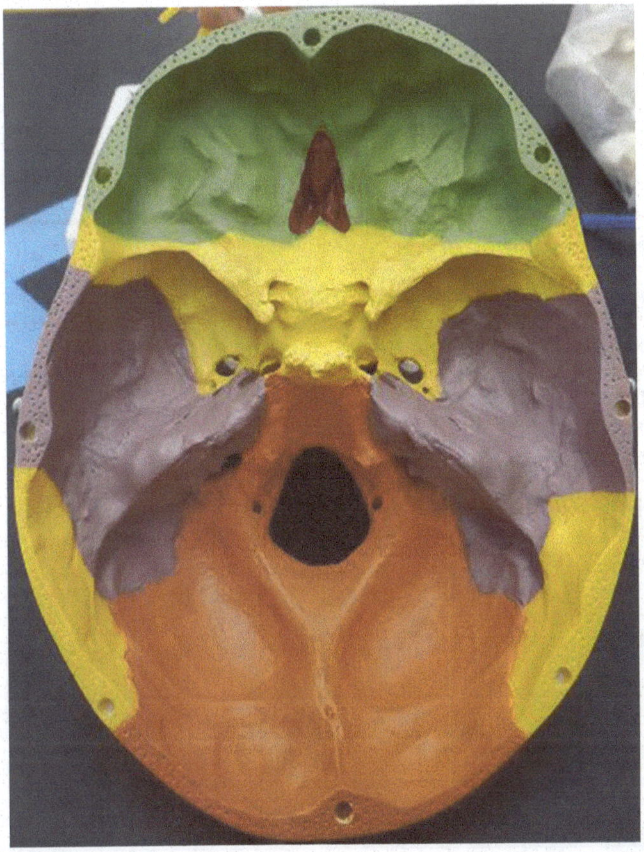

A20/2 © 3B Scientific GmbH, Germany, 2021 www.3bscientific.com

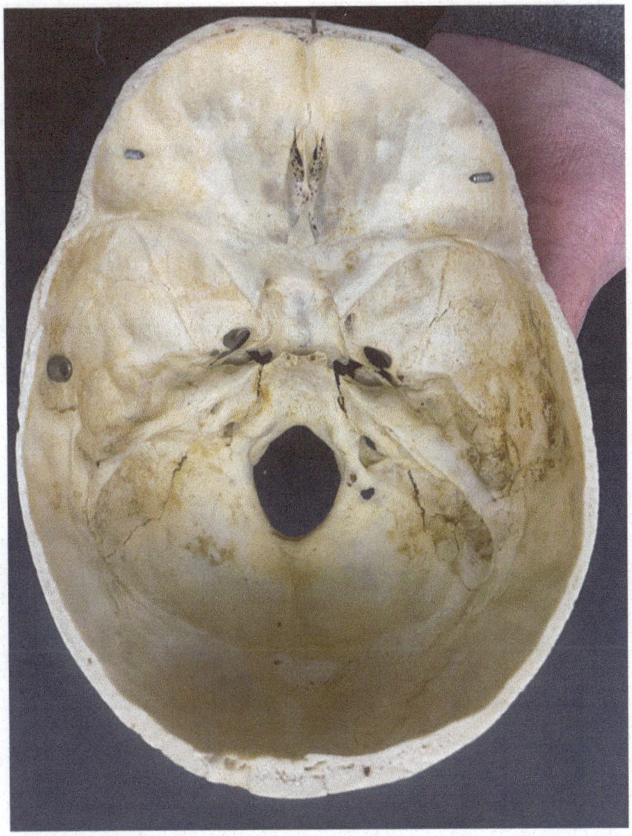

Source: Ben Jellen, 2021.

OUTCOME 6: Locate the foramina listed in the table. Describe their locations; name the structures that pass through them.

NAME	LOCATION	PASSAGE FOR
Carotid canal	temporal bone	internal carotid artery
Foramen magnum	occipital bone	continuation of medulla oblongata to spinal cord
Hypoglossal canal	occipital bone	hypoglossal n.—cranial nerve number twelve (CN XII)
Jugular foramen	between temporal and occipital bones	vagus n. (CN X), glossopharyngeal n. (CN IX), spinal accessory n. (CN XI)
Optic canal	sphenoid bone	optic n. (CN II)
Perforations of the cribriform plate	ethmoid	olfactory n. (CN I)
Internal acoustic meatus	temporal	vestibulocochlear n. (CN VIII)

OUTCOME 7: Locate the paranasal sinuses and the hyoid bone.

TOOLBOX

Sectional Anatomy of Skull
The Nasal Complex
The Hyoid Bone

sagittal head
frontal head
hyoid bone

■ ***Locate and identify the hyoid bone among the disarticulated bones before you and on the hanging skeleton.***

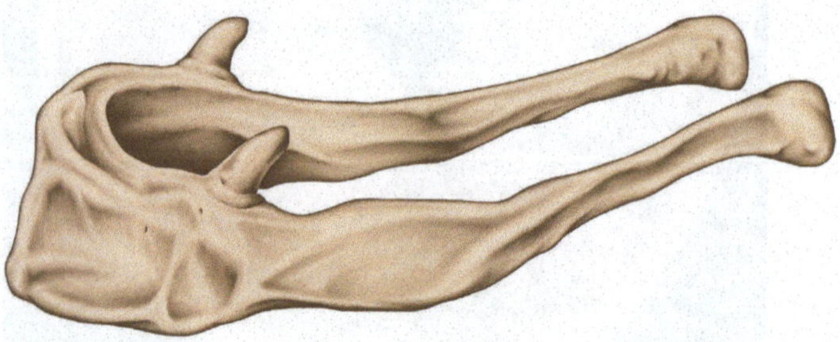

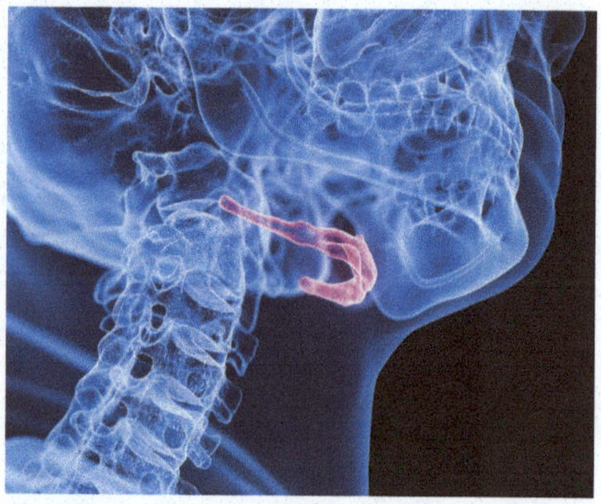

■ ***Locate the hollow spaces known as the paranasal sinuses.***

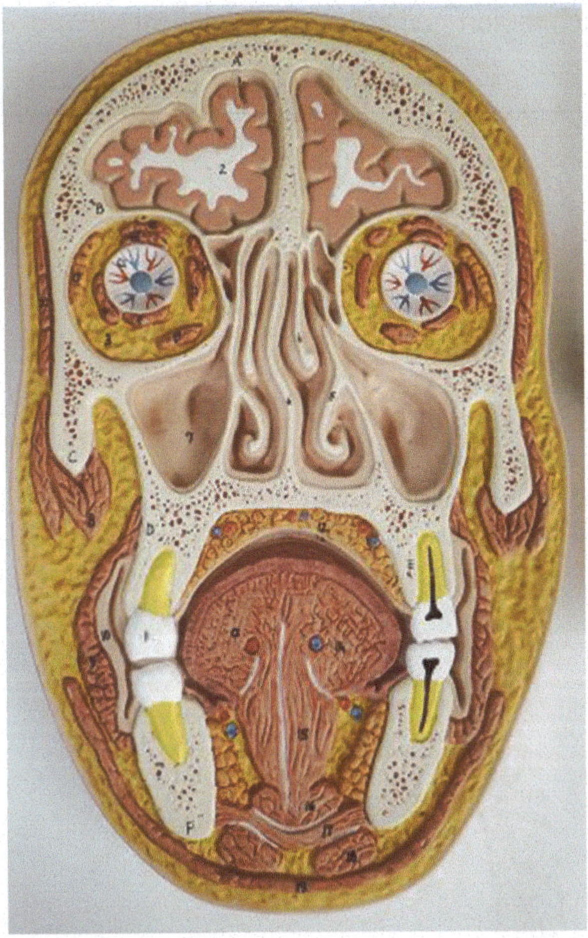

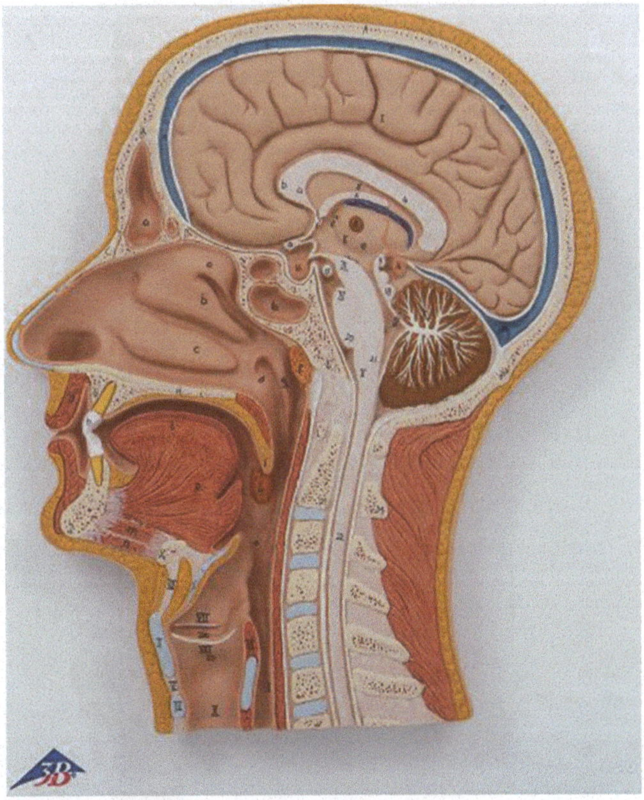

Photos C12/C11 © 3B Scientific GmbH, Germany, 2021
www.3bscientific.com

NAME OF SINUS	LOCATION
Frontal	Frontal bone
Sphenoid	Sphenoid bone
Ethmoid	Ethmoid bone
Maxillary	Maxillary bones left and right

OUTCOME 8: Name the five different regions of the vertebral column. Locate the regions on the articulated column. How many bones in each region? What is their numerical designation?

TOOLBOX

The Vertebral Column

vertebral column

cervical vertebrae ⟶ 7 bones, designated C1–C7
thoracic vertebrae ⟶ 12 bones, designated T1–T12
lumbar vertebrae ⟶ 5 bones, designated L1–L5
sacral region ⟶ 4–5 fused bones, designated sacrum
coccygeal region ⟶ 3–5 fused bones, designated coccyx

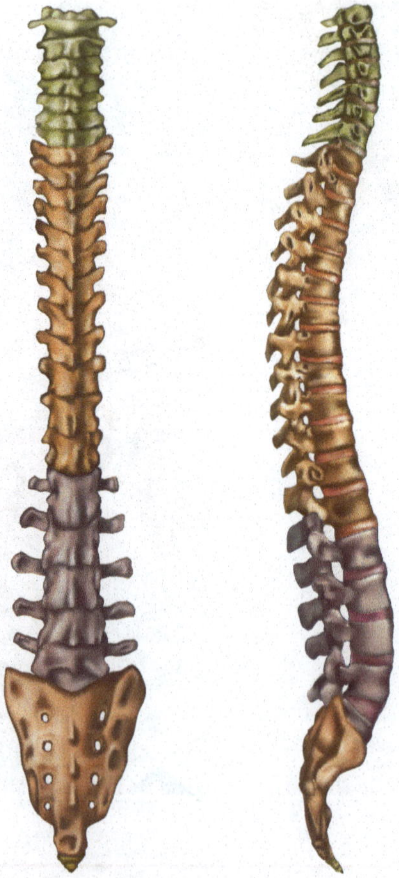

OUTCOME 9: Compare a single vertebra from each of the cervical, thoracic, and lumbar regions. Identify the structures that are common to each. Determine how you will distinguish one from another.

TOOLBOX

Cervical Vertebrae
Thoracic Vertebrae
Lumbar Vertebrae

individual vertebrae

cervical vertebra

lamina

body

thoracic vertebra

lumbar vertebra

body

transverse process

spinous process

vertebral (spinal) foramen

transverse costal facet
(thoracic vertebra only)

intervertebral foramen (IVF)

lateral view

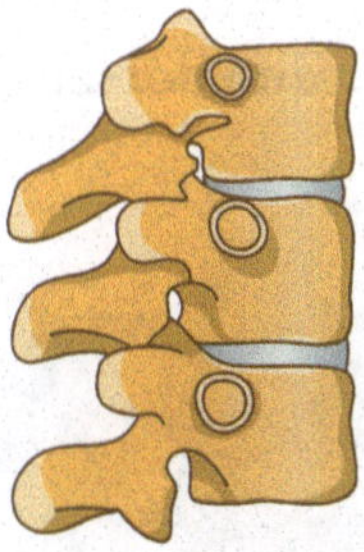

superior view

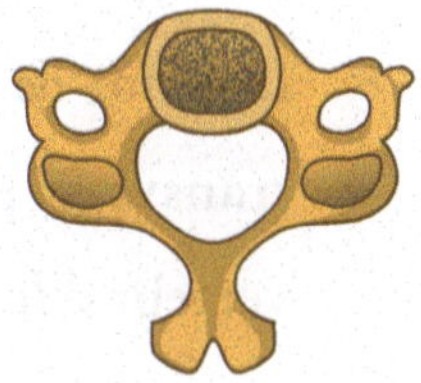

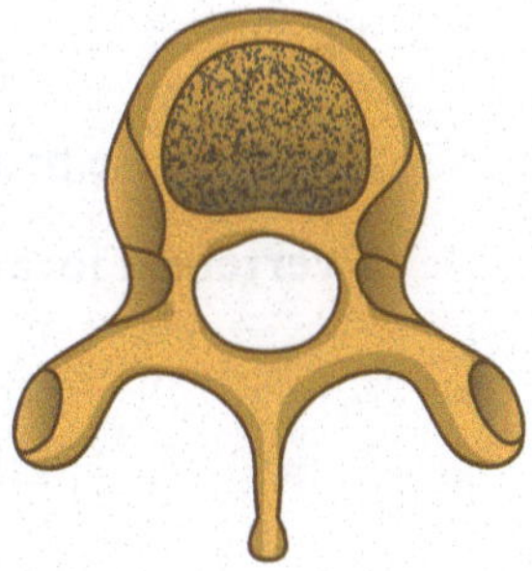

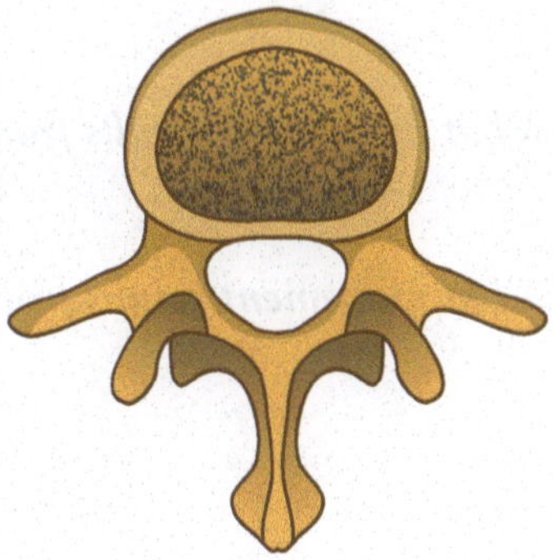

OUTCOME 10: Locate the first two cervical vertebrae: the ATLAS (C1) and the AXIS (C2). Identify them by name and numerical designation. Describe their articulation with the occipital condyles and with each other.

TOOLBOX

Atlas and Axis

atlas and axis bones

Atlas, C1:

facet for dens

transverse process

vertebral foramen

Axis, C2:

dens

spinous process

body

transverse process

vertebral foramen

______ superior and dorsal view

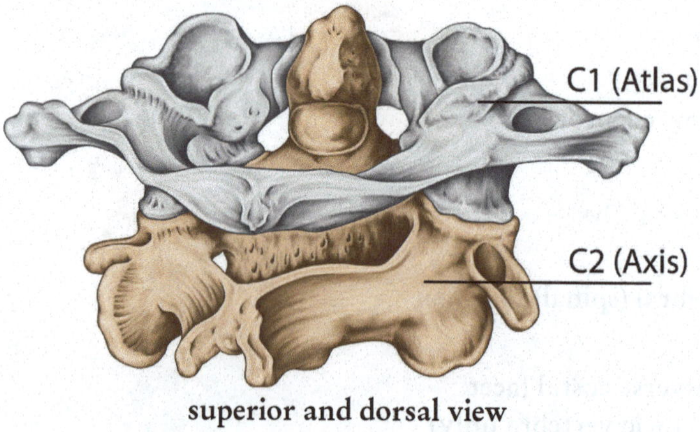

______ superior and dorsal view

■ *What movement results from the articulation of the atlas and the occipital condyles?*

■ *What movement results from the articulation of the axis and the atlas?*

OUTCOME 11: Identify the sacrum and coccyx.

TOOLBOX

The Sacrum and Coccyx

articulated pelvis
sacrum
coccyx

anterior surface of sacrum
posterior surface of sacrum
sacral canal
sacral hiatus

© SciePro/shutterstock.com

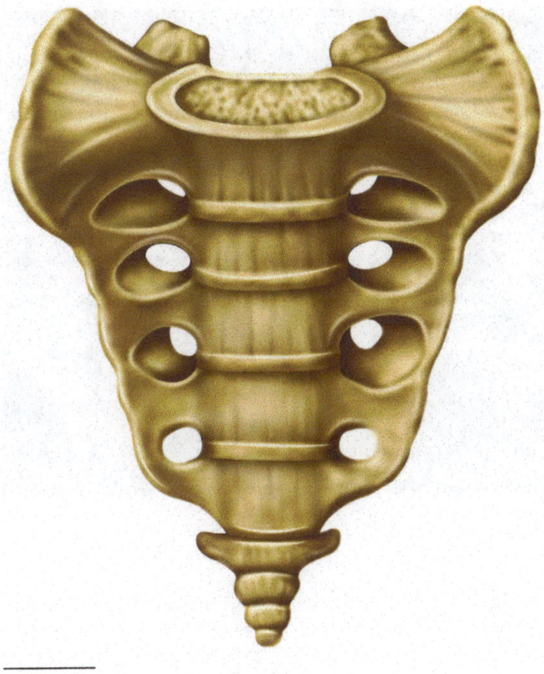

© ilusmedical/shutterstock.com

OUTCOME 12: Identify the parts of the thoracic cage. Locate the parts of the sternum. Determine which ribs are true, false, or floating.

TOOLBOX

The Thoracic Cage

disarticulated sternum and thoracic cage on Mr. Bones

Sternum

manubrium

body

xiphoid process

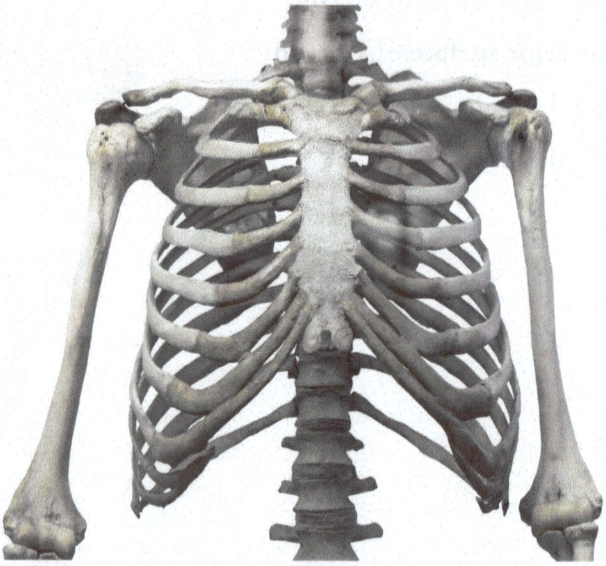

© aapsky/shutterstock.com

Ribs

true (1–7)

false (8–12)

vertebrochondral (8–10)
tubercle
shaft
costal groove
floating (11–12)

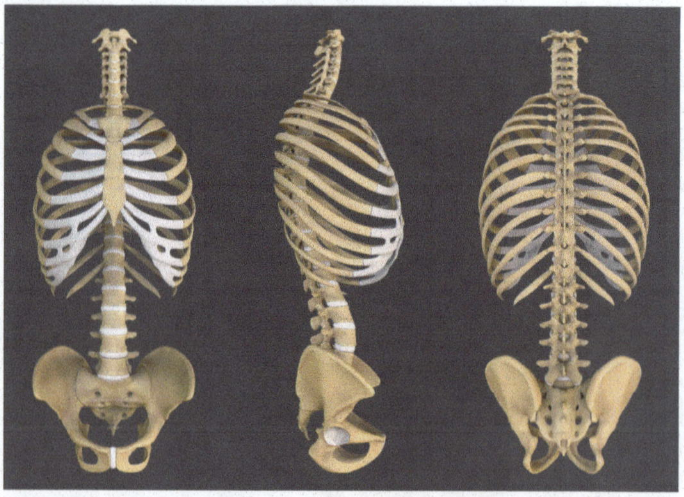

© mybox/shutterstock.com

© bogdan ionescu/shutterstock.com

G. *Details of the Appendicular Skeleton*

OUTCOME 13: Identify the details of the pectoral girdle.

TOOLBOX

The Clavicle

clavicle

sternal end
acromial end

inferior view

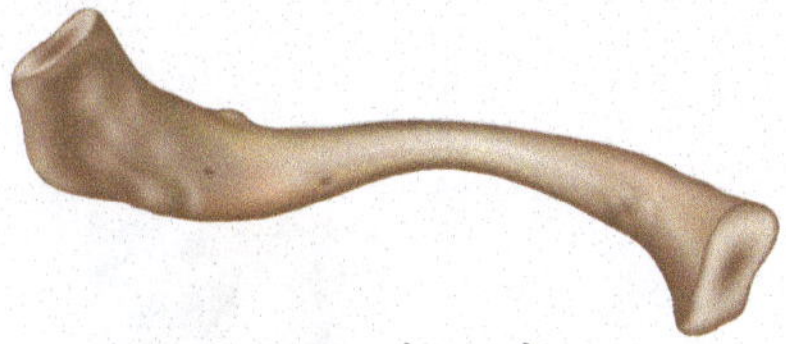

superior view

TOOLBOX

The Scapula

scapula

acromion process
spine
coracoid process

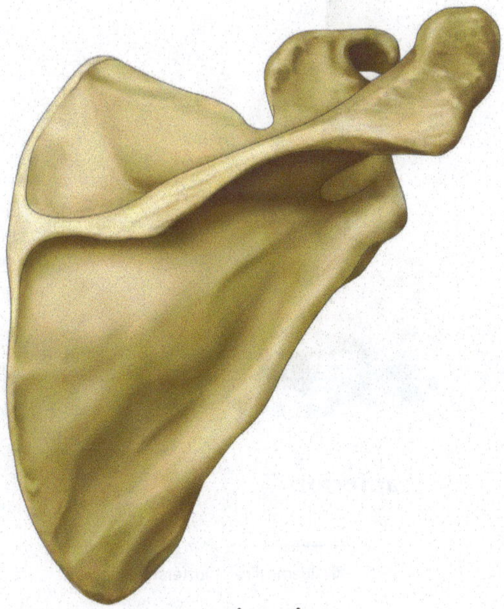

posterior view

OUTCOME 14: Identify the details of the bones of the upper appendages.

TOOLBOX

The Humerus

humerus

head
medial epicondyle

lateral epicondyle
olecranon fossa

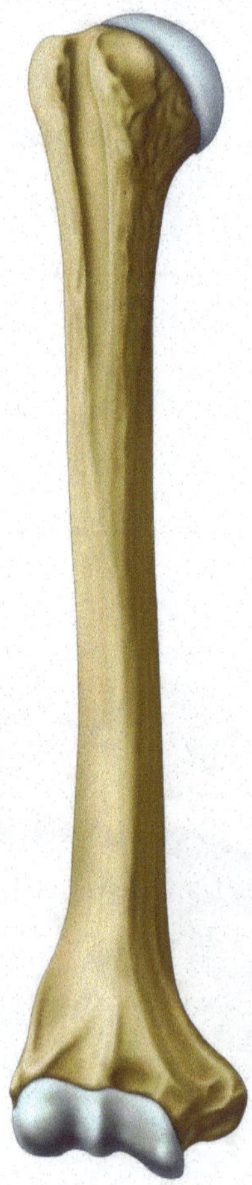

anterior

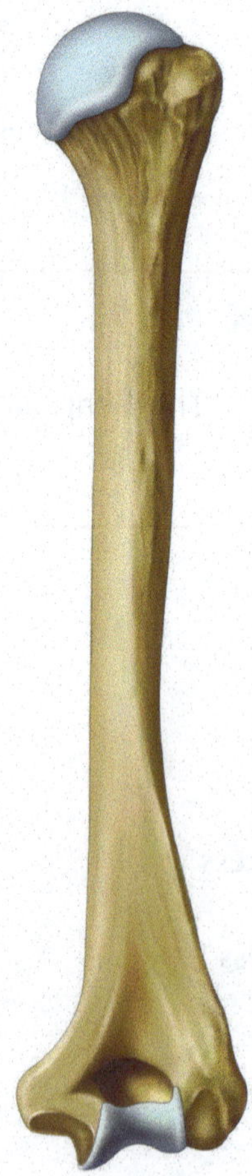

posterior

TOOLBOX

The Radius and the Ulna

radius and ulna

Radius
head of radius
radial styloid process
radial tuberosity

Ulna
head of ulna
trochlear notch
olecranon

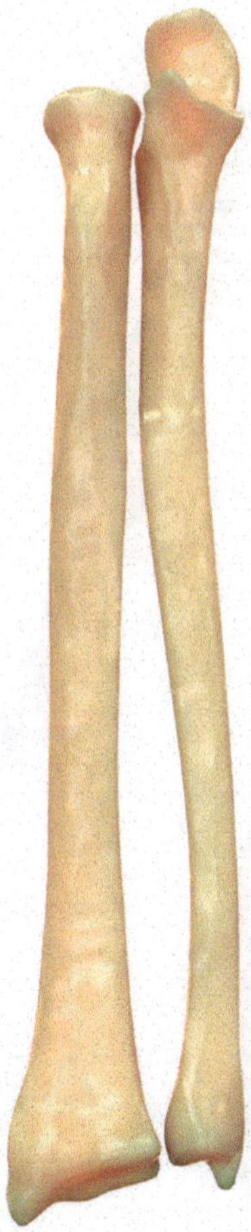

anterior view

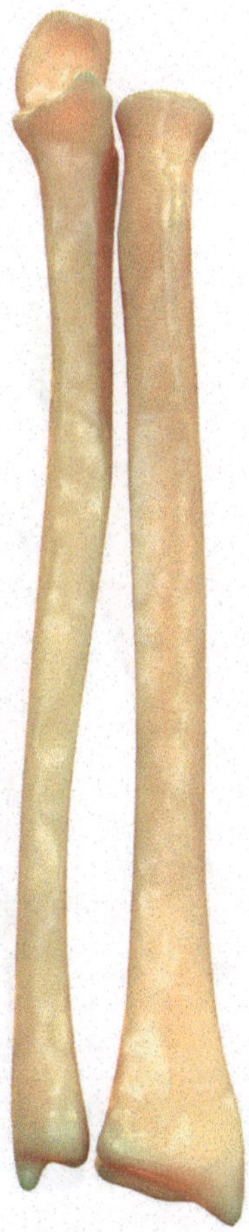

posterior view

TOOLBOX

The Bones of the Wrist and Hand

articulated hand

carpals, as a group
metacarpals
phalanx
phalanges

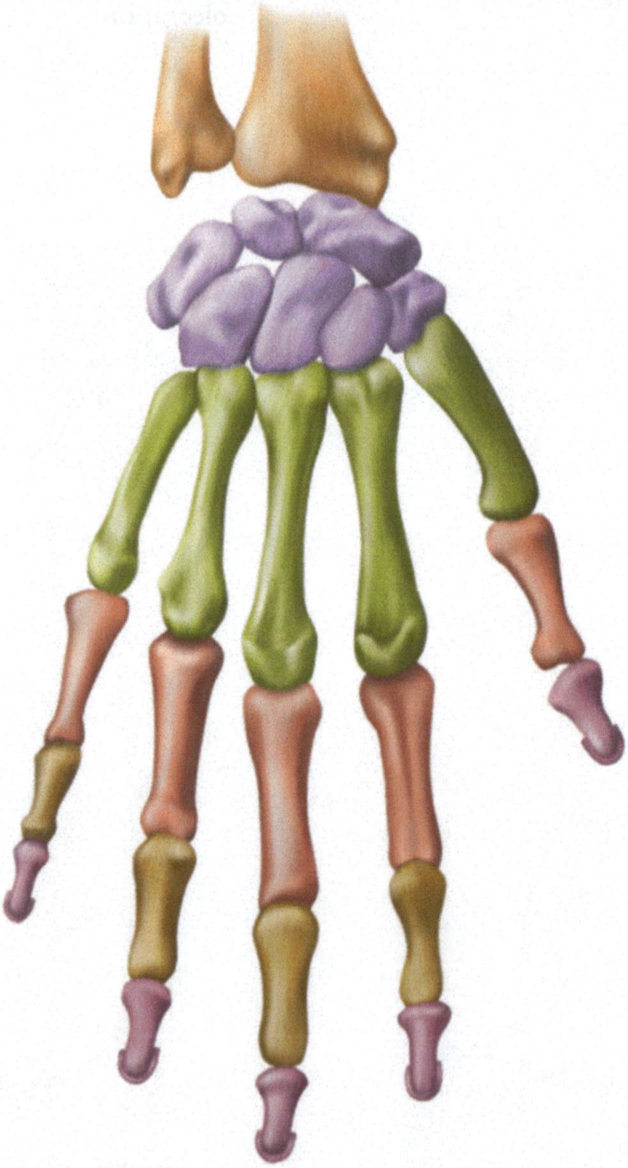

OUTCOME 15: Identify the details of the bones of the pelvic girdle.

TOOLBOX

The Pelvic Girdle

female, male os coxae

one os coxa
two os coxae (left, right)
ilium
pubis
ischium
iliac crest
pubic symphysis
acetabulum
iliac fossa
distinguish between female, male

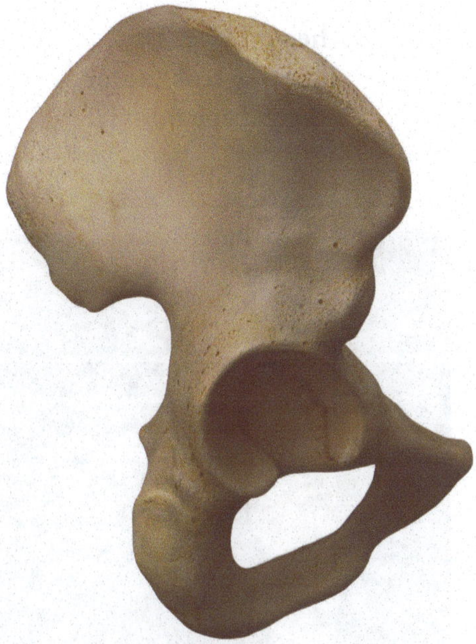

lateral view

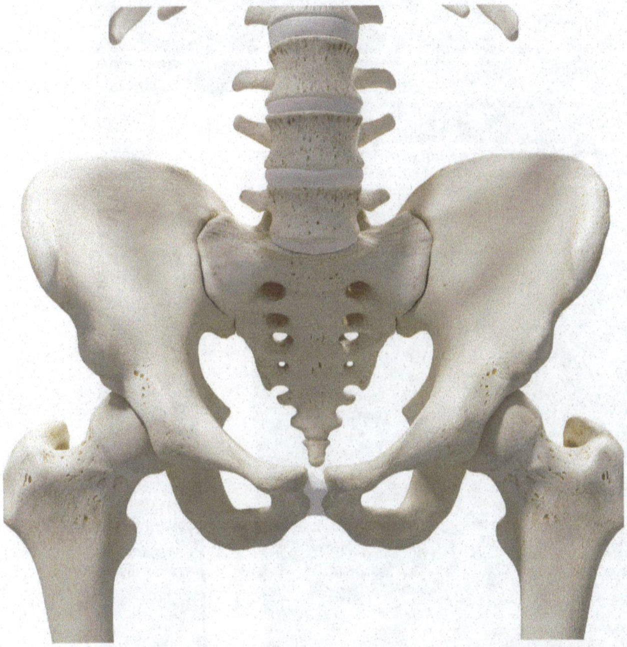

anterior view

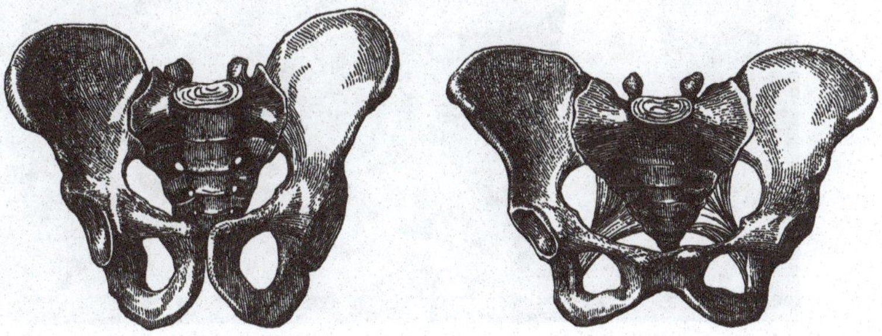

OUTCOME 16: Identify the details of the bones of the lower appendages.

TOOLBOX

The Femur

femur

head
greater trochanter
medial epicondyle
lateral epicondyle
medial condyle
lateral condyle

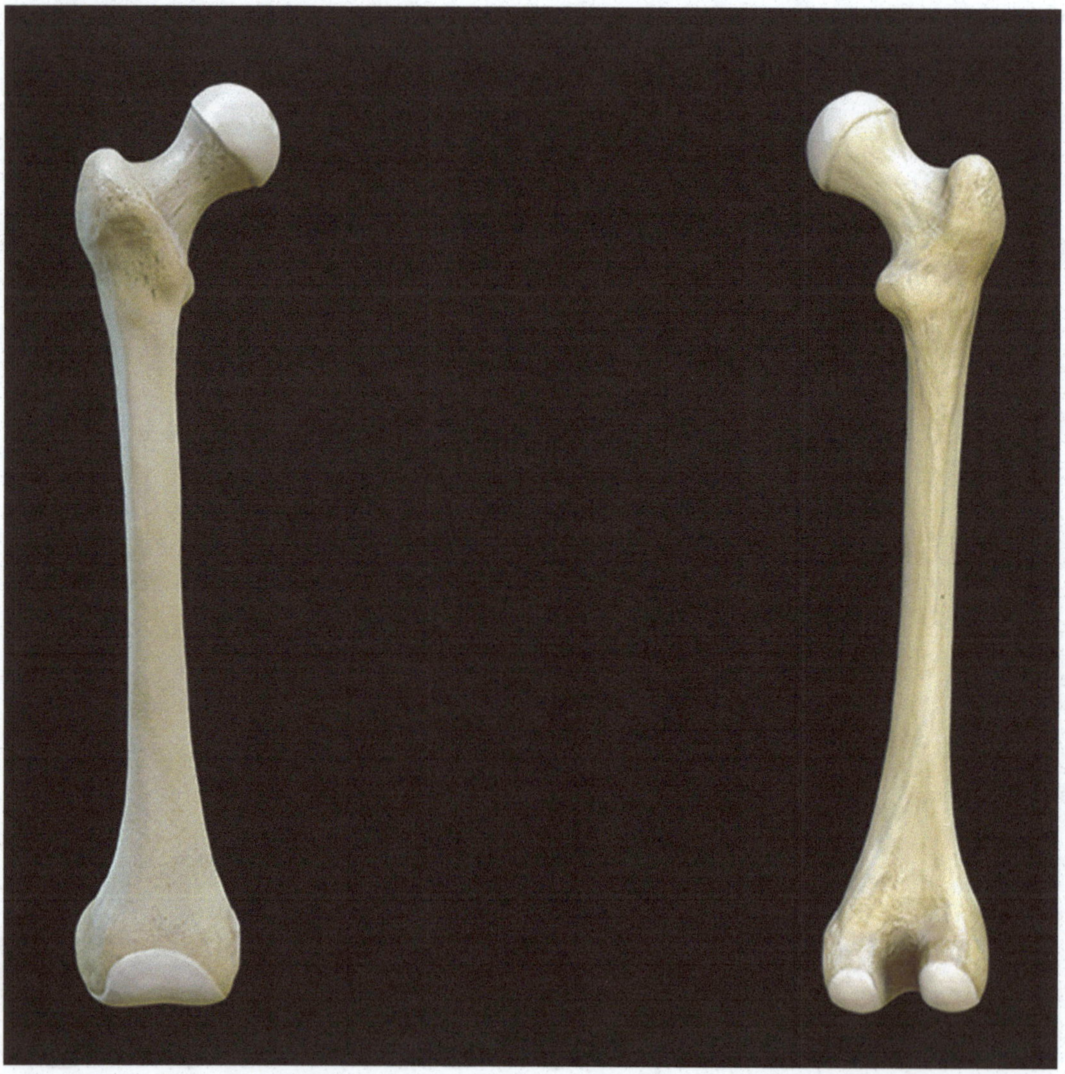

anterior view

posterior view

____

TOOLBOX

The tibia, fibula, and patella

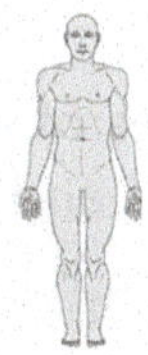

disarticulated tibia, fibula, and patella

Tibia
medial malleolus

Fibula
lateral malleolus

Patella
head
neck
shaft

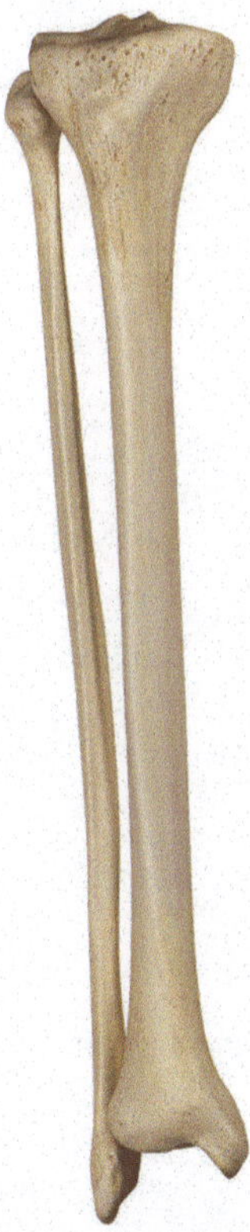

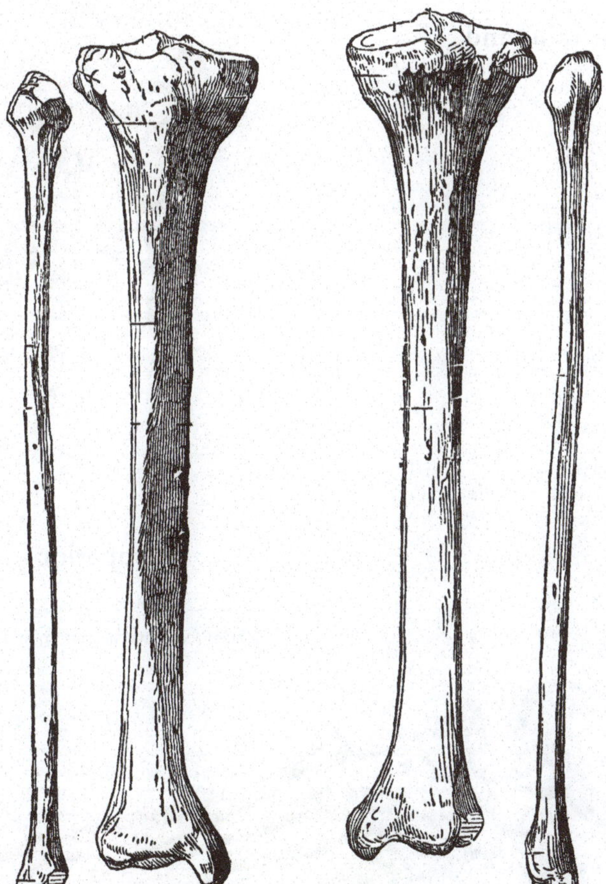

anterior view

posterior view

■ ***Examine the articulation of the patella at the knee.***

TOOLBOX

Bones of Ankle and Foot, Part I and II

articulated foot

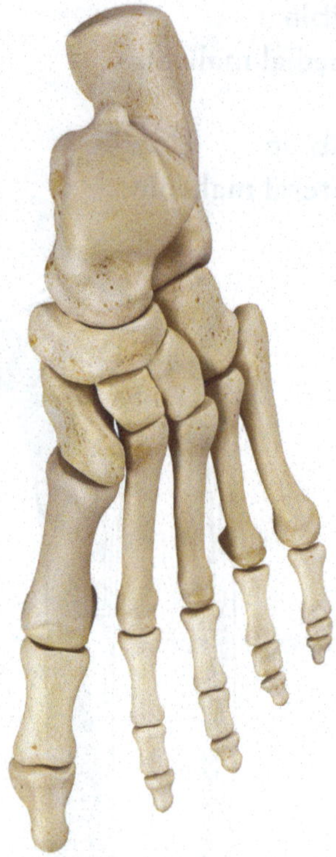

superior

tarsals (as a group), know the talus and calcaneus individually

metatarsals

phalanges

hallux

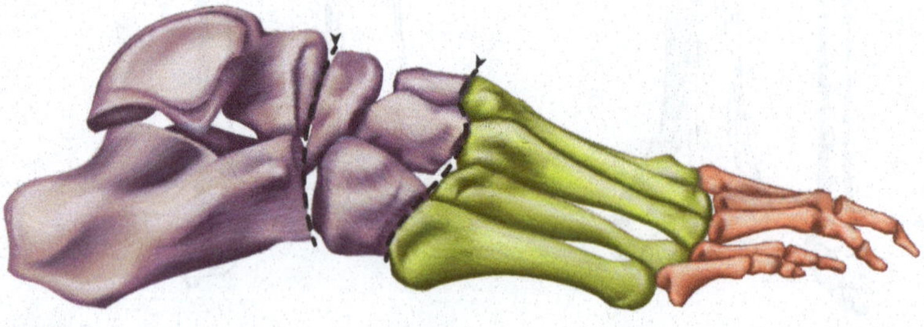

lateral

CHECKLIST: Summary Checklist Axial Skeleton

TERM/STRUCTURE	YOUR NOTES
Long Bone	
Proximal Epiphysis	
Distal Epiphysis	
Diaphysis	
Metaphysis	
Epiphyseal Plate/Line	
Medullary Cavity	
Trabeculae (Spongy Bone)	
Compact Bone	
Osteon	
Central Canal	
Concentric Lamellae	
Osteocyte	
Lacuna	
Canaliculi	
Lacuna	
Osteoclasts	
Osteoblasts	
Fibrous Layer of Periosteum	
Osteogenic Layer of Periosteum	
Circumferential Lamellae	
Interstitial Lamellae	
Canaliculi	
Frontal Sinus	
Sphenoid Sinus	

(*continued*)

TERM/STRUCTURE	YOUR NOTES
Ethmoid Sinus	
Maxillary Sinus	
Frontal Bone	
Superciliary Arch	
Parietal Bones	
Temporal Bones	
External Acoutis Meatus	
Internal Acoustic Meatus	
Carotid Canal	
Zygomatic Process	
Mastoid Process	
Styloid Process	
Mandibular Fossa	
Occipital Bone	
Foramen Magnum	
Jugular Foramen	
Occipital Condyle	
Hypoglossal Canal	
Maxilla	
Alveolar Processes	
Zygomatic Bones	
Temporal Processes	
Mandible	
Mandibular/Sigmoid Notch	
Ramus of Mandible	
Body of Mandible	

TERM/STRUCTURE	YOUR NOTES
Mental Protuberance	
Coronoid Process	
Mandibular Condyle/Head of the Mandible	
Nasal Bones	
Lacrimal Bones	
Lacrimal Groove	
Ethmoid Bone	
Cribriform Plate	
Perpendicular Plate	
Palatine Bone	
Sphenoid Bone	
Optic Foramen/Optic Canal	
Superior Orbital Fissure	
Inferior Orbital Fissure	
Sella Turcica	
Hypophyseal Fossa	
Foramen Ovale	
Vomer	
Wormian/Sutural Bone	
Nasal Septum	
Zygomatic Arch	
Coronal/Frontal Suture	
Sagittal/Longitudinal Suture	
Lambdoid Suture	
Cervical Vertebrae (7)	
Thoracic Vertebrae (12)	

(continued)

TERM/STRUCTURE	YOUR NOTES
Lumbar Vertebrae (5)	
Sacrum	
Coccyx	
Spinous Process	
Vertebral/Spinal Foramen	
Intervertebral Foramen	
Transverse Process	
Lamina	
Body of Vertebrae	
Costal Facet (On Thoracic Vertebrae)	
Facet for the Dens (On C1)	
Dens (On C2)	
Atlas (C1)	
Axis (C2)	
Anterior Surface of Sacrum	
Posterior Surface of Sacrum	
Sacral Canal	
Sacral Hiatus	
Sternum	
Manubrium	
Body of Sternum	
Xiphoid Process	
Tubercle of Rib	
Shaft of Rib	
Costal Groove of Rib	

CHECKLIST: Summary Checklist Appendicular Skeletal

TERM/STRUCTURE	YOUR NOTES
Clavicle	
Acromial End of Clavicle	
Sternal End of Clavicle	
Scapula	
Acromion Process of Scapula	
Coracoid Process of Scapula	
Spine of Scapula	
Humerus	
Head of Humerus	
Medial epicondyle of Humerus	
Lateral Epicondyle of Humerus	
Olecranon Fossa of Humerus	
Radius	
Head of Radius	
Radial Styloid Process	
Radial Tuberosity	
Ulna	
Head of Ulna	
Trochlear Notch of Ulna	
Olecranon	
Carpals	
Metacarpals	
Os Coxae	
Acetabulum	

(*continued*)

TERM/STRUCTURE	YOUR NOTES
Pubic Symphysis	
Ilium	
Iliac Crest	
Iliac Fossa	
Ischium	
Pubis	
Femur	
Head of Femur	
Greater Trochanter	
Medial Epicondyle of Femur	
Lateral Epicondyle of Femur	
Medial Condyle of Femur	
Lateral Condyle of Femur	
Patella	
Tibia	
Medial Malleolus of Tibia	
Fibula	
Head of Fibula	
Neck of Fibula	
Shaft of Fibula	
Lateral Malleolus of Fibula	
Tarsals	
Metatarsals	
Phalanges	
Hallux	

EXERCISE 5

Musculoskeletal System

A. *Introduction*

This chapter focuses on the muscles attached to our bones that allow us to move: our skeletal muscles. These are the **organs** of the **muscle organ system.** Skeletal muscles are striated and under voluntary control. In this Exercise, you will learn the location of these muscles as well as their respective actions either individually or in a group.

Muscle Actions

The action of a muscle describes the movement of a joint and limb controlled by that muscle. Two muscles, or a group of muscles with a coordinated action, are **synergists.** Synergistic muscles tend to be grouped together in **compartments.** Two muscles, or a group of muscles with opposite actions, are **antagonists.** It is important to remember that muscles can only pull. The muscles use the bones as levers to achieve different movements. **One single muscle cannot perform antagonistic actions.** For example, the masseter muscle is responsible for closing your mouth by pulling on the body of the mandible and elevating it at the temporomandibular joint (TMJ). Some students mistakenly describe the action of the masseter as the muscle responsible for chewing, but chewing involves opening *and* closing the mouth. This one muscle, however, cannot be responsible for these two antagonistic actions.

Origin and Insertion

The action of any muscle is easier visualized by observing the two attachment sites of that muscle. The **origin** is the attachment of a muscle that does not move during muscle contraction and is typically proximal. The **insertion** is the attachment site that moves during contraction and is typically distal. Try to imagine a marionette whenever you consider the action of a muscle. The string attaches at the origin and insertion. As you picture the attachments, visualize what would happen if you were to pull on the string. For example, you are probably already familiar with the biceps brachii. The name of this muscle comes from the fact that it has two heads and sits on the anterior surface of the brachium. Its origin is on the scapula, and its insertion is on the radius. Imagine the muscle as a string attached in these places. Now imagine that you pull on that string. What happens at your forearm? What happens at your elbow?

B. *Outcomes for Exercise 5*

1. Define the terms of movement. If provided with a definition, provide the term. Illustrate the actions with your own body. As you apply the terminology, always include the joint being acted upon. Explain why one muscle cannot perform antagonistic actions. What is meant by the origin and insertion of a muscle?

2. Identify the superficial skeletal muscles of the human body, anterior and posterior views.

3. Study the actions of the muscles. Apply the terms of movement as you learn the actions of each muscle or muscle group. Make sure to include the joint being acted upon.

C. *Muscle Actions*

OUTCOME 1: Define the terms of movement. If provided with a definition, give the term. Illustrate the actions with your own body. As you apply the terminology, always include the joint being acted upon. Explain why one muscle cannot perform antagonistic actions. What is meant by the origin and insertion of a muscle?

TOOLBOX

Types of Movements

Specific terms describe the action of a muscle on a joint. Learn the terms and be able to apply them appropriately. Read the definitions and act out the examples in the anatomical position.

ALWAYS INCLUDE THE JOINT BEING ACTED UPON.

ACTION	DESCRIPTION
flexion	a decrease in the angle of a joint. Bending the elbow, bending the knee, bending at the waist. Flex your head at your neck to look at your toes. Occurs in anterior-posterior plane.
extension	the return from a flexed position. An increase in the angle of a joint. Return your head to anatomical position by extending your head at the neck. When you swing your leg back to kick a ball, you are extending your leg at the hip.
hyperextension	an increase in the angle of a joint beyond its normal range of motion, usually resulting in joint damage.
abduction	movement of a limb in the coronal plane away from the body or midline. When you abduct a limb, you are "taking it away" from your body. Always refers to movements of the appendicular skeleton.
adduction	movement of a limb in the coronal plane toward the body or midline. When you adduct a limb, you are "adding it" to your body. Always refers to movements of the appendicular skeleton.
rotation	a circular movement around a fixed, longitudinal axis. You rotate your head at the neck when you move it left and right to shake your head "no." May be internal/medial or external/lateral.
circumduction	a circular movement that maintains a fixed proximal position, while drawing an imaginary circle in the air at the distal end. Try to circumduct your head at the neck, and see how it is different than rotation.
elevation	a move in a superior direction. Shrugging your shoulders.
depression	a move in an inferior direction.
protraction	a horizontal move in an anterior direction. Jutting your chin anteriorly.
retraction	a horizontal move in a posterior direction.

The following special actions have very specific meanings:

ACTION	DESCRIPTION
inversion	movement of the sole of the foot medially
eversion	movement of the sole of the foot laterally
supination	rotation of the forearm to position the palm anteriorly or superiorly. Imagine the position of your palm when you hold a bowl of "soup."
pronation	rotation of the forearm to position the palm posteriorly or inferiorly
plantar flexion	pointing the toes so that the plantar surface of the foot is flexed. You "plant" your toes on the ground.
dorsiflexion	flexing the dorsum. Elevating the toes off the floor the way you might if you walked on your heels.

D. *Overview of Human Musculature*

OUTCOME 2: Identify the skeletal muscles of the human body, anterior and posterior views.

Orbicularis oculi

Orbicularis oris

Massater

Sternocleidomastoid

Deltoid

Pectoralis major

External oblique

Rectus abdominis

Linea alba

Biceps brachii

Brachialis

Pronator teres

Flexor carpi radialis

Adductor longus

Adductor magnus

Tensor fascia latae

Gracilis

Rectus femoris

Vastus lateralis

Vastus medialis

Sartorius

Tibialis anterior

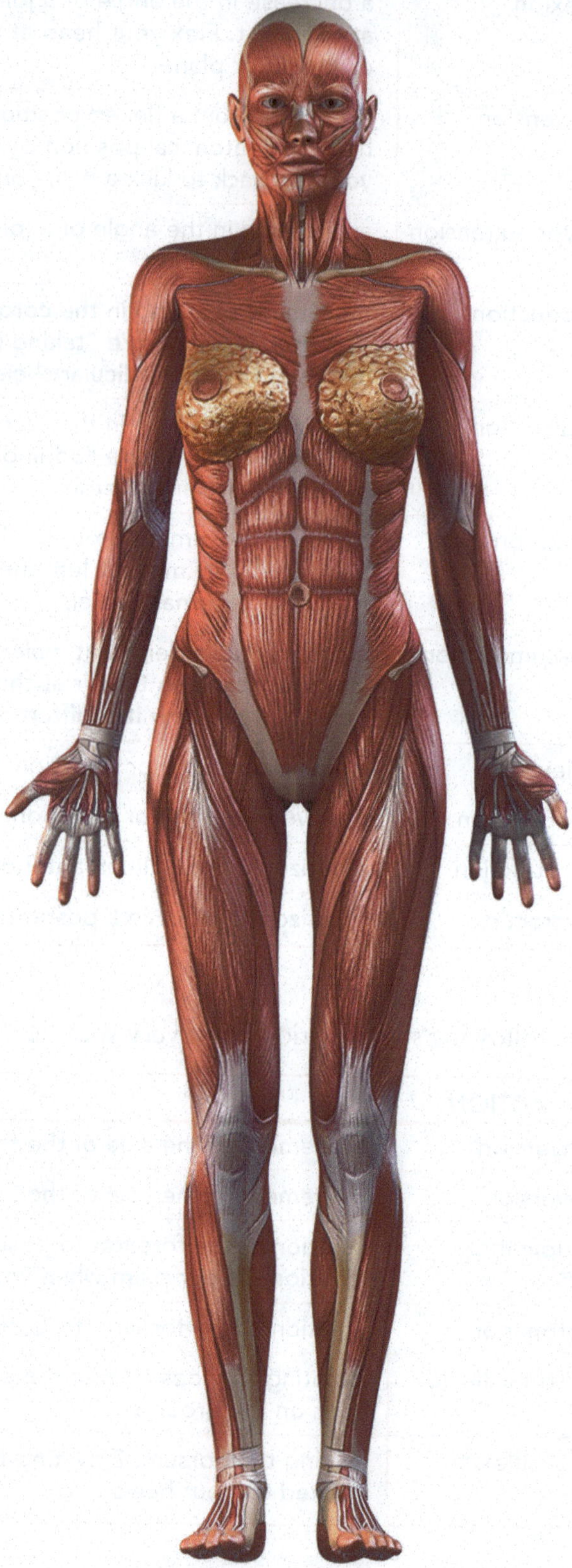

anterior view

TOOLBOX

Superficial Skeletal Muscles, Posterior View

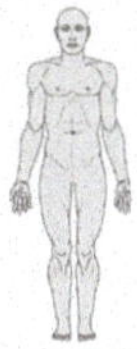

Mr. Muscle
arm model
leg model

Trapezius

Deltoid

Supraspinatus

Infraspinatus

Latissmus dorsi

Triceps brachii

Gluteus maximus

Iliotibial band

Semitendinosis

Semimembranosis

Gastrocnemius

Calcaneal/Achilles Tendon

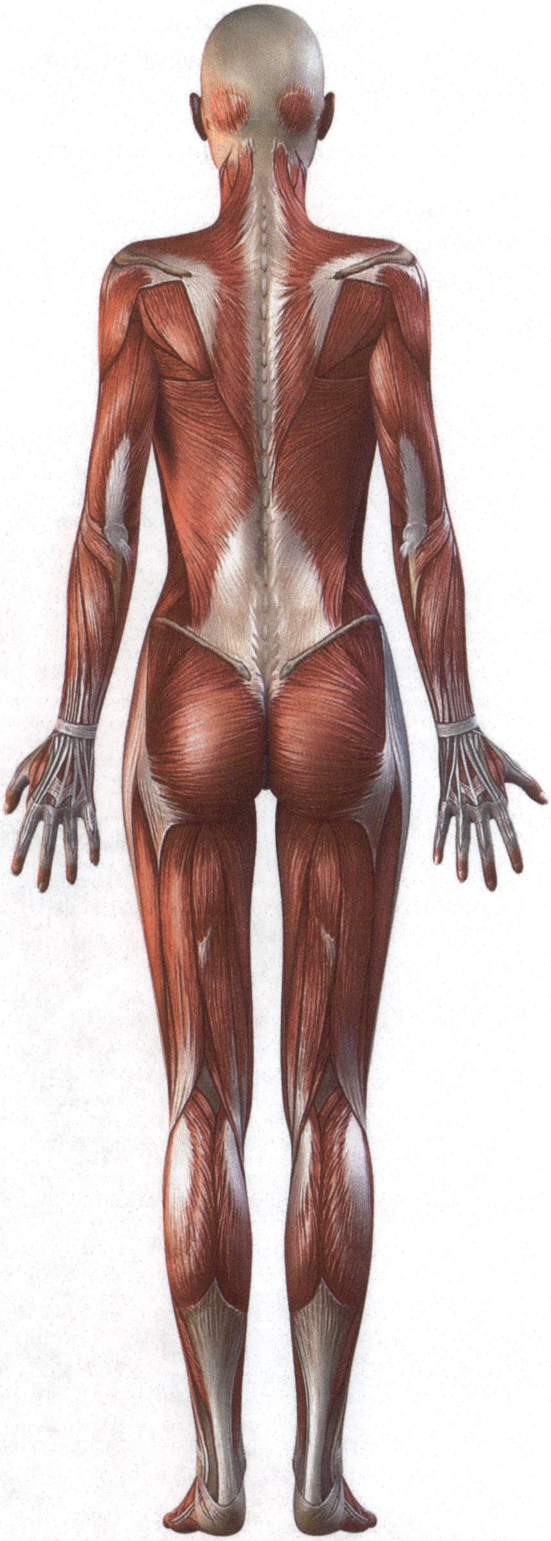

posterior view

TOOLBOX

Use your lecture text to locate and identify the following muscles.

Mr. Muscle
Miniature Human Model
Mr. Flip Chart

■ ***Identify the muscles of the face and neck.***

Facial Muscles
- Orbicularis oculi
- Orbicularis oris
- Masseter

Neck and Shoulder Muscles
- Sternocleidomastoid

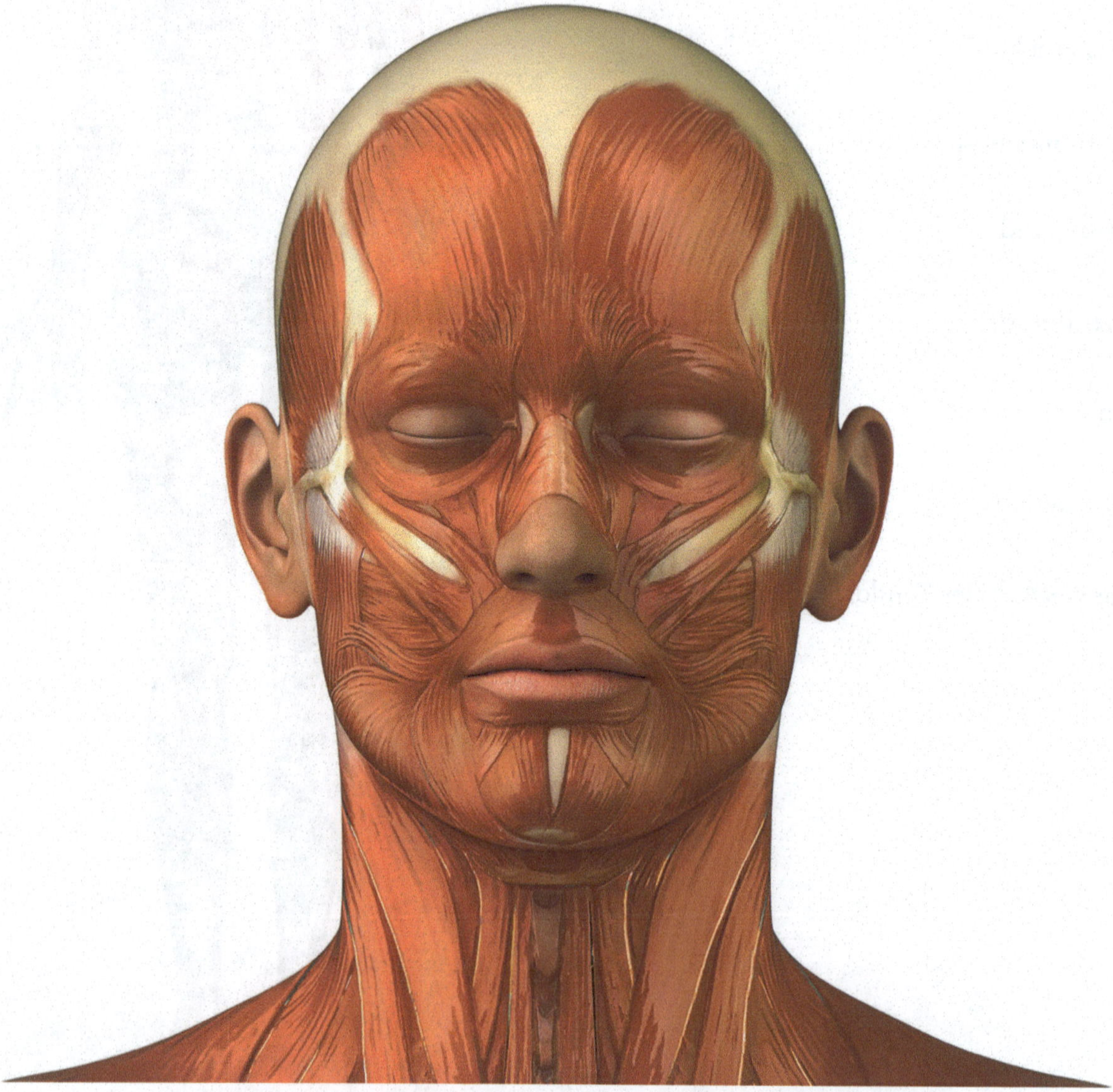

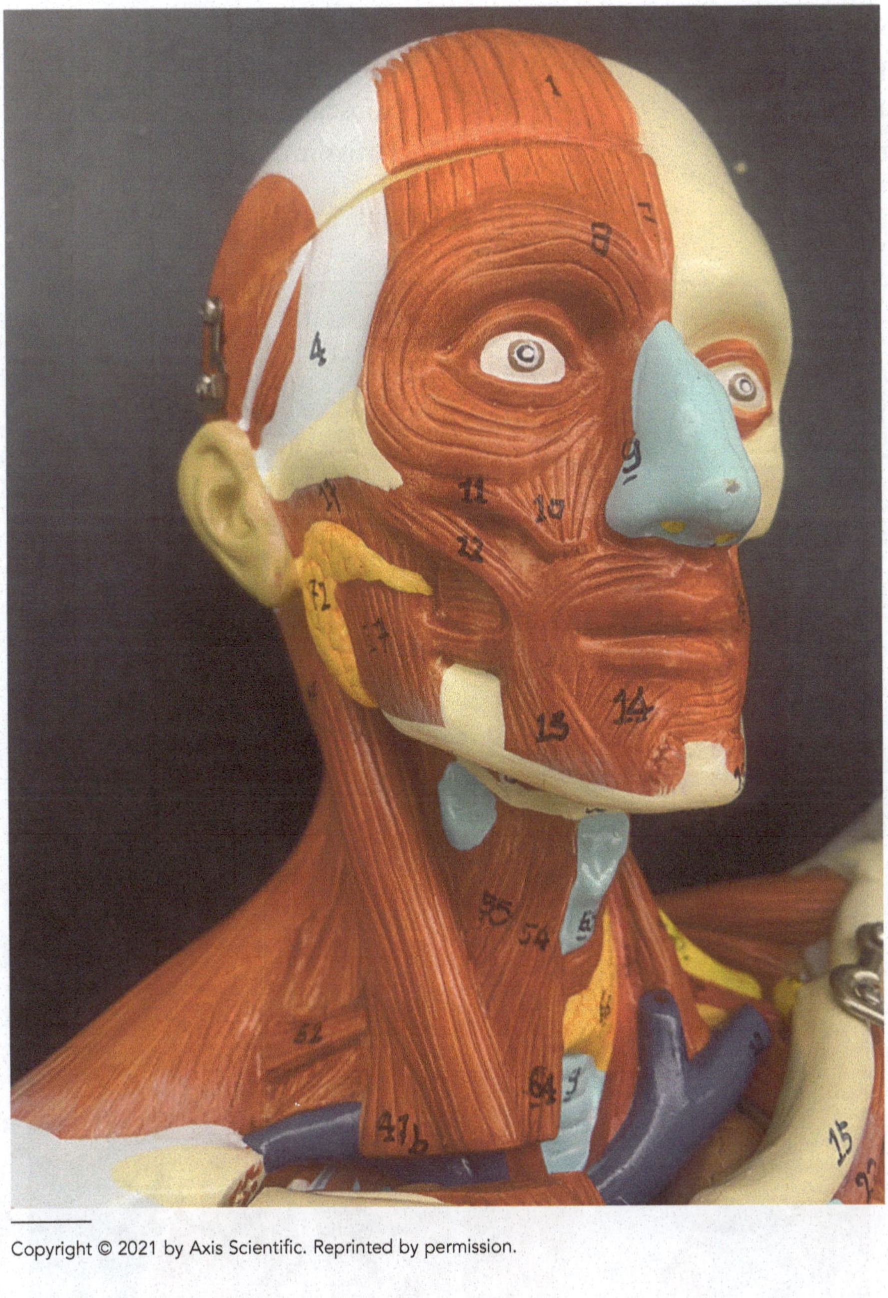

■ ***Identify the muscles of the trunk.***

Trapezius
Deltoid
External intercostal muscles
Internal intercostal muscles
External oblique muscles
Internal oblique muscles
Transversus abdominis
Rectus abdominis
Diaphragm
Linea alba (nonmuscular tissue)
Pectoralis major
Pectoralis minor
Serratus anterior
Supraspinatus
Infraspinatus
Subscapularis
Latissimus dorsi

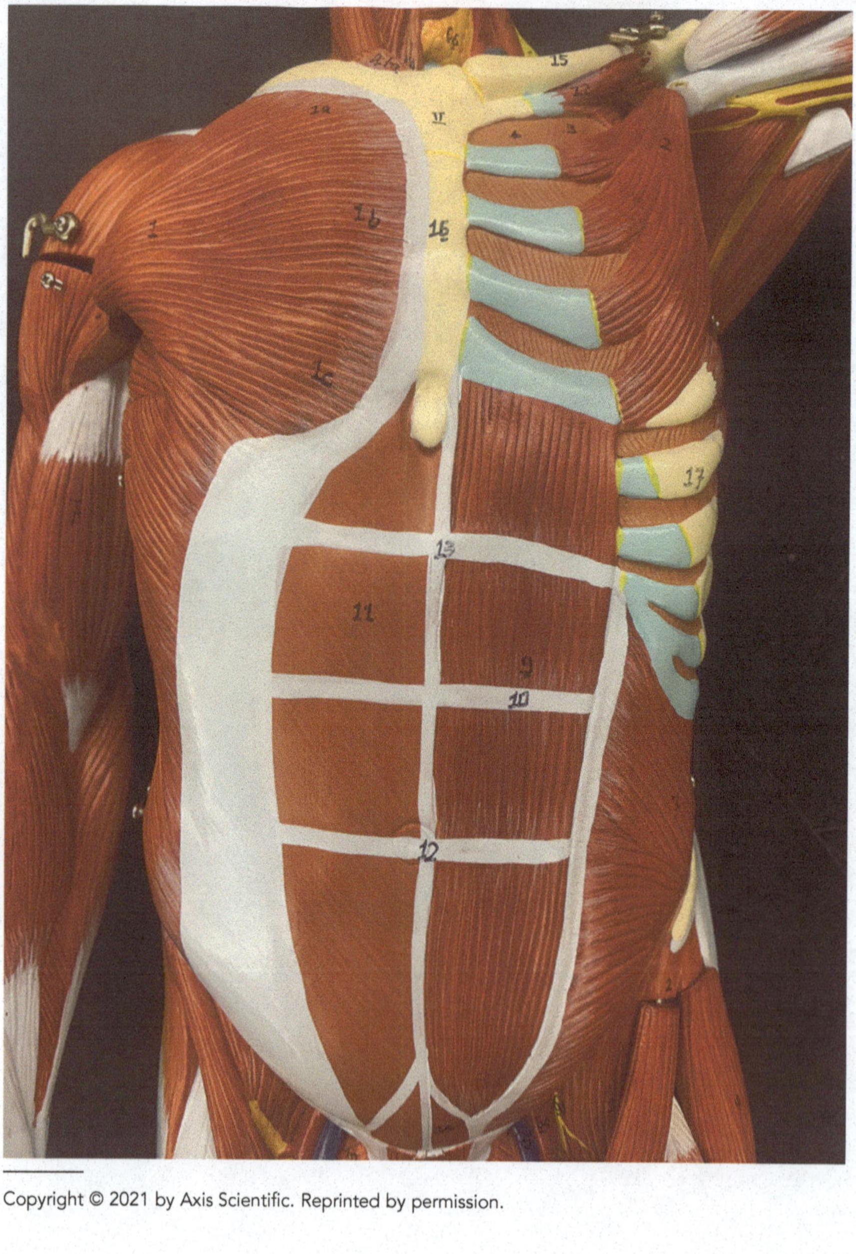

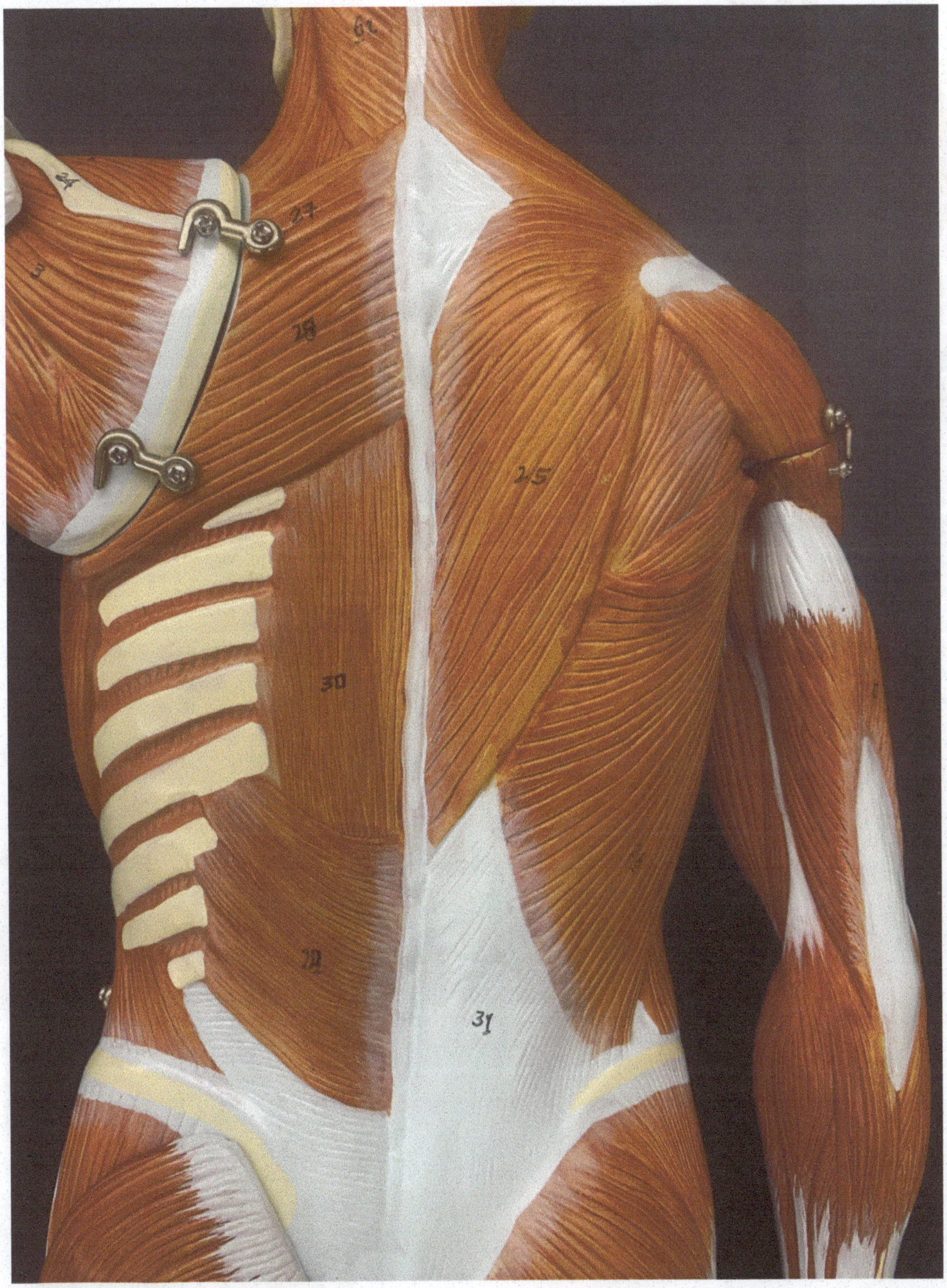
24
27
28
25
30
31

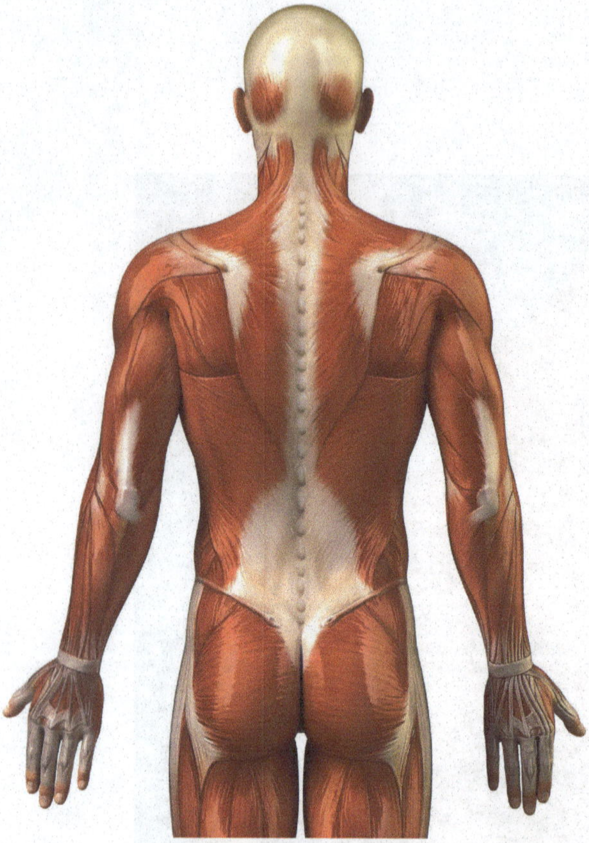

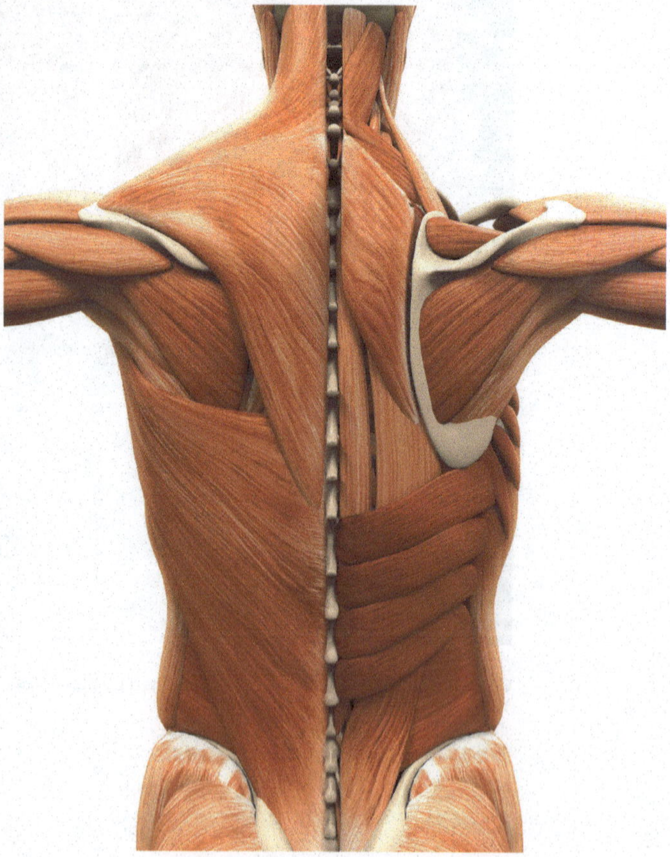

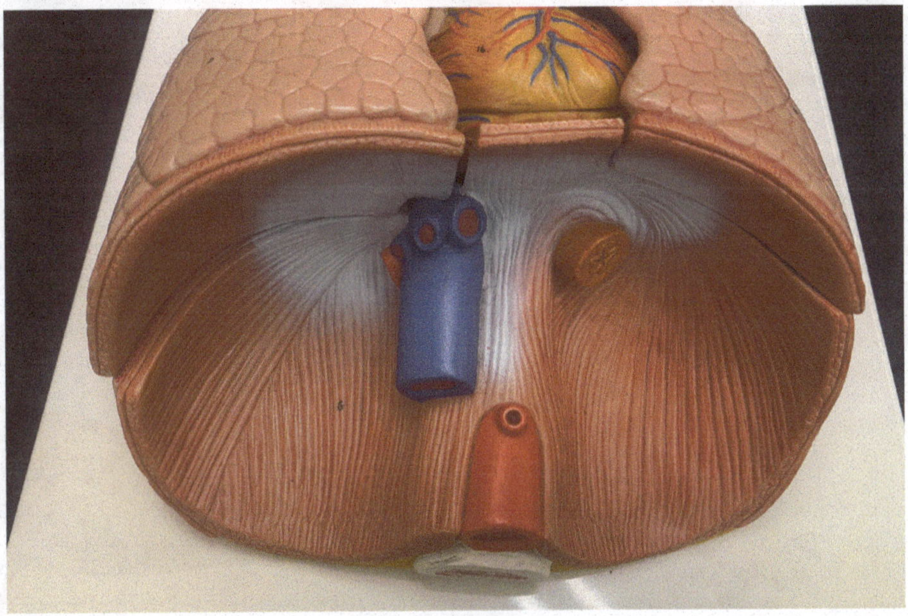

G15 © 3B Scientific GmbH, Germany, 2021 www.3bscientific.com

■ *Identify the muscles of the arm.*

Biceps brachii—long head and short head
Coracobrachialis
Triceps brachii—lateral, long, and medial heads
Brachialis
Pronator teres
Supinator
Flexor carpi radialis
Extensor carpi radialis

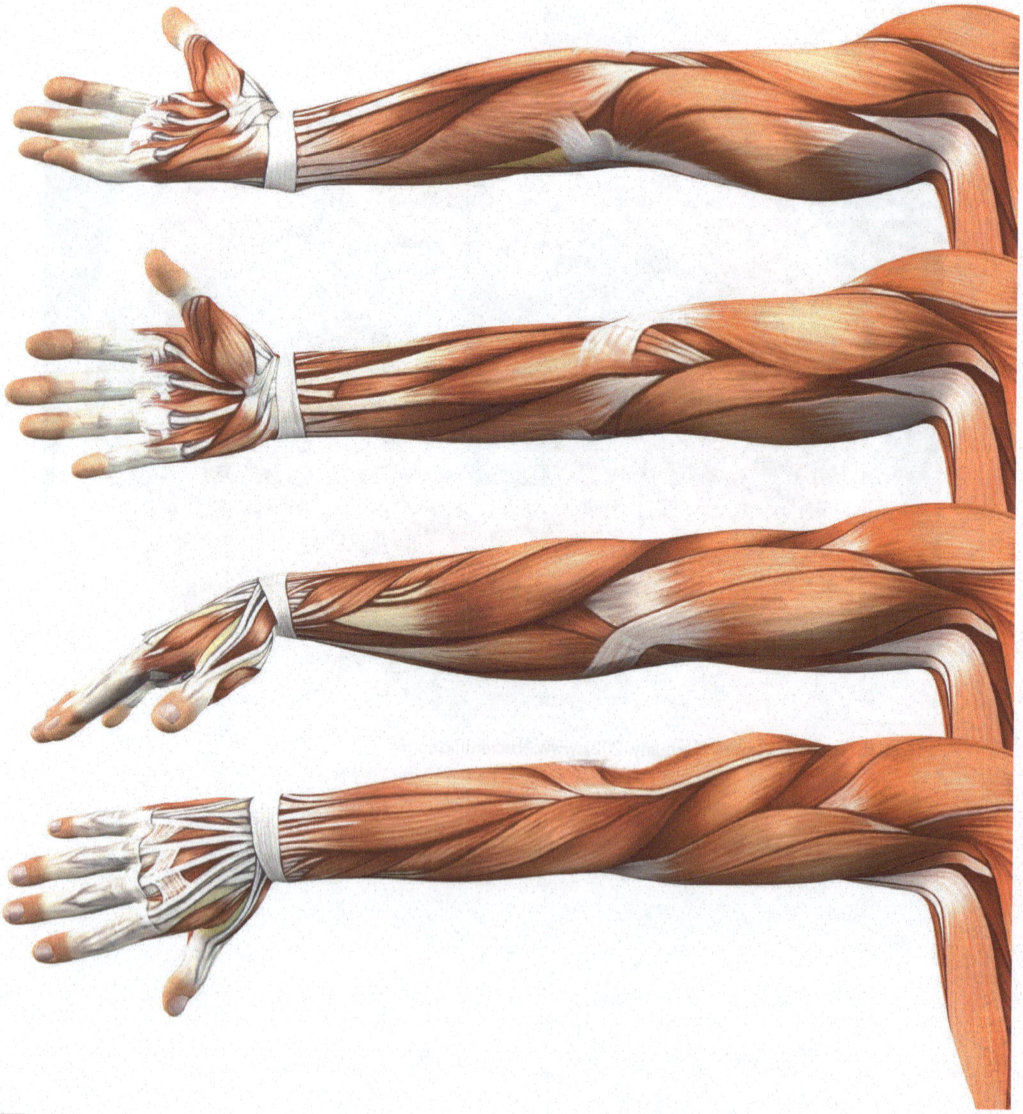

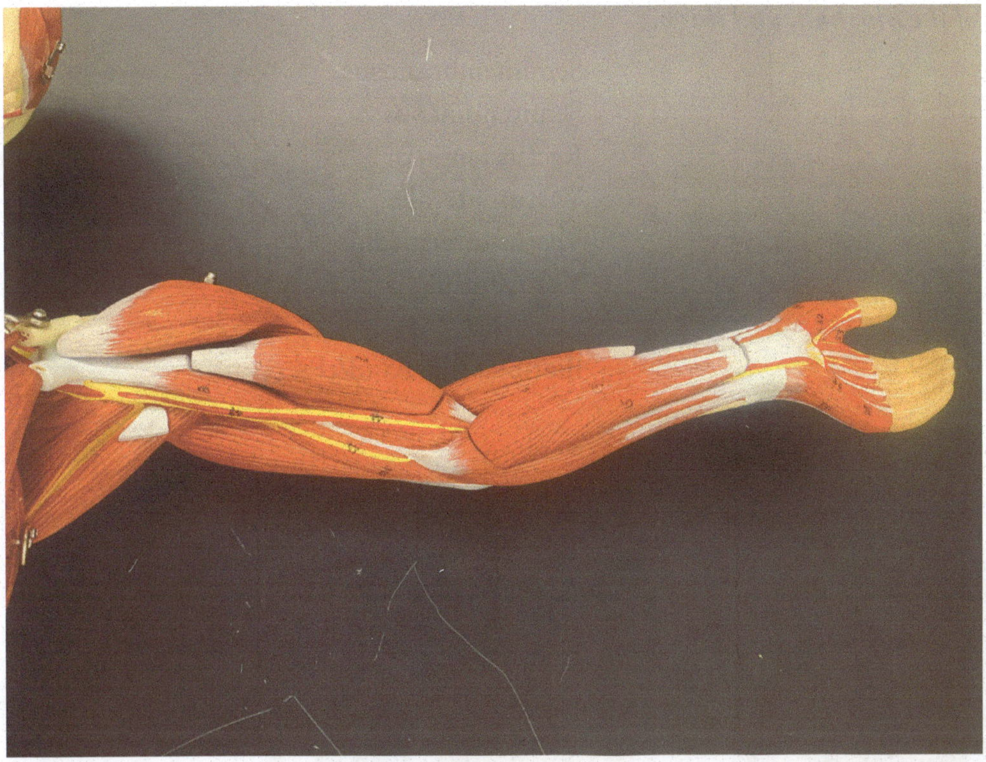

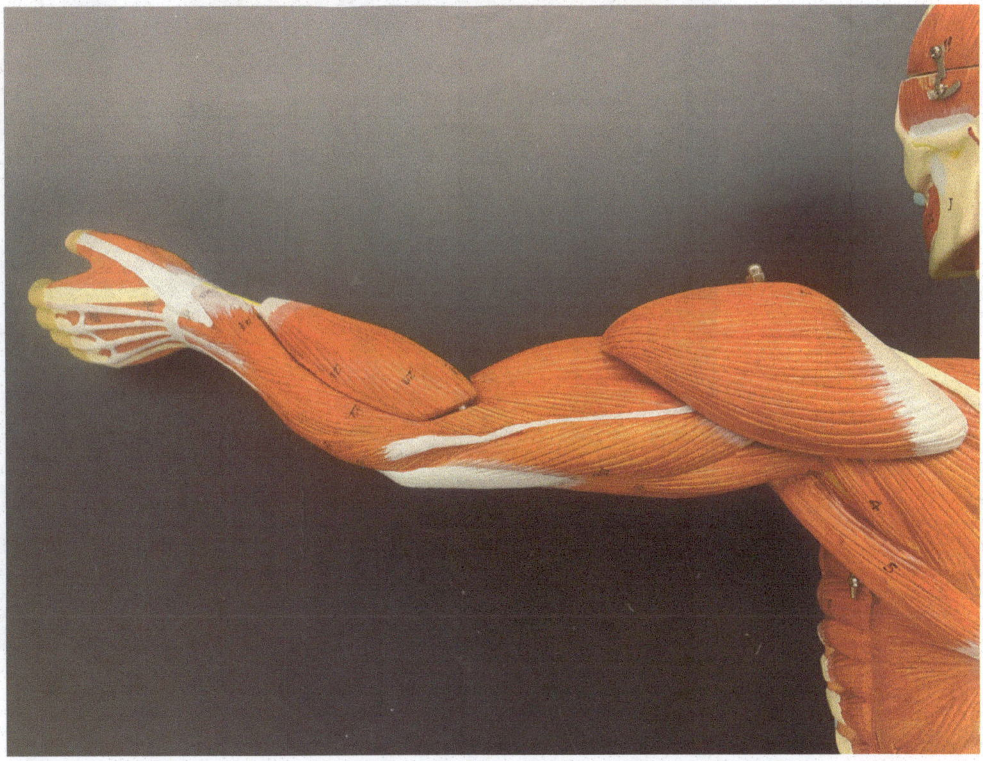

■ ***Identify the muscles of the hip, butt, and thigh.***

Gluteus maximus
Gluteus medius
Tensor fasciae latae

Muscles of the Thigh
Adductor longus
Adductor magnus
Gracilis
Biceps femoris
Semimembranosus
Semitendinosus
Rectus femoris
Vastus lateralis
Vastus intermedius
Vastus medialis
Sartorius
Iliotibial (IT) band/tract

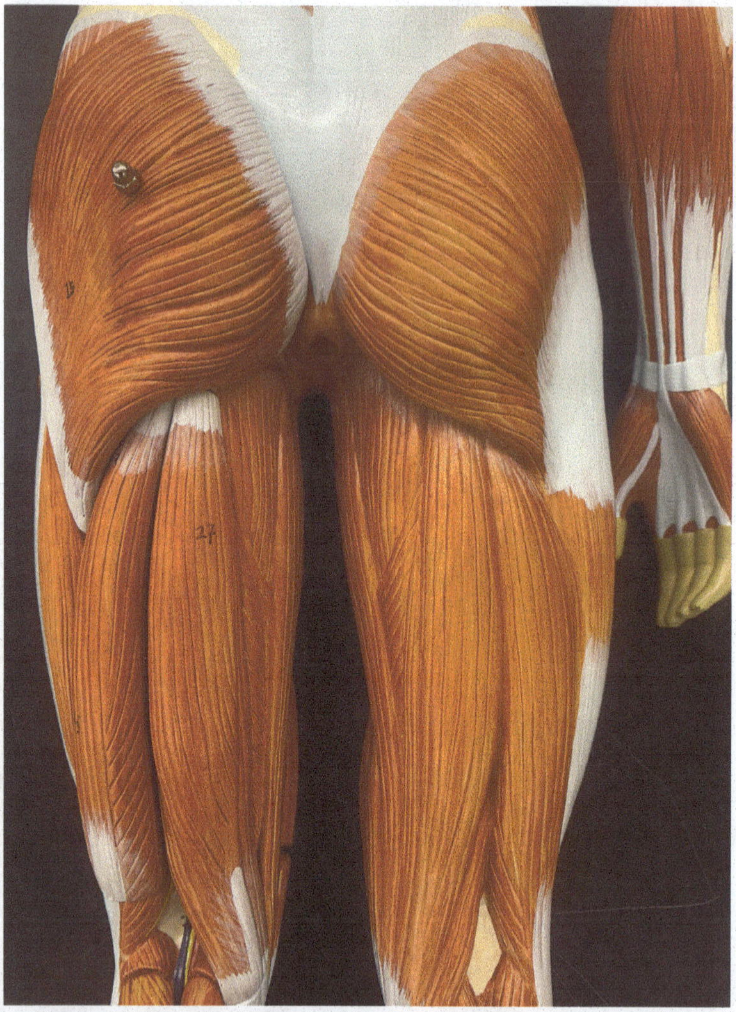

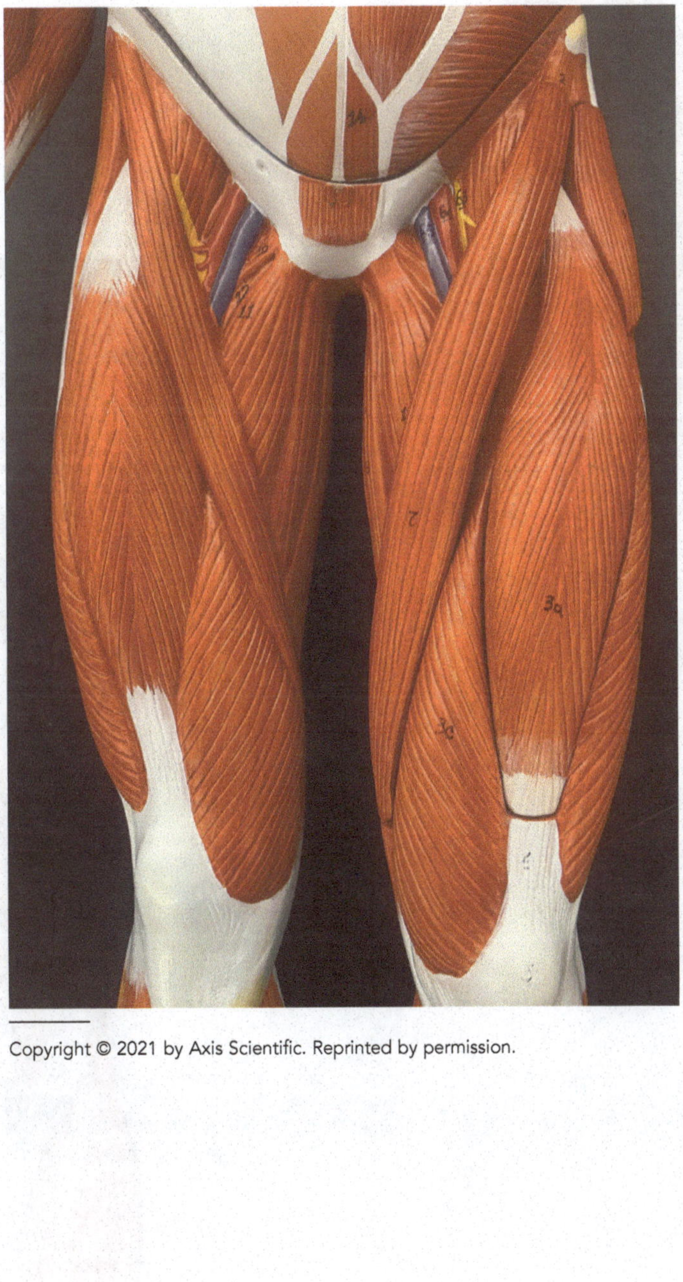

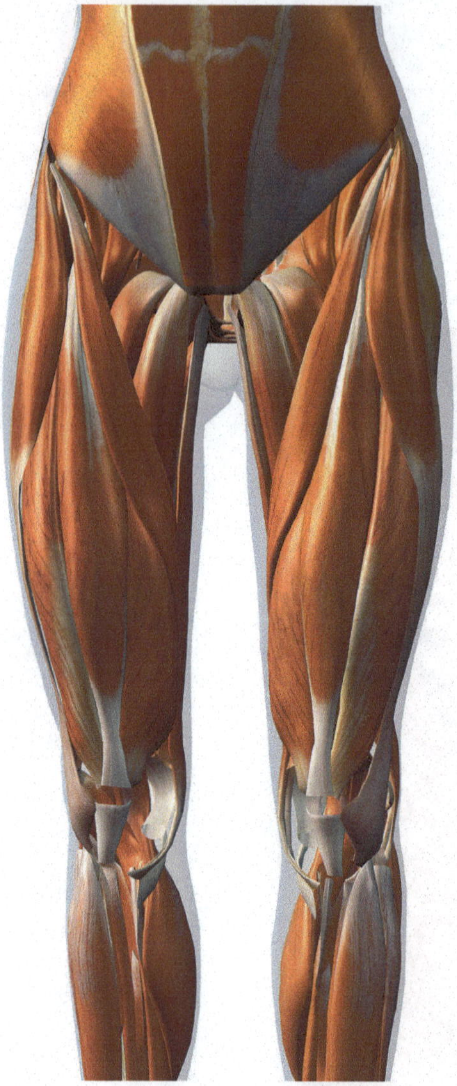

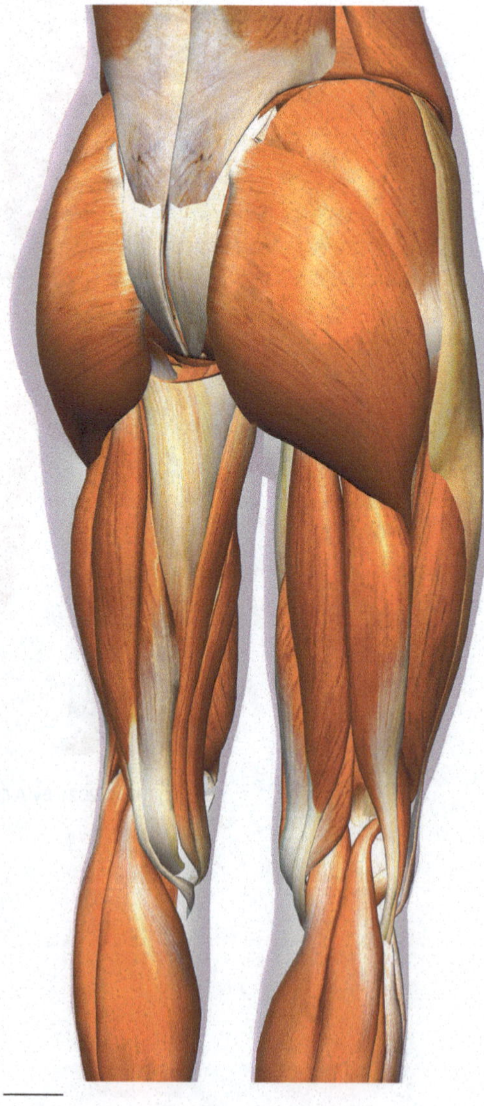

■ ***Identify the muscles of the leg.***

Tibialis anterior
Tibialis posterior
Gastrocnemius
Soleus
Calcaneal/Gastrocnemius/Achilles Tendon (nonmuscular tissue)

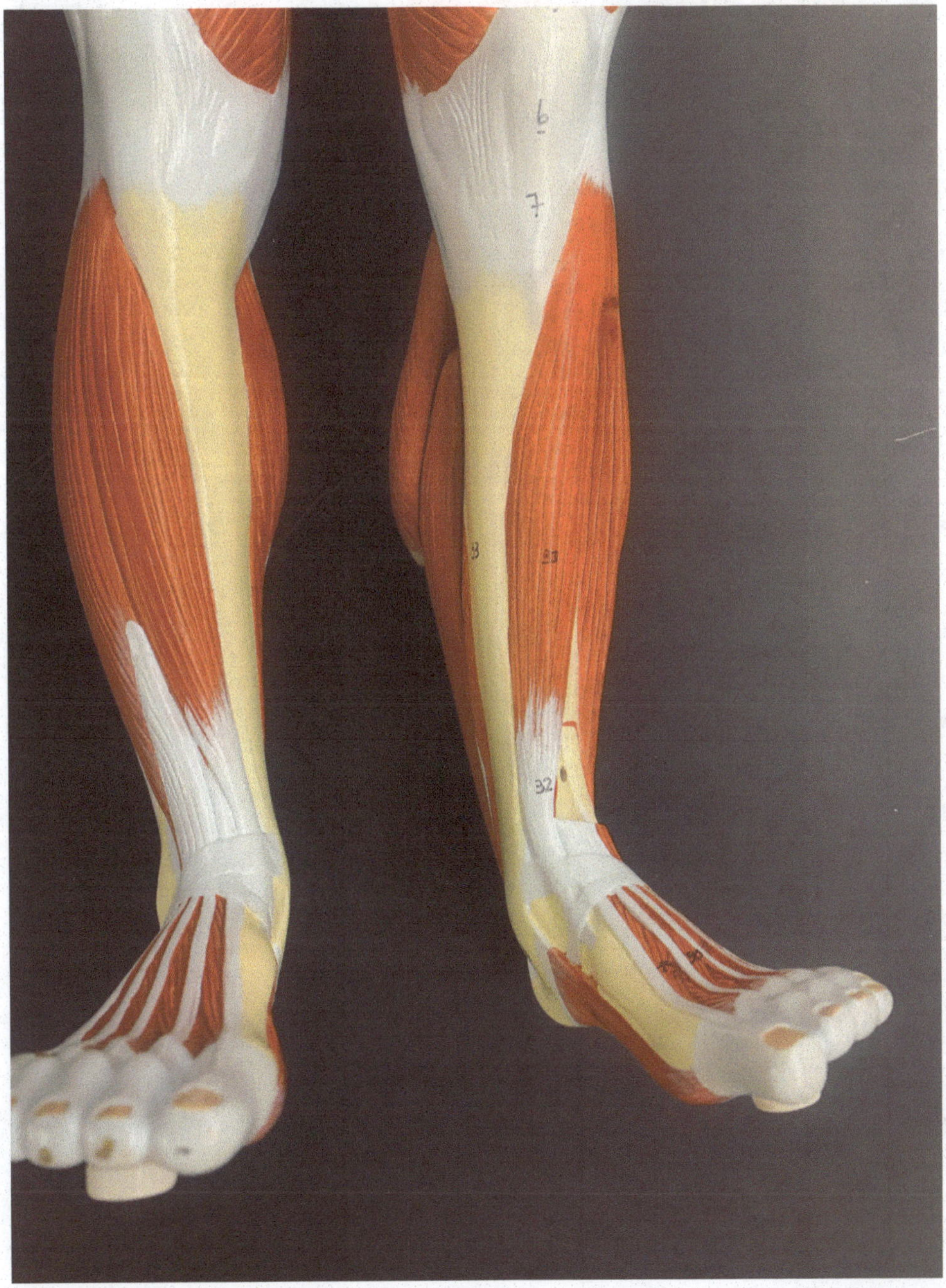

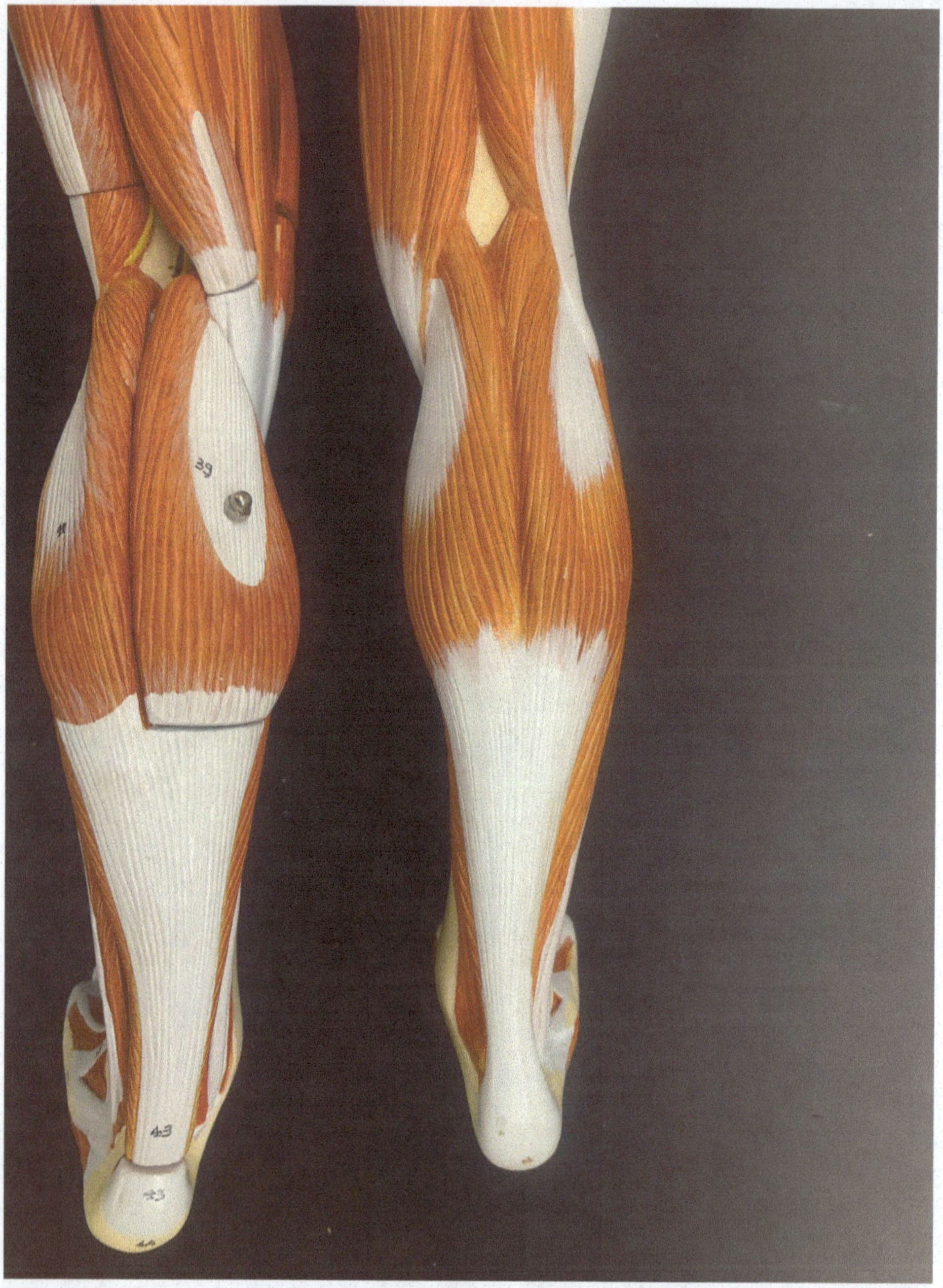

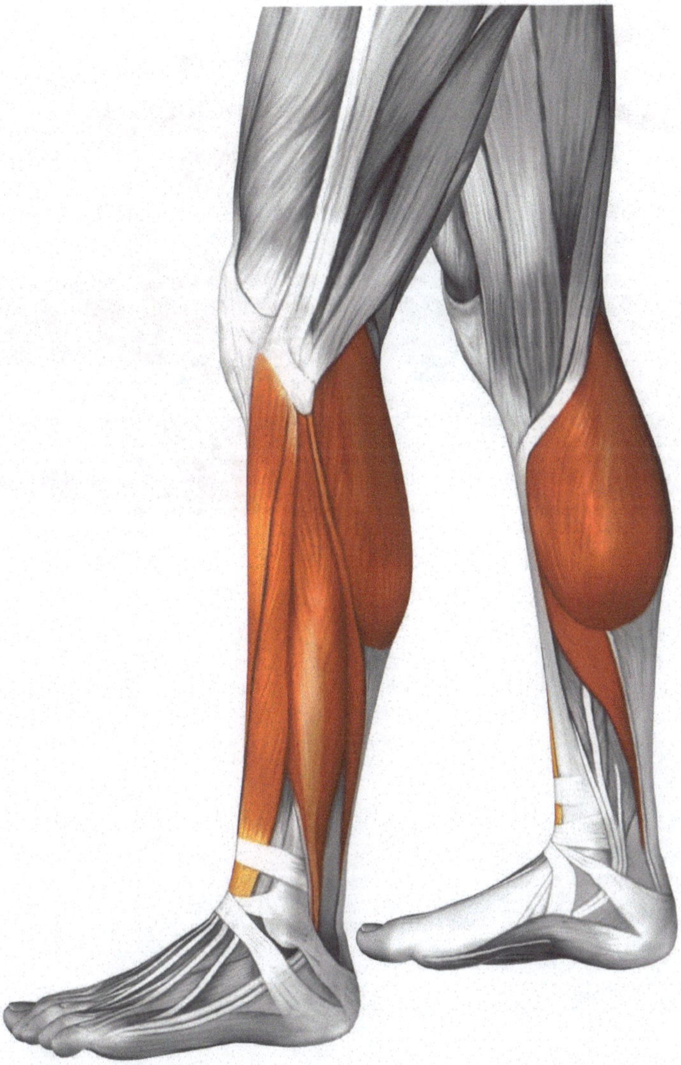

■ ***Identify the extrinsic muscles of the eye.***

Superior oblique
Inferior oblique
Superior rectus
Medial rectus
Inferior rectus
Lateral rectus

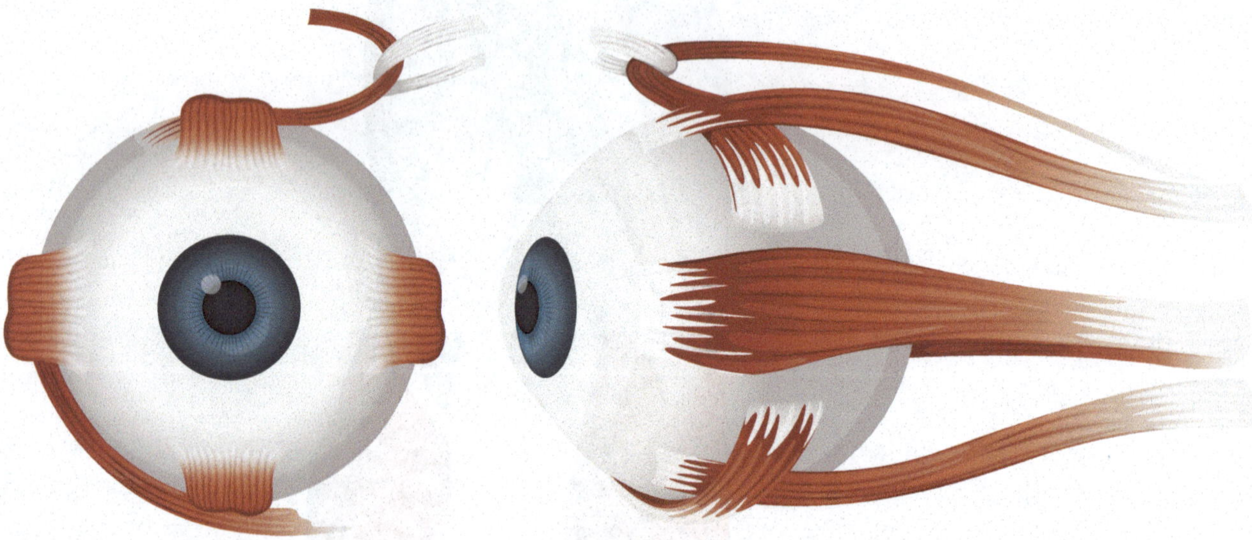

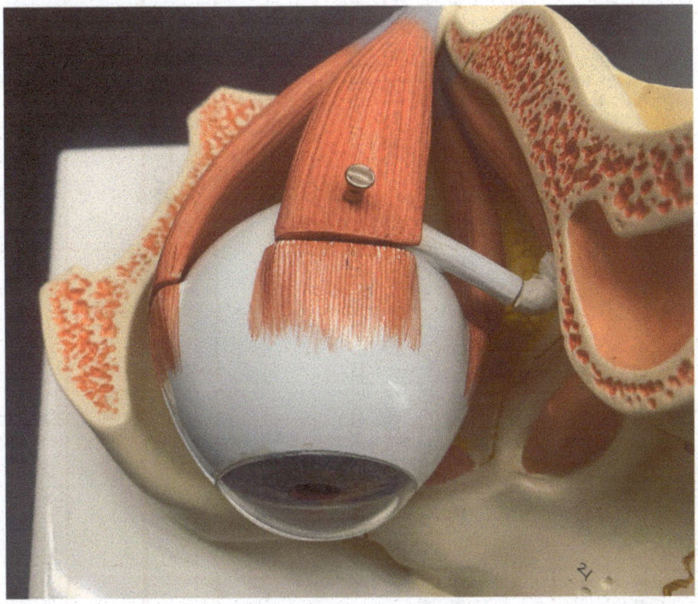

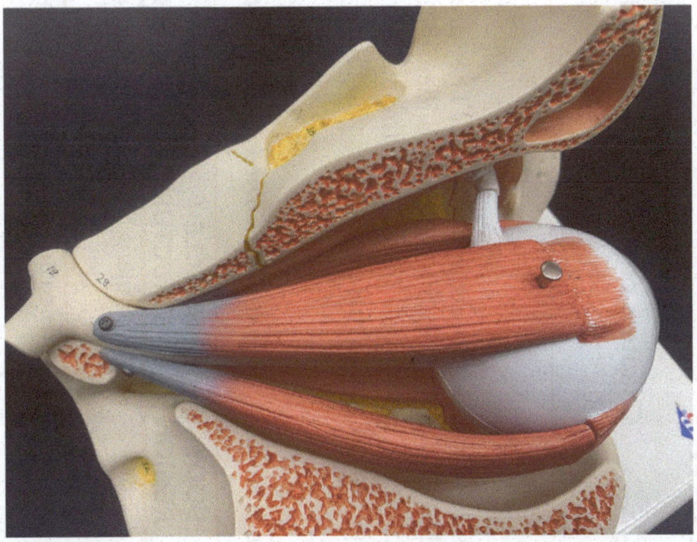

F13 © 3B Scientific GmbH, Germany, 2021 www.3bscientific.com

E. *Examination of Human Muscles and Their Actions*

OUTCOME 3: Study the actions of the muscles. Apply the terms of movement as you learn the actions of each muscle or muscle group.

TOOLBOX

Chapters on the Muscle System

arm, leg, Mr. Muscle, Mr. Flip Chart

The table lists the muscles you must locate and identify. They are listed by body region. The muscles of the leg are listed by synergistic compartments. In addition, the muscles that you are to identify are illustrated in the diagrams, photographs, and images of your textbook. Locate each muscle on the models of the arm, the leg, Mr. Muscle, and Mr. Flip Chart. Use your textbook's labeled diagrams, photographs, and images to aid you in this process. As you locate the muscles, make any notes that will help you remember information about the muscle.

■ ***Make sure to use the action words that you learned and name the affected joint.***

AXIAL/ APPENDICULAR	BODY REGION	MUSCLE / STRUCTURE	ACTION	NOTES
Axial	Facial	Orbicularis oculi	Closes eyelids	
		Orbicularis oris	Compresses/Purses lips; "kiss"	
		Masseter	Elevates the mandible to close mouth	
Axial	Neck	Sternocleidomastoid	Acting together, both flex the neck and draw the head forward; acting alone, each rotates the face up and to the opposite side	
Appendicular	Shoulder	Deltoid	Contraction of the entire muscle abducts the shoulder; contraction of lateral fibers abduct humerus; contraction of anterior fibers flex and medially rotate humerus; posterior fibers extend and laterally rotate humerus	

AXIAL/ APPENDICULAR	BODY REGION	MUSCLE / STRUCTURE	ACTION	NOTES
Axial	Trunk	Trapezius	Elevation of scapula, retraction of scapula, upward rotation of scapula and/or clavicle, and abduction & extension of neck	
Axial	Trunk	External intercostals	Elevates ribs to expand thoracic cavity	
Axial	Trunk	Internal intercostals	Depresses ribs to compress thoracic cavity	
Axial	Trunk	Transversus thoracis	Depresses ribs	
Axial	Trunk	External obliques	Together, both compress the abdomen and depress the ribs; acting alone, each one laterally flexes or rotates the vertebral column to the opposite side	
Axial	Trunk	Internal obliques	Together, both compress the abdomen and depress the ribs; acting alone, each one laterally flexes or rotates the vertebral column to the same side	
Axial	Trunk	Transversus abdominis	Compresses abdomen	
Axial	Trunk	Rectus abdominis	Depresses ribs; compresses abdomen	
Axial	Trunk	Diaphragm	Expansion of thoracic cavity and compression of abdominopelvic cavity	
Axial	Trunk	Linea alba	Midsagittal nonmuscular tissue	
Axial	Trunk	Pectoralis major	Shoulder flexion; adduction and medial rotation of humerus	
Axial	Trunk	Pectoralis minor	Depresses and protracts scapula when ribs are fixed; elevates ribs when scapula is fixed	
Axial	Trunk	Serratus anterior	Protracts shoulder; rotates scapula, abducts and elevates arm	
Axial	Trunk	Supraspinatus	Abduction of humerus	
Axial	Trunk	Infraspinatus	Extends and laterally rotates humerus	
Axial	Trunk	Subscapularis	Medial rotation of humerus	

(continued)

AXIAL/ APPENDICULAR	BODY REGION	MUSCLE / STRUCTURE	ACTION	NOTES
Axial	Trunk	Latissimus dorsi	Extension, adduction, and medial rotation at shoulder	
Appendicular	Arm	Biceps brachii (long & short heads)	Flexes elbow, abducts arm, supinates forearm	
Appendicular	Arm	Coracobrachialis	Adducts arm, flexes shoulder	
Appendicular	Arm	Triceps brachii (lateral, long, medial heads)	Extend elbow, long head adducts humerus	
Appendicular	Arm	Brachialis	Flexes elbow	
Appendicular	Arm	Pronator teres	Pronation of forearm	
Appendicular	Arm	Supinator	Supination of forearm	
Appendicular	Arm	Flexor carpi radialis	Flexion of wrist, abduction of hand	
Appendicular	Arm	Extensor carpi radialis	Extension and abduction at wrist	
Axial	Butt	Gluteus maximus	Extension of hip, lateral rotation of femur	
Axial	Butt	Gluteus medius	Abduction and medial rotation of femur	
Axial	Hip	Tensor fasciae latae	Flexion of hip joint; abducts and medially rotates femur	
Appendicular	Thigh	Adductor longus	Adduction and lateral rotation of femur	
Appendicular	Thigh	Adductor magnus	Anterior portion adducts and laterally rotates femur and flexes hip; posterior portion extends hip	
Appendicular	Thigh	Gracilis	Adducts femur, extends knee	
Appendicular	"Hamstring"	Biceps femoris	Flexion at knee, extension at hip	
Appendicular	"Hamstring"	Semimembranosus	Flexion at knee; extension at hip	
Appendicular	"Hamstring"	Semitendinosus	Flexion at knee; extension at hip	

AXIAL/ APPENDICULAR	BODY REGION	MUSCLE / STRUCTURE	ACTION	NOTES
Appendicular	"Quadriceps femoris"	Rectus femoris	Extension at the knee; flexion at the hip	
Appendicular	"Quadriceps femoris"	Vastus lateralis	Extension at the knee	
Appendicular	"Quadriceps femoris"	Vastus intermedius	Extension at the knee	
Appendicular	"Quadriceps femoris"	Vastus medialis	Extension at the knee	
Appendicular	Thigh	Sartorius	Flexion at knee and hip; abduction, flexion, & lateral rotation at hip	
Appendicular	Thigh	Iliotibial (IT) Band/Tract		
Appendicular	Leg	Tibialis anterior	Dorsiflexion and inversion of foot	
Appendicular	Leg	Tibialis posterior	Plantar flexion and inversion of foot	
Appendicular	Leg	Gastrocnemius	Plantar flexion at ankle; flexion at knee	
Appendicular	Leg	Soleus	Plantar flexion of foot	
Appendicular	Leg	Calcaneal / Gastrocnemius / Achilles Tendon		
Axial	Eye	Superior oblique	Rotates eye inferiorly and laterally, look down and out	
Axial	Eye	Inferior oblique	Rotates eye superiorly and laterally, look up and out	
Axial	Eye	Superior rectus	Elevates eye, pulls eye upward	
Axial	Eye	Medial rectus	Adducts eye, pulls eye towards nose	
Axial	Eye	Inferior rectus	Depresses eye, pulls eye downward	
Axial	Eye	Lateral rectus	Abducts eye, pulls eye away from nose	

CHECKLIST: Summary Checklist Skeletomucsular

TERM/STRUCTURE	YOUR NOTES
Orbicularis Oculi	
Orbicularis Oris	
Masseter	
Sternocleidomastoid	
Trapezius	
Deltoid	
External Intercostal Muscles	
Internal Intercostal Muscles	
Transverse Thoracis	
External Oblique Muscles	
Internal Oblique Muscles	
Transversus Abdominis	
Rectus Abdominis	
Diaphragm	
Superior Oblique (of the Eye)	
Inferior Oblique (of the Eye)	
Superior Rectus (of the Eye)	
Medial Rectus (of the Eye)	
Inferior Rectus (of the Eye)	
Lateral Rectus (of the Eye)	
Pectoralis Major	
Pectoralis Minor	
Serratus Anterior	
Supraspinatus	
Infraspinatus	
Subscapularis	
Latissimus Dorsi	
Biceps Brachii—Long and Short Head	

TERM/STRUCTURE	YOUR NOTES
Coracobrachialis	
Triceps Brachii—Lateral, Long, and Medial Head	
Brachialis	
Pronator Teres	
Supinator	
Flexor Carpi Radialis	
Extensor Carpi Radialis	
Gluteus Maximus	
Gluteus Medius	
Tensor Fasciae Latae	
Adductor Longus	
Adductor Magnus	
Gracilis	
Biceps Femoris	
Semimembranosus	
Semitendinosus	
Rectus Femoris	
Vastus Lateralis	
Vastus Intermedius	
Vastus Medialis	
Sartorius	
Tibialis Anterior	
Tibialis Posterior	
Gastrocnemius	
Soleus	
Iliotibial (IT) Band/Tract	
Calcaneal/Achilles Tendon	
Linea Alba	

EXERCISE 6

Brain and Cranial Nerves

A. *Introduction*

The **brain, spinal cord** and **meninges** are the elements of the Central Nervous System. Review the pattern of a **reflex arc:**

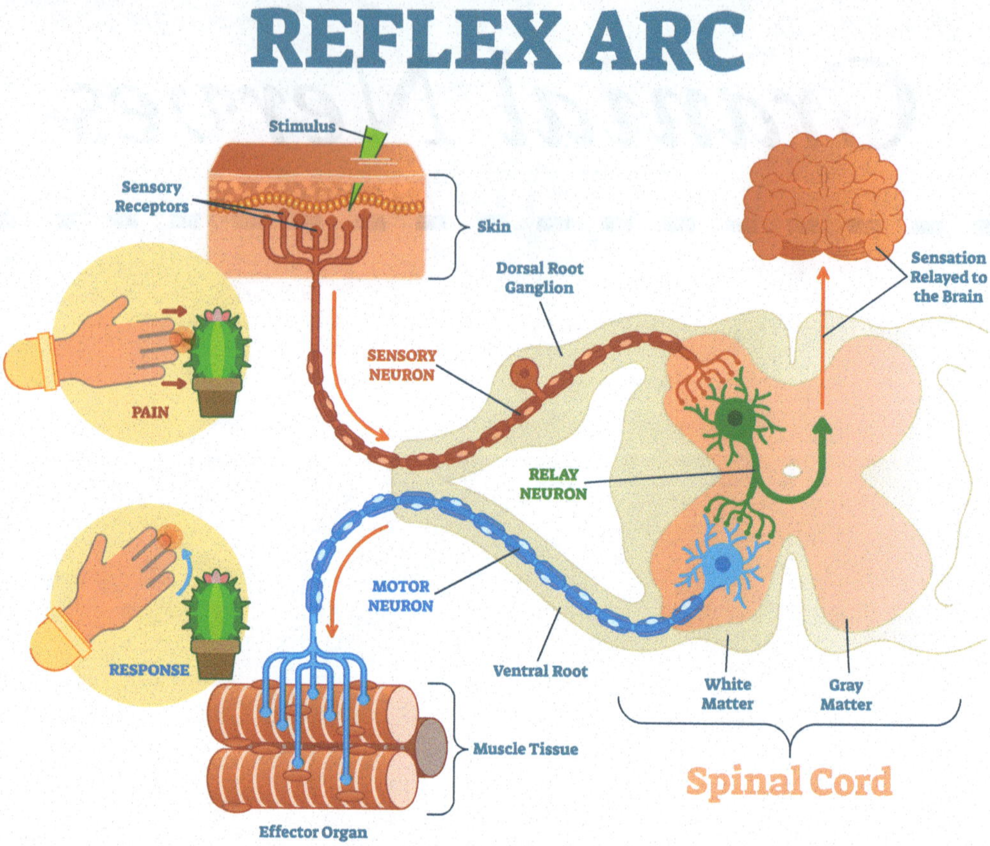

As you examine the anatomy of these elements, remember that the CNS is the control center of a reflex arc—its function is to process input information and to plan an output action. The input pathway is the sensory / afferent pathway and the output pathway is the motor / efferent pathway. The spinal and cranial nerves are the connecting structures within the Peripheral Nervous System. The spinal cord and brain are the integrating and coordinating structures within the Central Nervous System.

B. *Outcomes for Exercise 6*

1. Locate, label, and identify the ventricles of the brain and some related anatomical features.

2. Locate, label, and identify the cranial meninges.

3. Use a flowchart to help describe the pathway of CSF by listing the structures it passes through in the correct order. What structures produce the CSF? Be able to determine where the fluid was immediately before a specific location and where it will be immediately after a specific location.

4. Locate, label, and identify the external and internal structures of the brain.

5. Locate, label, and identify the cranial nerves by name and Roman numeral designation (CNI–CNXII). For each nerve, determine its general modality (sensory, motor, or both) and its general function.

6. Dissect the sheep brain and locate the structures listed. Be able to identify all structures. Identify the following cranial nerves by name, specific function, and Roman numeral designation: CNI, CNII, CNIII, CNV.

C. *Human Brain*

OUTCOME 1: Locate, label, and identify the ventricles of the brain and some related anatomical features.

TOOLBOX

Ventricles of the Brain

human brain models

ventricle brain model

cerebral hemispheres

cerebellum

medulla oblongata

pons

sagittal/longitudinal fissure

left and right lateral ventricles

- **anterior horns**
- **posterior horns**
- **inferior horns**

left and right interventricular foramina

3rd ventricle

aqueduct of mid-brain (cerebral aqueduct)

4th ventricle

central canal

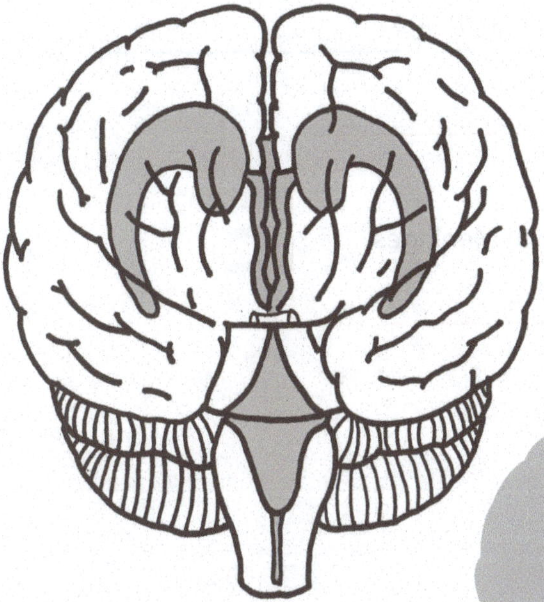

Anterior View

Illustration by Mark Neilsen, © 2015.

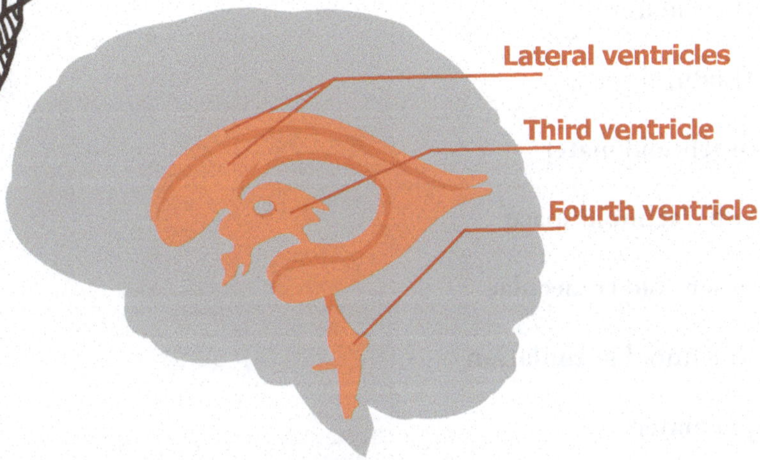

© Systemoff/shutterstock.com

Lateral View

VH410 © 3B Scientific GmbH, Germany, 2021 www.3bscientific.com

OUTCOME 2: Locate, label, and identify the cranial meninges.

TOOLBOX

The Cranial Meninges

dura mater

subdural space

arachnoid mater

subarachnoid space

arachnoid trabeculae

arachnoid granulation

pia mater

cerebral cortex

cerebral vein

superior sagittal sinus (a vein)

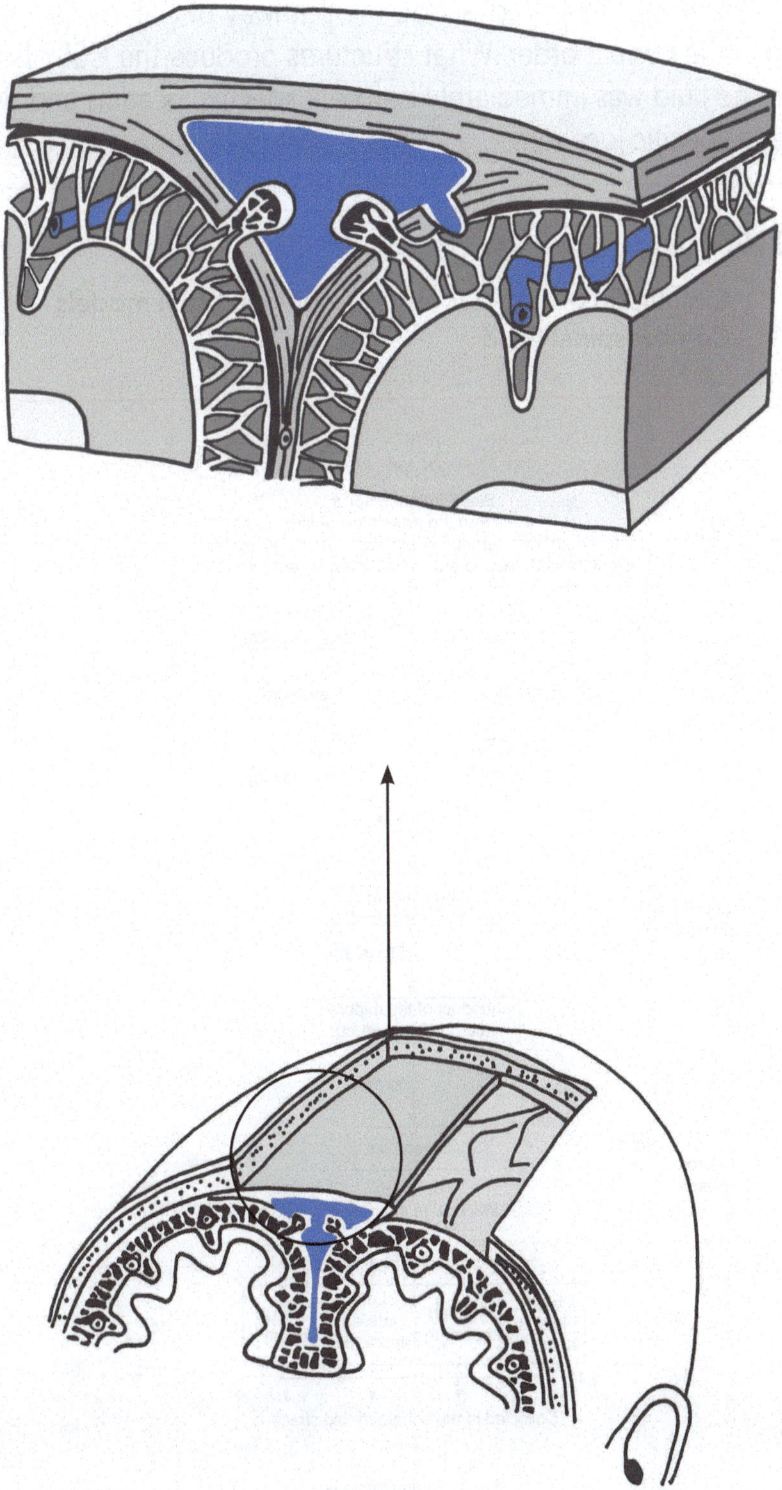

Illustration by Lucia Tranel, © 2015.

OUTCOME 3: Use flowchart to help describe the pathway of CSF by listing the structures it passes through in the correct order. What structures produce the CSF? Be able to determine where the fluid was immediately before a specific location and where it will be immediately after a specific location.

TOOLBOX

Circulation of Cerebrospinal Fluid

human brain models

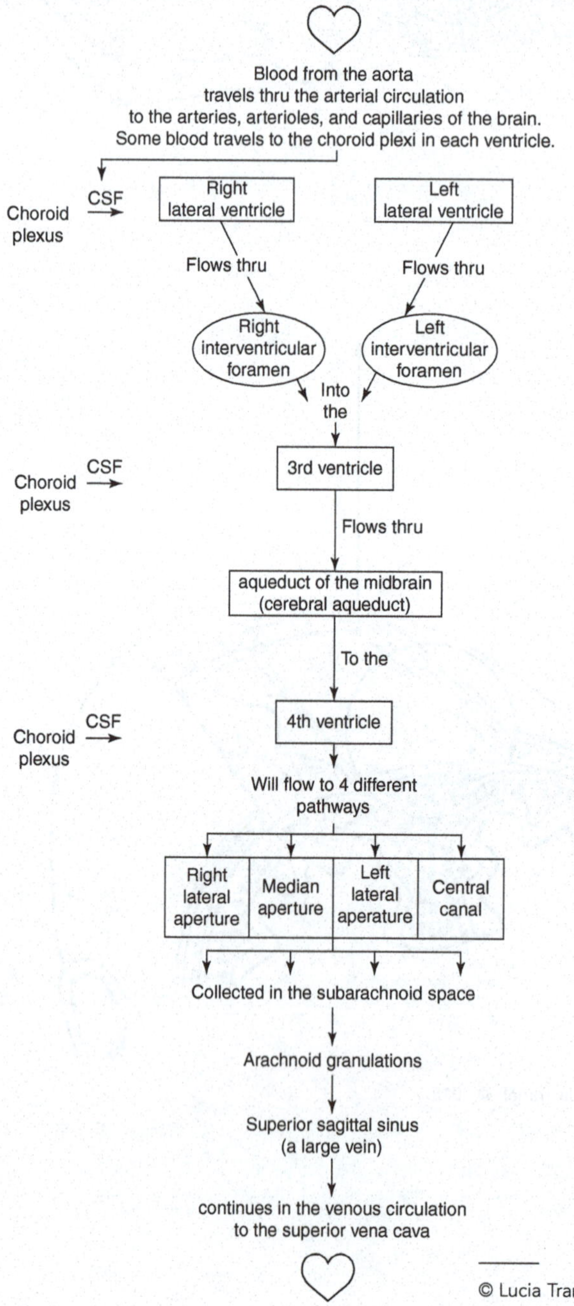

- ***Use the drawing to locate the structures of the CSF flow pattern (only the right side of the CNS is represented).***
- ***Draw arrows in the image to indicate the flow pattern.***

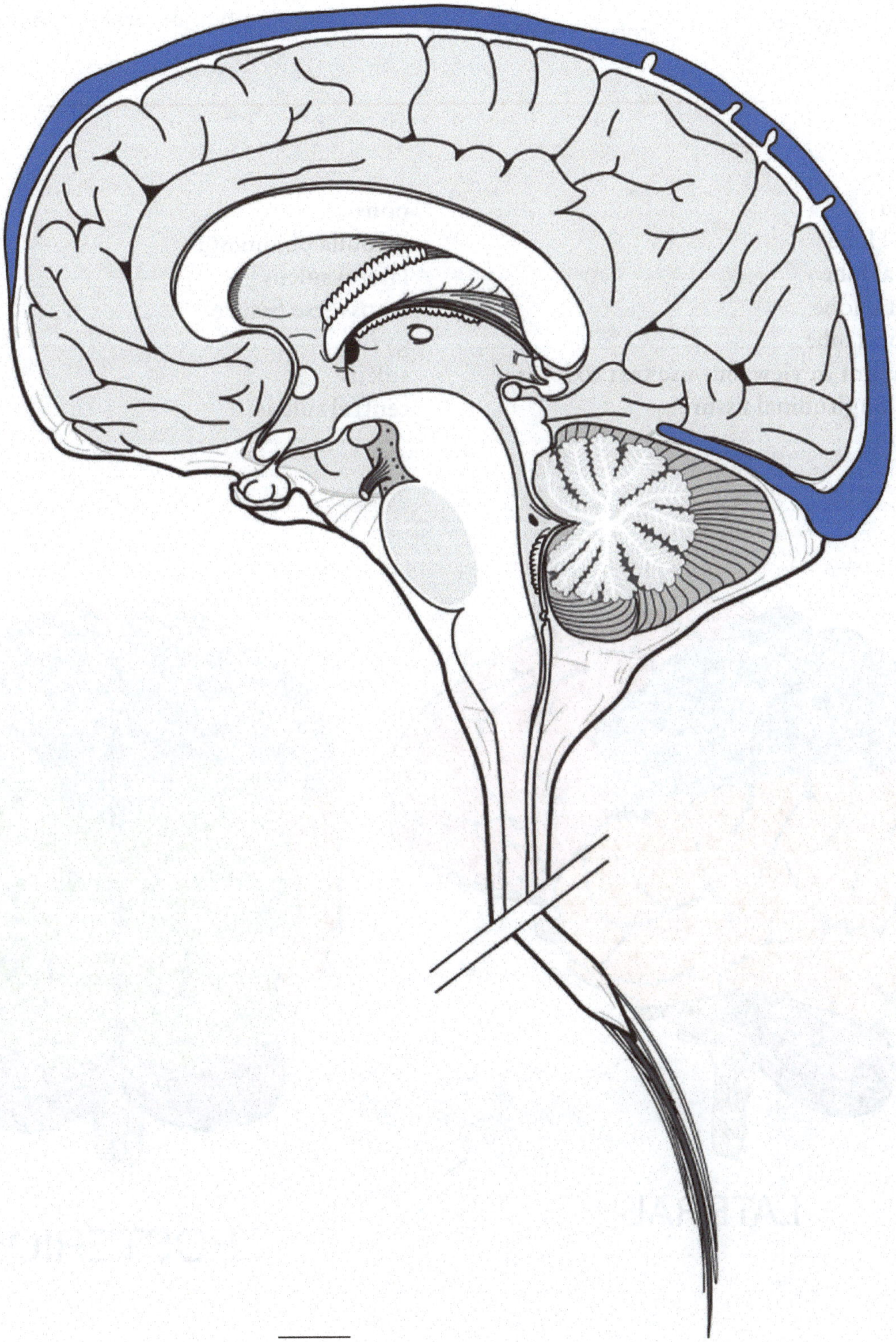

Illustration by Mark Neilsen, © 2015.

OUTCOME 4: Locate, label, and identify the external and internal structures of the brain.

TOOLBOX

Cerebral Hemishperes

human brain models

cerebrum
- frontal lobe
- parietal lobe
- occipital lobe
- temporal lobe
- insula (not in view, but use text to locate)

sagittal/longitudinal fissure
cerebellum
pons
medulla oblongata
lateral sulcus
transverse fissure
gyrus
sulcus
central sulcus

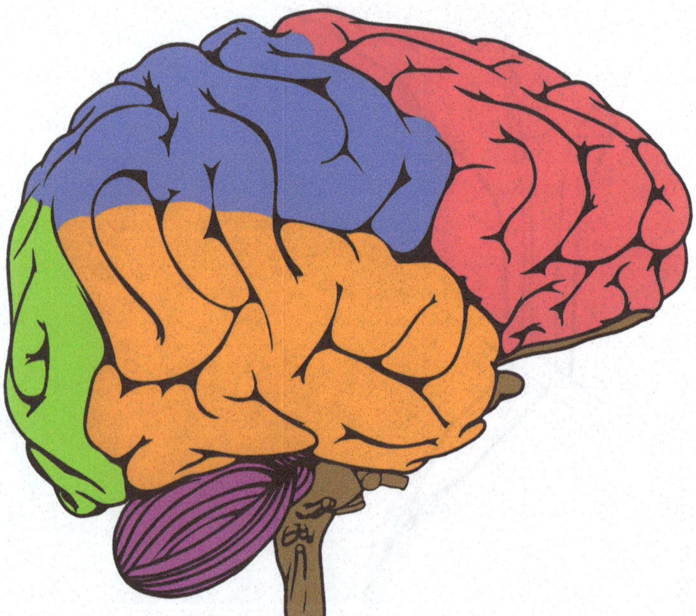

LATERAL

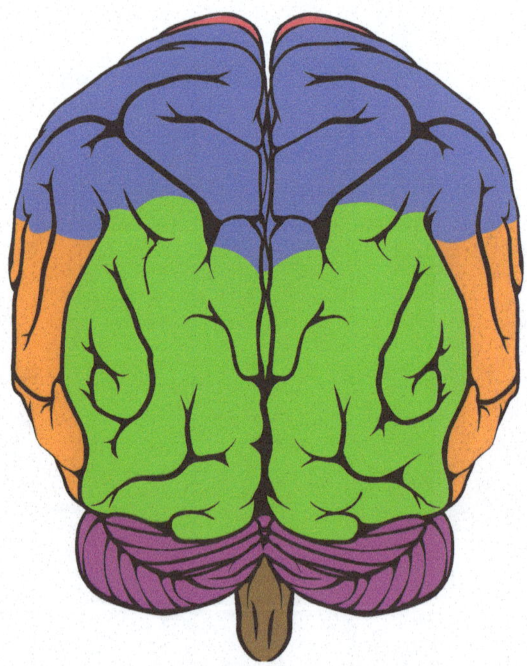

POSTERIOR

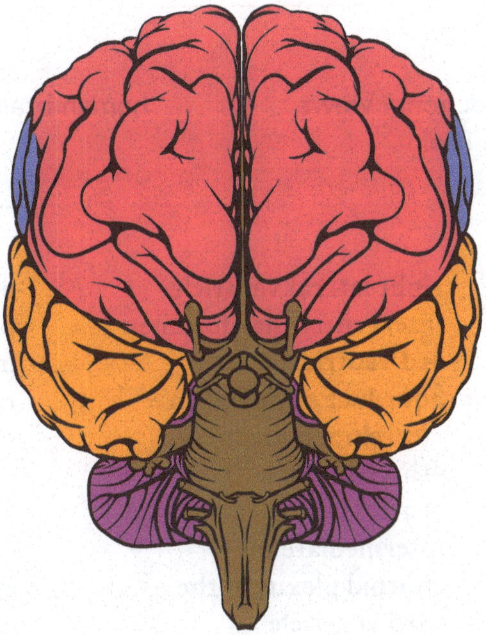

ANTERIOR

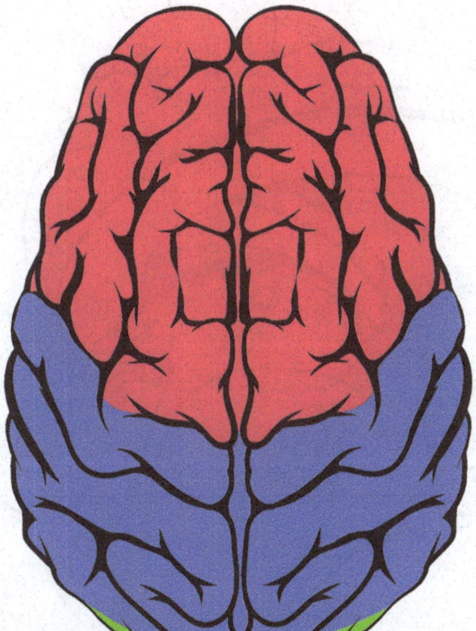

SUPERIOR

TOOLBOX

The Sectional Views of the Brain

human brain models

cerebrum
- frontal lobe
- temporal lobe
- parietal lobe
- occipital lobe

cerebellum (arbor vitae)
pons
medulla oblongata
central sulcus
corpus callosum
fornix
right lateral ventricle
interventricular foramen
3rd ventricle
aqueduct of midbrain (cerebral aqueduct)
4th ventricle
central canal
intermediate mass
choroid plexus of the 3rd ventricle
optic chiasma
pituitary (hypophysis)
infundibulum
mammillary body
pineal body
corpora quadrigemina
- superior colliculus
- inferior colliculus

cerebral peduncle (midbrain)

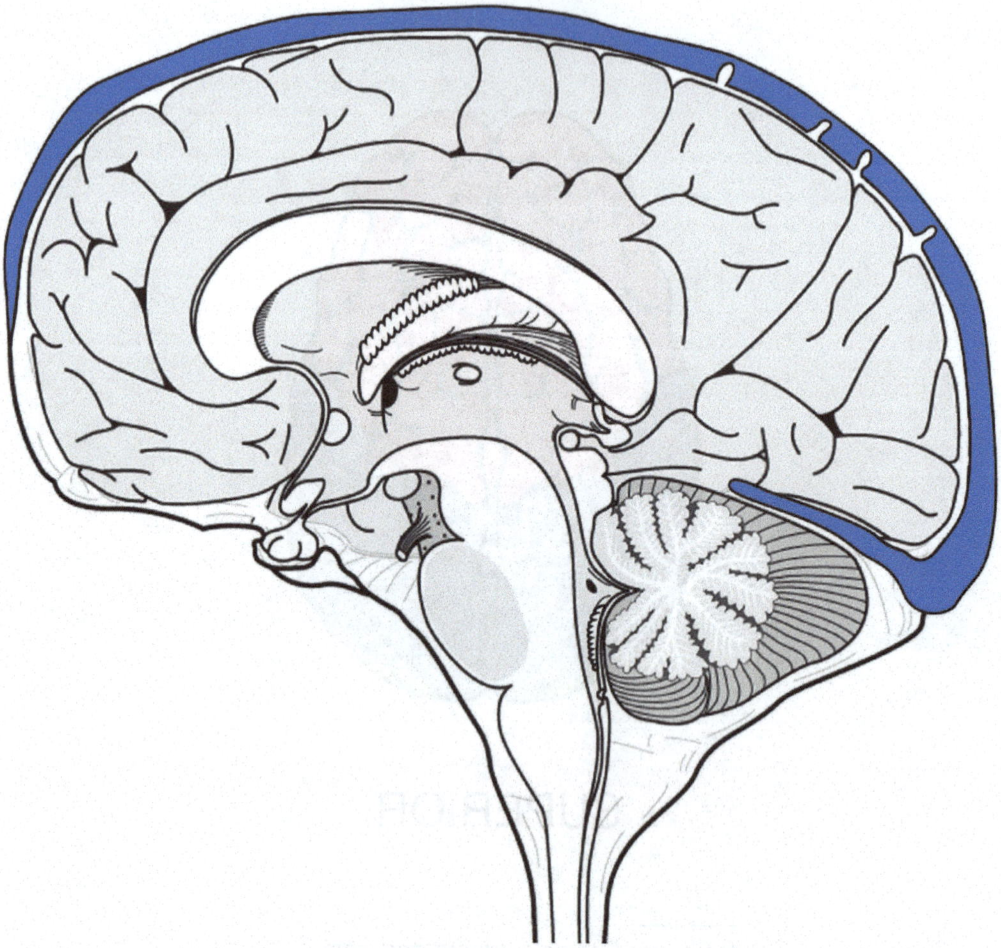

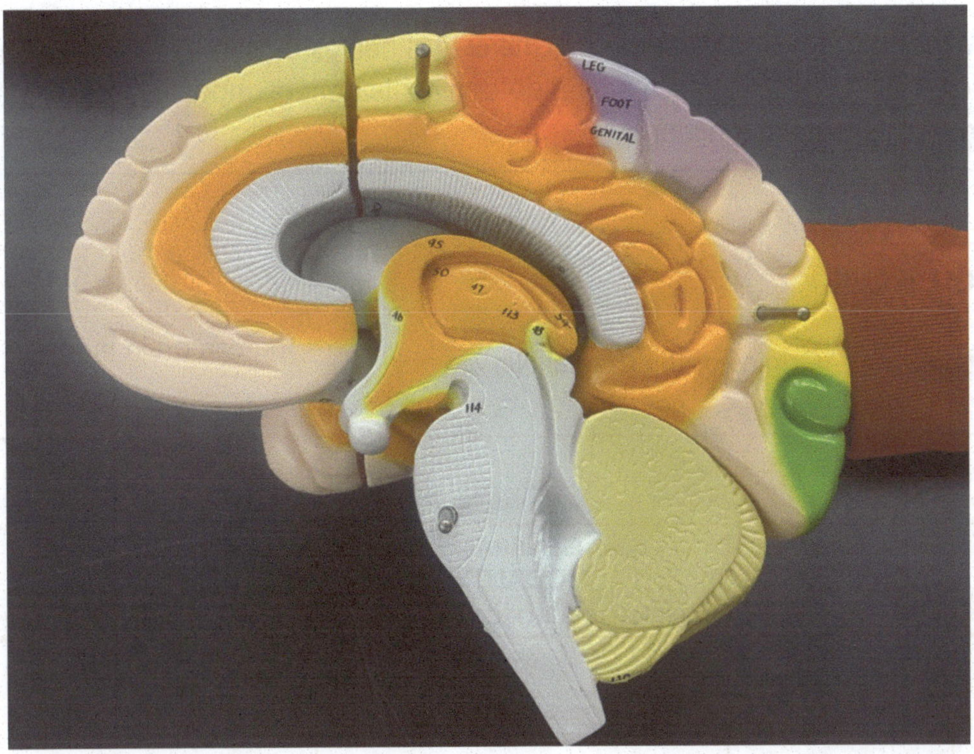
LEG
FOOT
GENITAL

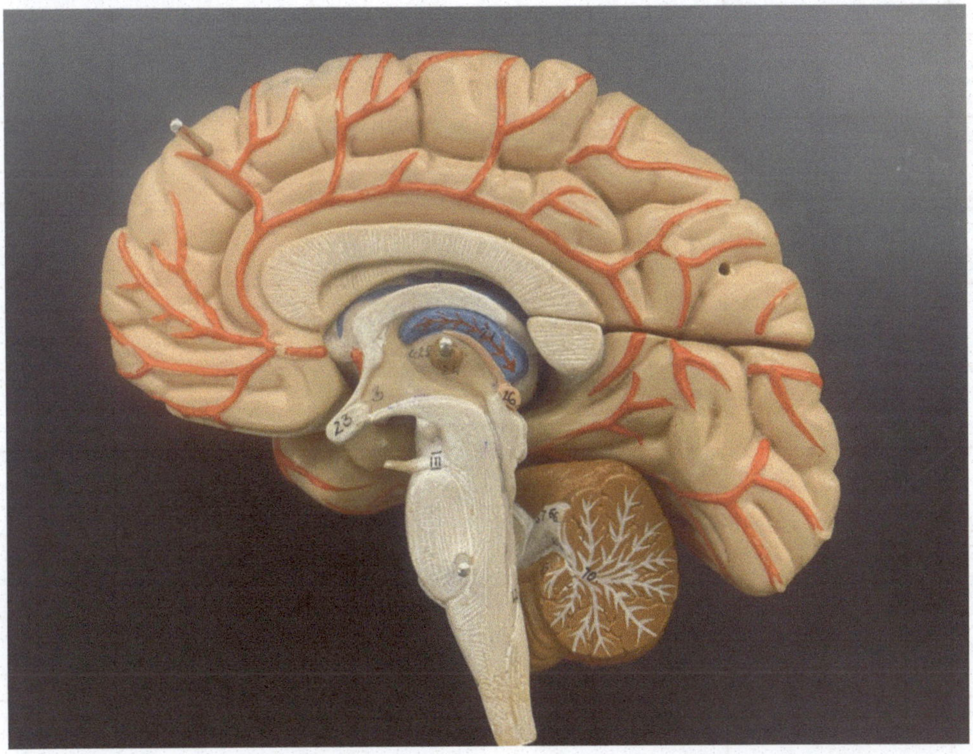

OUTCOME 5: Locate, label, and identify the cranial nerves by name and Roman numeral designation (CNI–CNXII). For each nerve, determine its general modality (sensory, motor, or bother) and its general function.

TOOLBOX

Origins of the Cranial Nerves

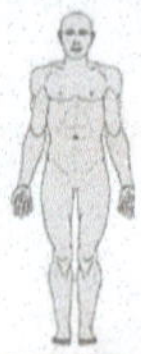

human brain models

NUMBER	NAME	GENERAL MODALITY (SENSORY, MOTOR, BOTH)	GENERAL FUNCTION
CN I			
CN II			
CN III			
CN IV			
CN V			
CN VI			
CN VII			
CN VIII			
CN IX			
CN X			
CN XI			
CN XII			

- ***Write the name of each nerve at the designated line on the image.***
- ***Color the nerves as follows:*** *YELLOW = SENSORY*, *GREEN = MOTOR*, *PURPLE = BOTH*

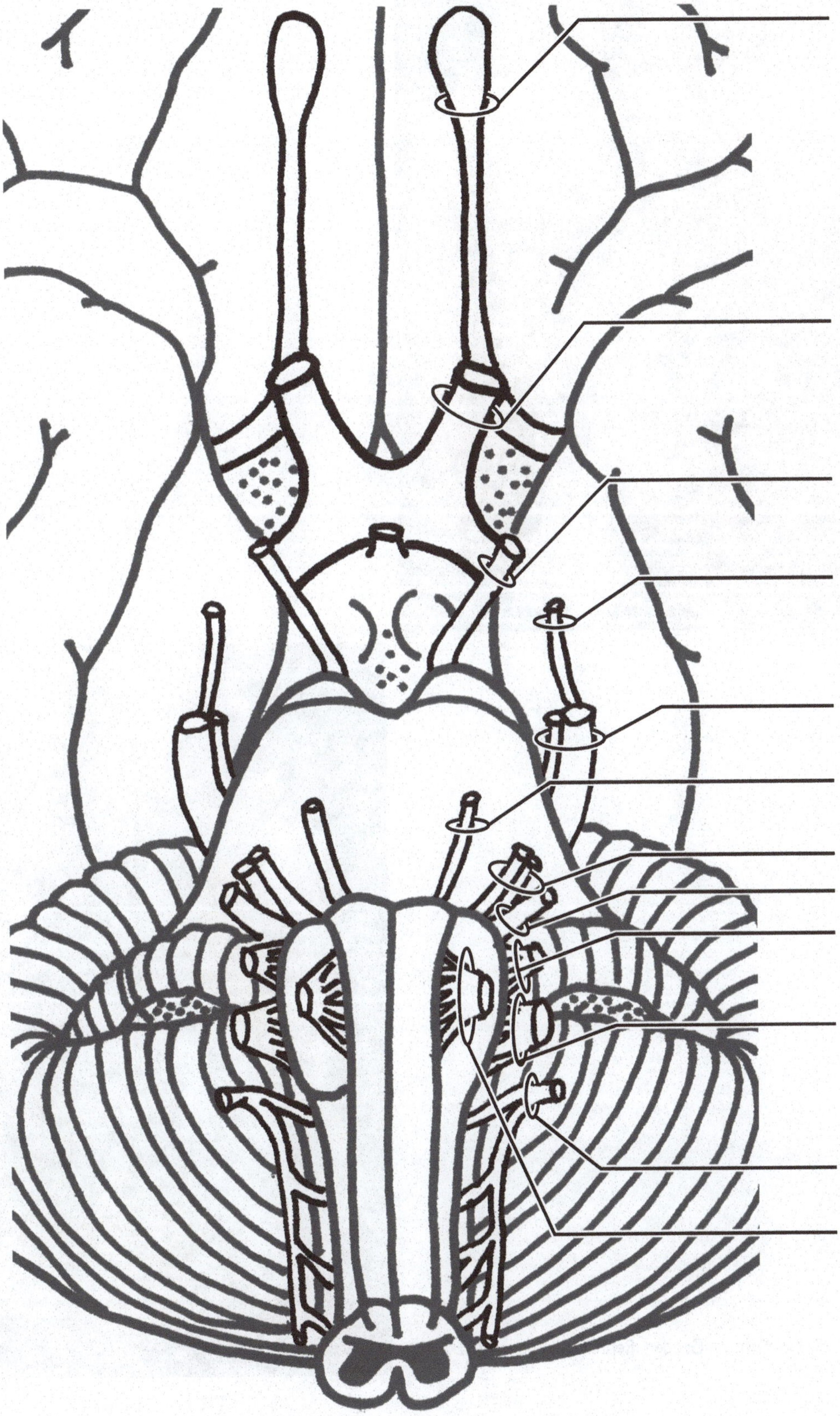

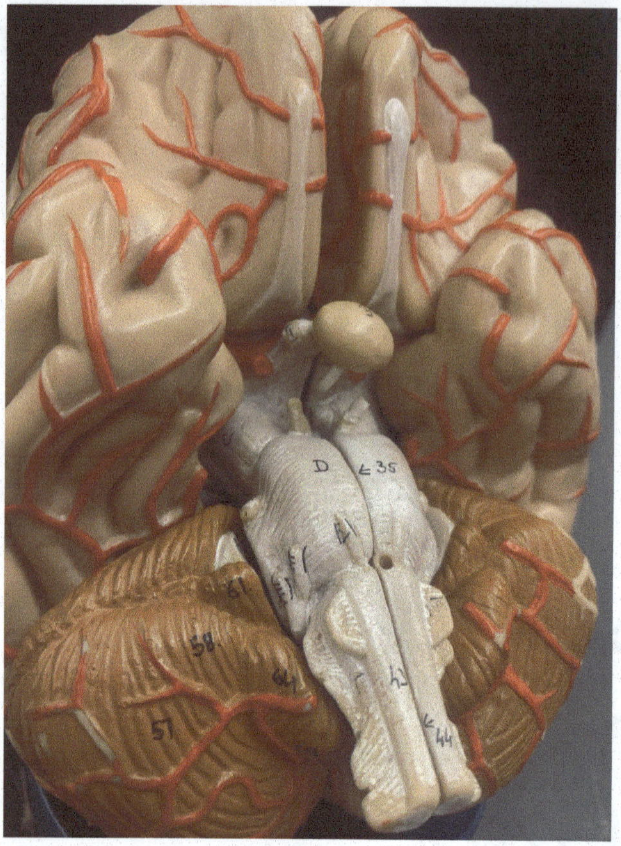
D
35
61
58
64
57
42
44

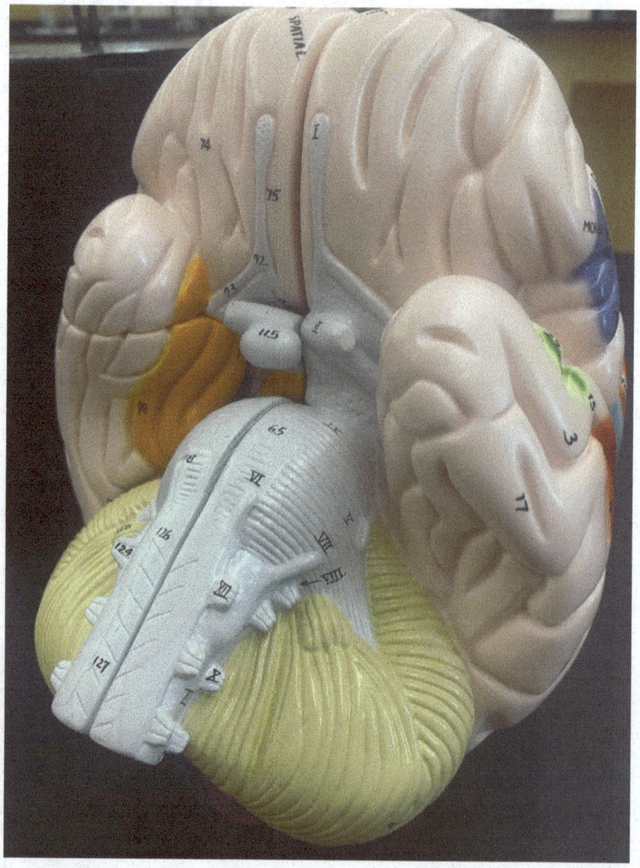
SPATIAL
74
I
75
115
65
VI
126
124
127
3
17

E. *Comparative Anatomy—Dissection*

OUTCOME 6: Dissect the sheep brain and locate the structures listed. Be able to identify all structures. Identify the following cranial nerves by name, specific function, and Roman numeral designation: CNI, CNII, CNIII, CNV.

We will use a sheep brain for the comparative anatomy study of the brain, because it is very similar to human brains, and unlike a human brain, it is readily available and relatively inexpensive. You will not need to perform any dissection; the specimens are dissected for you. All you will need is a probe to aid in pointing out structures. Each specimen is in an appropriately labeled container and submerged in preservative. *Please make sure to put all specimens back in the correct container and tightly close all lids.*

TOOLBOX

✂

Brain with Dura Mater

Obtain the brain from the specimen jar.

1. Three layers of membranes known as the **meninges** enclose the brain and spinal cord. Identify the outermost meningeal layer, the **dura mater.** It is a thick, white, fibrous covering. The **dura mater encephali** covers the brain; the **dura mater spinalis** covers the spinal cord. Can you understand why this membrane has a name that means "tough mother"? If the dura mater was removed, the **arachnoid mater** would be visible. The **pia mater** lies directly on the surface of the brain, beneath the arachnoid mater.

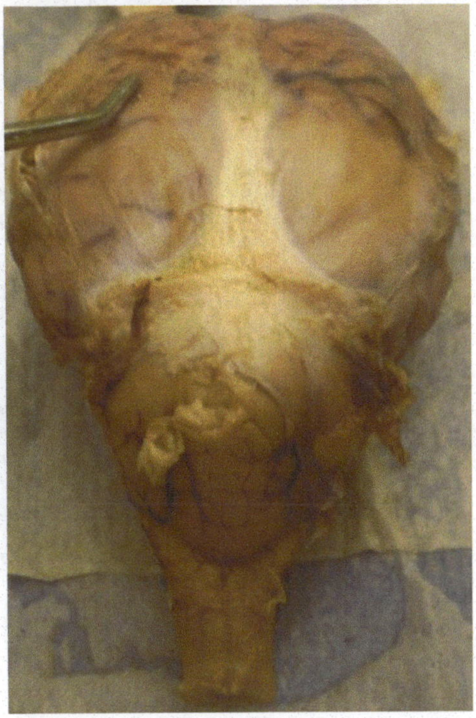

Source: Ben Jellen, 2021.

2. Use your textbook to find images of the Cranial Meninges.

3. The dura mater obscures most of the detail surface anatomy of the brain, but you should be able to identify the three major regions of the brain:

   - **cerebrum**—the largest and most superior/anterior section of the brain. This is the region that develops from the **telencephalon.** The **diencephalon** and **mesencephalon (mid-brain)** develop into structures deep within the cerebrum.
   - **cerebellum**—"small brain." Develops from the **metencephalon.**
   - **medulla oblongata**—The region that is continuous with the spinal cord as it passes through the foramen magnum. It derives from the **myelencephalon.**

4. Examine the ventral view of the brain. Find the **pituitary gland, optic nerves, optic chiasma,** and **trigeminal nerves.** Make sure to identify these structures now, as they are usually removed or destroyed on the next specimen to be examined: the brain without the dura mater.

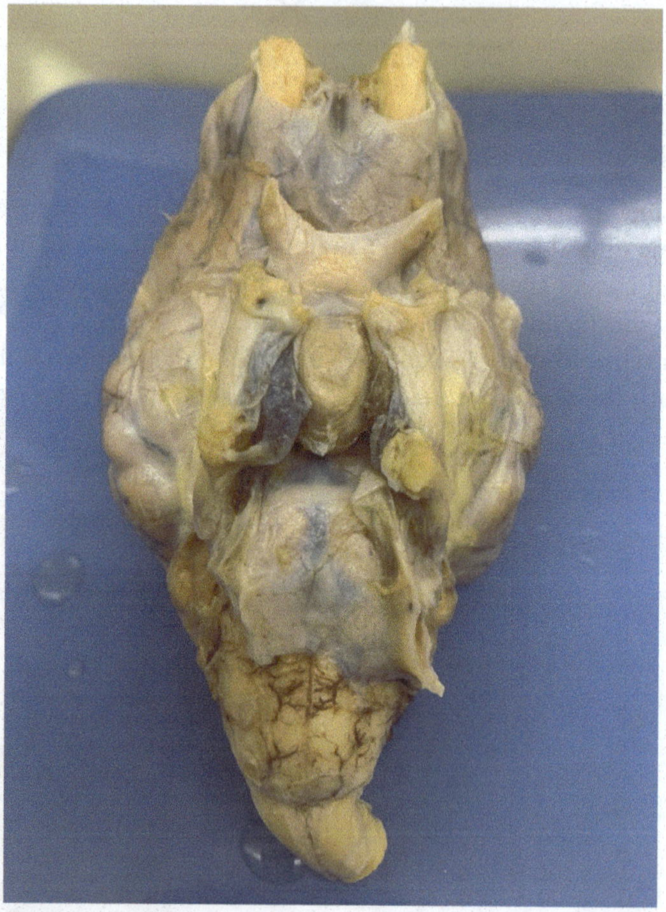

Source: Ben Jellen, 2021.

Brain, without Dura Mater

Obtain the brain from the specimen jar.

5. Locate:

- **cerebral hemispheres**—paired halves of the cerebrum. Regions of the cerebrum are designated based upon the bone under which the region sits: **frontal lobe, occipital lobe, parietal lobe, temporal lobe.**
- **longitudinal fissure**—deep groove that separates the two hemispheres.
- **corpus callosum**—Gently spread the cerebral hemispheres apart and observe the thick, white band that connects the two hemispheres.
- **gyrus/sulcus**—Notice that the surface area of the cerebrum is increased with **gyri** and **sulci.** A gyrus is a raised convolution; a sulcus is a shallow groove. What is the function of these structures?
- **transverse fissure**—separates the cerebrum from the cerebellum.

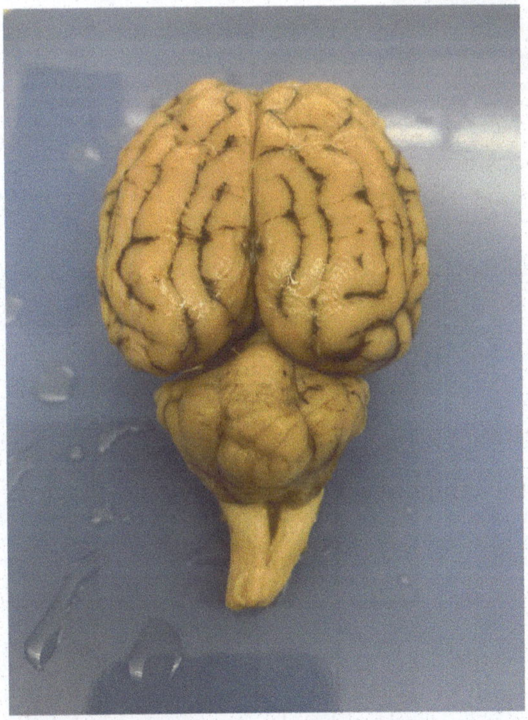

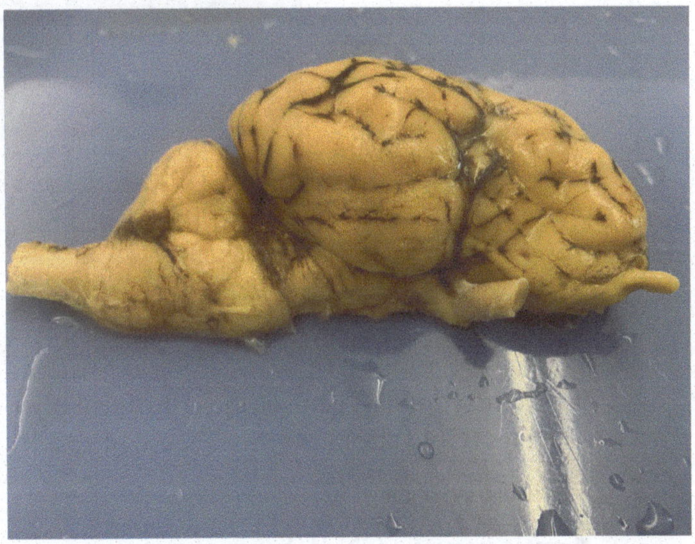

Source: Ben Jellen, 2021.

6. The roof of the mid-brain can be seen by spreading the cerebrum and the cerebellum apart at the transverse fissure.
   - **corpora quadrigemina**—four prominent lumps. The larger, superior pair is the **superior colliculi.** The smaller, posterior pair is the **inferior colliculi.**
   - **pineal body**—seen very deep where the longitudinal fissure and transverse fissure meet, just superior to the superior colliculi.
   - **cerebellum**—on the dorsal surface, posterior to the cerebrum.

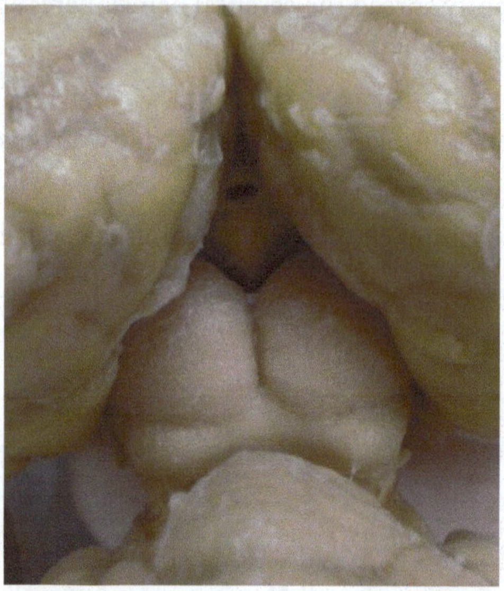

7. Observe the ventral surface of the sheep brain. Locate:
   - **olfactory bulbs**—lie over the cribriform plate of the ethmoid bone and receive the olfactory nerves (CN I) from the nose. Sense of smell.
   - **olfactory tract**
   - **optic nerves**—CN II. Sense of sight.
   - **optic chiasma**—some of the fibers of the left and right optic nerves cross at this point before proceeding into the **optic tract.** The optic chiasma is located in the chiasmatic groove of the skull.
   - **infundibulum**—the stalk of the **pituitary (hypophysis).** The pituitary came off with the removal of the dura mater. It sits in the hypophyseal fossa.

- **mammillary body**
- **cerebral peduncle**—at the mid-brain, the trunk on which the cerebrum sits. The **oculomotor nerve (CN III)** emerges from the cerebral peduncle. This nerve controls a number of muscles in the eye.
- **pons**—on either side of the pons, you might be able to locate a stub of the **trigeminal nerve (CN V).** This nerve receives sensations from the face and controls some motor activity of chewing.
- **medulla oblongata**
- **spinal cord**

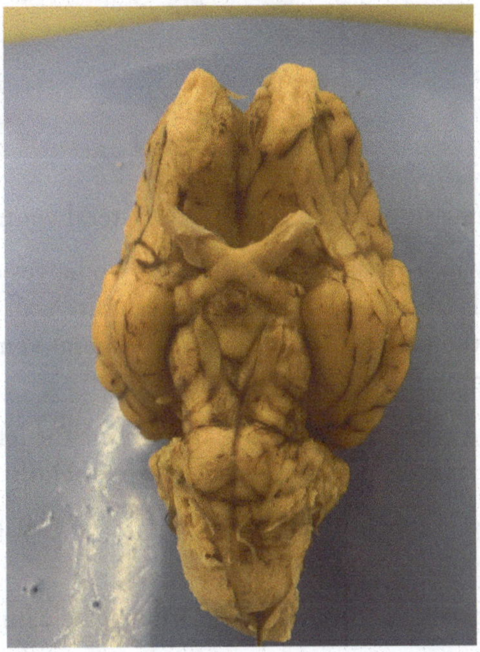

Source: Ben Jellen, 2021.

Brain, Sagittal Section

Obtain the brain from the specimen jar.

8. Locate:
   - **cerebrum**
   - **cerebellum**
   - **medulla oblongata**
   - **spinal cord**
   - **corpus callosum**—white matter composed of myelinated commissural fibers.
   - **lateral ventricle**—a thin wall of tissue, the **septum pellucidum,** usually separates lateral ventricles. Since the lateral ventricles are symmetrical, their names are not ventricles 1 and 2. Refer to them simply as **left lateral ventricle** and **right lateral ventricle.**

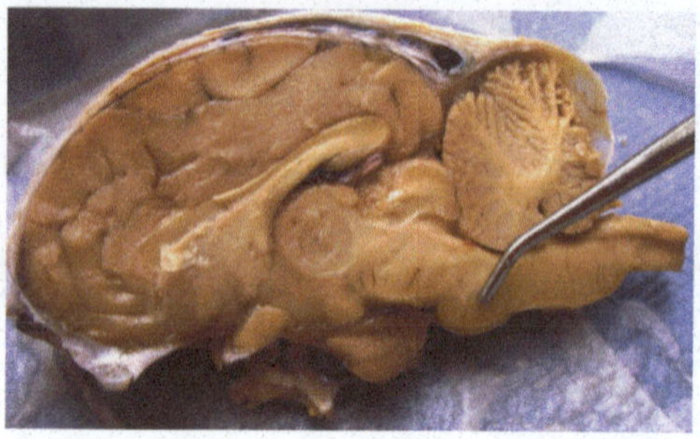

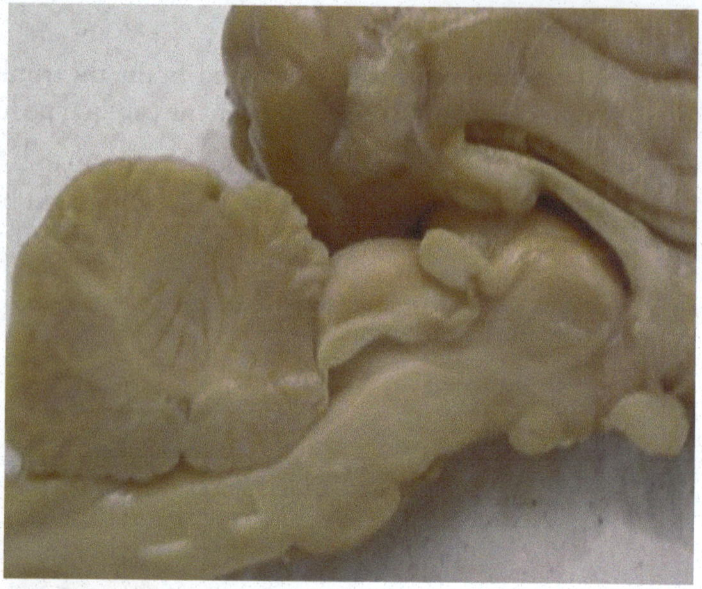

- **fornix**—a white commissure on the floor of the lateral ventricles.
- **3rd ventricle**—a very narrow space between the two halves of the brain. An epithelial membrane covers the walls of the 3rd ventricle, making its wall appear smooth and shiny. Very gently, try to locate the connection between the lateral ventricle and the 3rd ventricle, the **interventricular foramen.**
- **intermediate mass**—a gray commissure that connects the left and right thalamic nuclei. Will appear as a dull circular area, not covered by the epithelium.
- **thalamus**—nucleus *in the wall* of the 3rd ventricle. The thalamus is a relay station for most sensory information coming into the brain.
- **hypothalamus**—a nucleus in the wall of the 3rd ventricle immediately superior to the infundibulum and pituitary.
- **optic chiasma**
- **infundibulum**
- **mammillary body**
- **cerebral aqueduct**—connects the 3rd and 4th ventricles
- **pineal body**

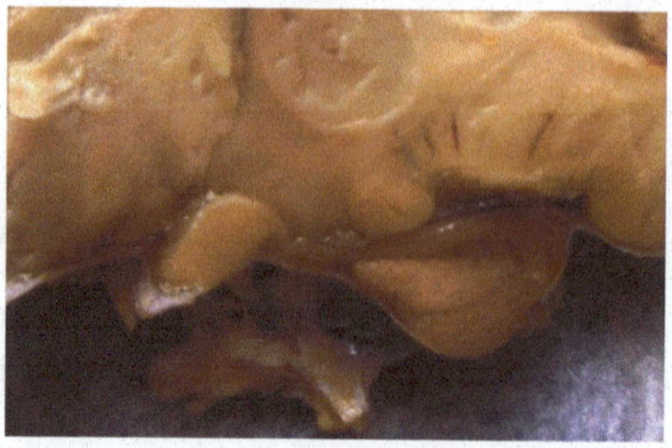

- **corpora quadrigemina**—on the roof of the cerebral aqueduct.
- **cerebral peduncle (mid-brain)**—this structure marks the beginning of the "brain stem"
- **cerebellum**—note the pattern of gray and white matter known as the **arbor vitae** ("tree of life").
- **4th ventricle**
- **central canal**—continuation of the 4th ventricle into the spinal cord.
- **pons**

Brain, Coronal Section

Obtain the brain from the specimen jar.

9. Locate:
   - **cerebral cortex**—gray matter on the surface of the cerebrum
   - **left and right lateral ventricles**
   - **corpus callosum**
   - **fornix**

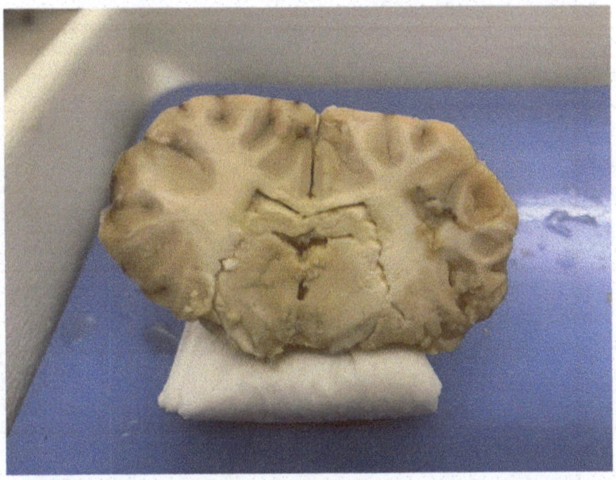

Source: Ben Jellen, 2021.

   - **3rd ventricle**
   - **thalamus**
   - **hypothalamus**

*** The assessment over the sheep brain will include all structures listed. It will also include the location, name, number and function of CN I, II, III, and V.**

CHECKLIST: Summary Checklist Nervous I

TERM/STRUCTURE	YOUR NOTES
Meninges	
Dura Mater	
Arachnoid Mater	
Pia Mater	
Cerebrum	
Cerebellum	
Medulla Oblongata	
Cerebral Hemispheres	
Frontal Lobe	
Parietal Lobe	
Temporal Lobe	
Occipital Lobe	
Arbor Vitae	
Olfactory Bulb	
Olfactory Tract	
Optic Chiasm	
Optic Tracts	
Pituitary Gland/Hypophysis	
Midbrain	
Cerebral Peduncle	
Corpora Quadrigemina	

TERM/STRUCTURE	YOUR NOTES
Pineal Gland/Body Superior and Inferior Colliculi	
Pineal Gland/Body	
Corpus Callosum	
Septum Pellucidum	
Lateral Ventricles	
Fornix	
Mammillary Body	
Third Ventricle	
Fourth Ventricle	
Diencephalon	
Thalamus/Thalamic Nuclei	
Hypothalamus	
Mesencephalon	
Cerebral Aqueduct/Aqueduct of the Midbrain	
Insula	
Pons	
Intermediate Mass/ Interthalamic Adhesion	
Interventricular Foramine	
Arachnoid Trabeculae	
Arachnoid Granulations/Villi	
Lateral and Median Aperture	
Central Canal of Spinal Cord	

(continued)

TERM/STRUCTURE	YOUR NOTES
Periosteal Layer of Dura Mater	
Meningeal Layer of Dura Mater	
Subarachnoid Space	
Subdural Space	
Superior Sagittal Sinus	
Cerebral Cortex	
Cranium	
Choroid Plexus (In Each Ventricle)	
Infundibulum	
Frontal Sinus	
Sphenoid Sinus	
CN I	
CN II	
CN III	
CN IV	
CN V	
CN VI	
CN VII	
CN VIII	
CN IX	
CN X	
CN XI	
CN XII	

EXERCISE 7

The Spinal Cord and Special Senses

A. Introduction

The first portion of Exercise 7 addresses the anatomy of the spinal cord of the central nervous system (CNS). The second portion addresses the anaomty of the eye and the ear, which are **special somatic sensory structures.**

You will learn about the nervous control for the musculature of the eye in addition to learning the external and internal anatomy of the eye. This includes: **extrinsic (skeletal) muscles** for controlling the movement of the eyeball, which you began in Exercise 5, and **intrinsic (smooth) muscles** for changing the **diameter of the pupil** and the **shape of the lens** to focus on objects at varying distances (a process called **accommodation**). Understanding the nervous control of these functions is important, because there are many drugs that can interfere with the nervous control of these activities.

B. *Outcomes for Exercise 7*

1. Locate, label, and identify the structures within the spinal cord and spinal nerves. Review the histology of the nervous tissue from Exercise 2.
2. Locate, label, and identify the external and internal structures of the eye.
3. Locate, label, and identify the extrinsic (extra-ocular) muscles of the eye. Describe their insertions, actions, and innervations. Are these muscles voluntary or involuntary? Made of smooth muscle or skeletal muscle tissue?
4. Locate, label, and identify the antagonistic muscles that control the diameter of the pupil. When each contract, what is the effect on the pupil? What nerves control each? Are these muscles voluntary or involuntary? Made of smooth muscle or skeletal muscle tissue?
5. Define accommodation. Describe the action of the ciliary muscle on the suspensory ligaments during accommodation. When the ciliary muscle contracts, what is the resulting shape of the lens? Is that action suited for near or far vision? When the ciliary muscle relaxes, what is the resulting shape of the lens? Is that action suited for near or far vision?
6. Describe the pathway of lacrimal (extrinsic) fluid by listing the structures it passes through/over in the correct order. Be able to determine where the fluid was immediately before a specific location and where it will be immediately after a specific location.
7. Describe the pathway of aqueous (intrinsic) fluid by listing the structures it passes through in the correct order. What structure produces the aqueous fluid? Be able to determine where the fluid was immediately before a specific location and where it will be immediately after a specific location.
8. Name the three layers of cells that compose the retina. Which cells are more internal (inner)? Which cells are more external (outer)? Describe the pathway of light as it travels through the cells of the retina. Describe the pathway of the action potential.
9. Locate, label, and identify the external and internal structures of the ear.
10. Examine the retina and cochlea under the microscope. Identify the structures listed in the tables. Draw and label what you see.
11. Dissect a beef eye. Locate and identify the structures listed below.

C. *Anatomy of the Spinal Cord and Spinal Nerves*

OUTCOME 1: Locate, label, and identify the structures within the spinal cord and spinal nerves. Review the histology of the nervous tissue from Exercise 2.

TOOLBOX

Sectional Organization of the Spinal Cord

spinal cord

spinal nerve
dorsal root
dorsal root ganglion
ventral root
posterior horn
lateral horn
anterior horn
anterior median fissure
posterior median sulcus
posterior column
lateral column
anterior column
posterior white commissure
posterior gray commissure
anterior gray commissure
anterior white commissure
central canal
white ramus
gray ramus
sympathetic ganglion

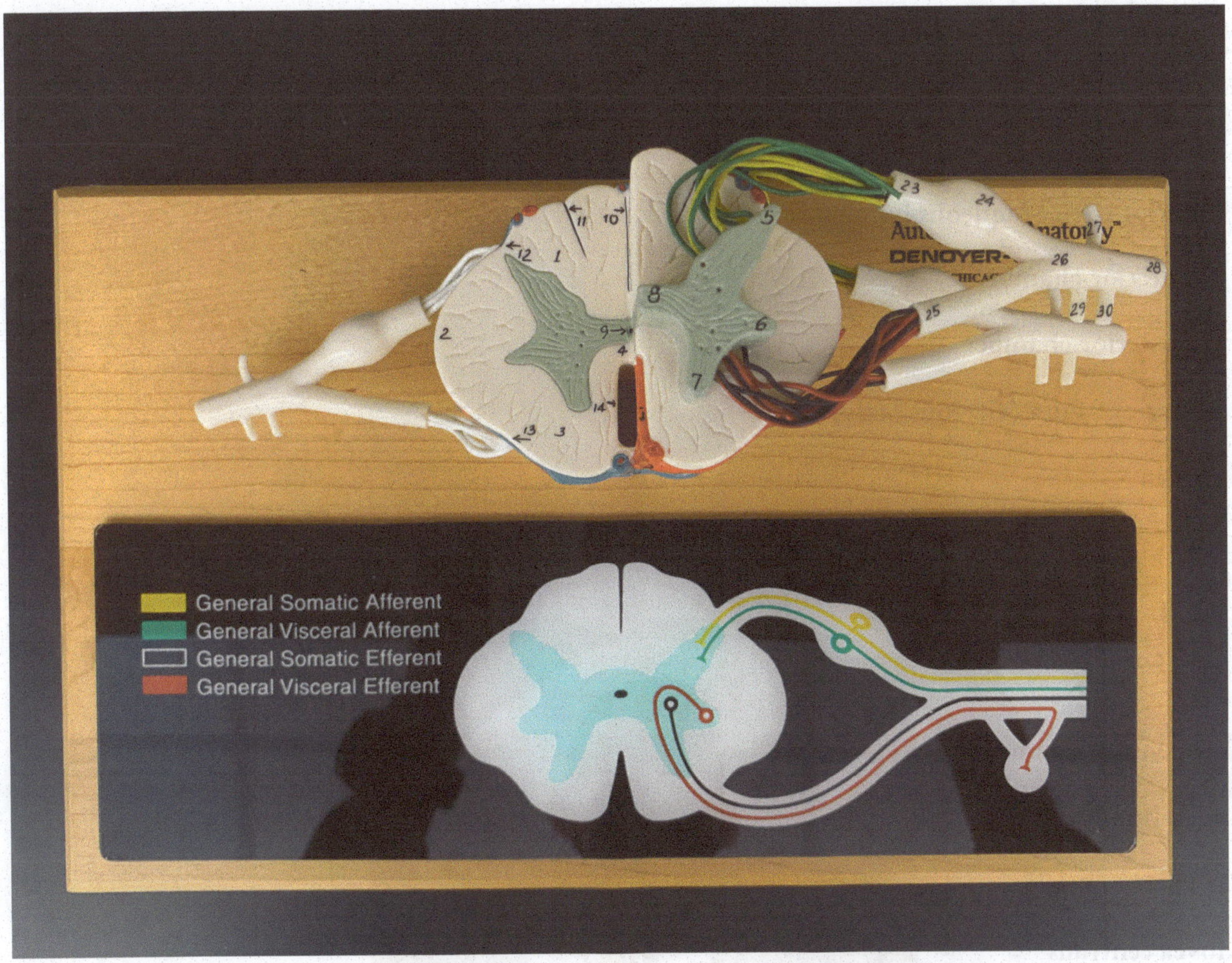

D. *The Eye*

OUTCOME 2: Locate, label, and identify the external and internal structures of the eye.

TOOLBOX

Sectional Anatomy of the Eye

eye model

sclera

choroid

retina

optic nerve

optic disc

ocular conjunctiva

anterior cavity (aqueous humor)

- anterior chamber
- posterior chamber

posterior cavity (vitreous body)

cornea

iris

pupil

lens

suspensory ligaments

ciliary body

- ciliary muscle

fovea centralis

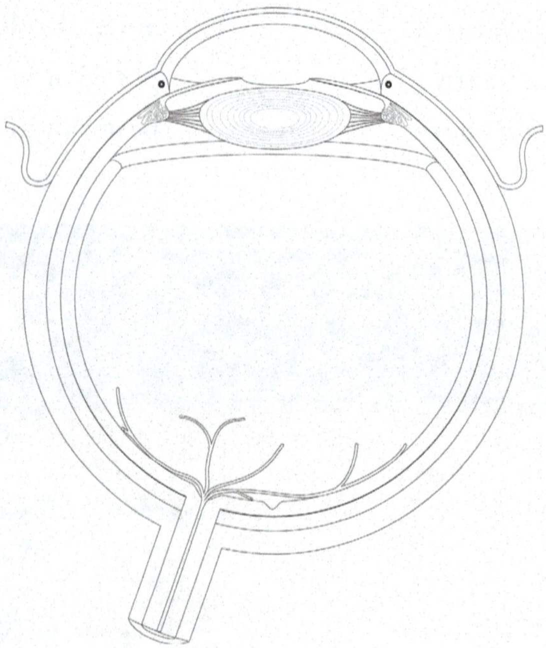

Superior View of Right Eye

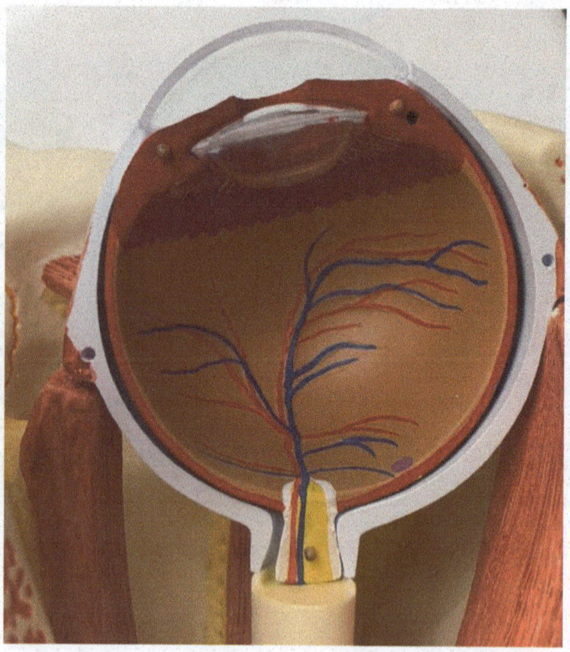

F13 © 3B Scientific GmbH, Germany, 2021 www.3bscientific.com

OUTCOME 3: Locate, label, and identify the extrinsic (extra-ocular) muscles of the eye. Describe their insertions, actions, and innervations. Are these muscles voluntary or involuntary? Made of smooth muscle or skeletal muscle tissue?

TOOLBOX

Sectional Anatomy of the Eye
The Lens and Chambers of the Eye

eye model

MUSCLE	INSERTION	ACTION	INNERVATION
superior rectus	superior surface	eye looks up	CN III Oculomotor n.
inferior rectus	inferior surface	eye looks down	CN III Oculomotor n.
medial rectus	medial surface	eye looks medially	CN III Oculomotor n.
lateral rectus	lateral surface	eye looks laterally	CN VI Abducens n.
superior oblique	through trochlea to superior-lateral surface	eye rolls, looks down & laterally ("down & out")	CN IV Trochlear n
inferior oblique	inferior-lateral surface	eye rolls, looks up & laterally ("up & out")	CN III Oculomotor n.

■ ***Locate, label, and identify the extrinsic muscles of the eye. (Left eye shown)***

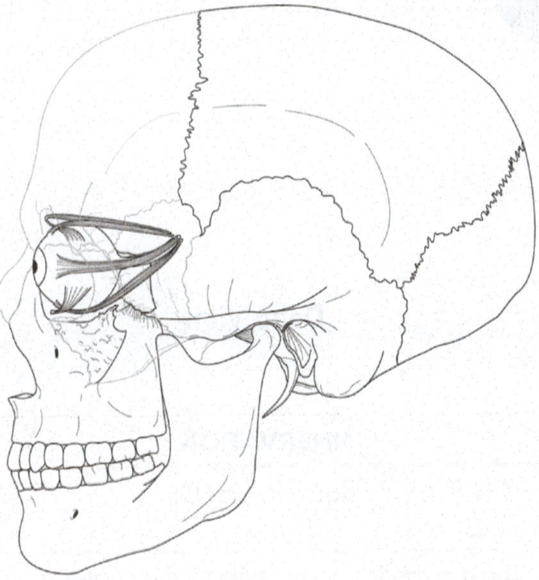

Illustration by Jamey Garbett. © 2003 Mark Nielsen.

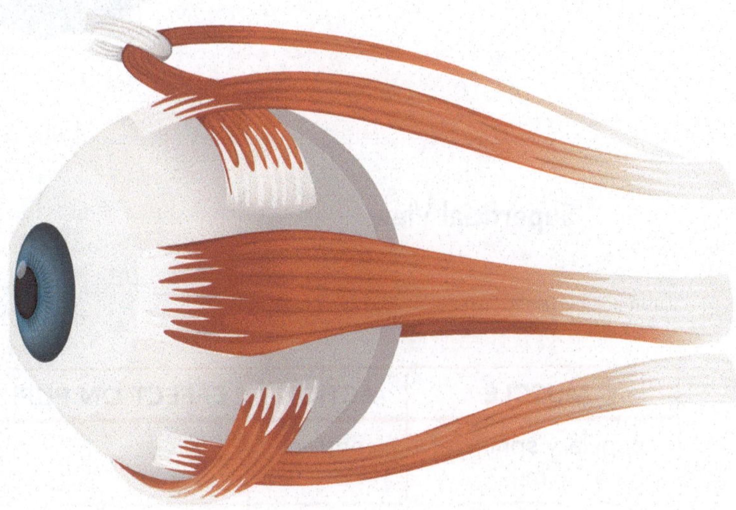

© BlueRingMedia/Shutterstock.com

OUTCOME 4: Locate, label, and identify the antagonistic muscles that control the diameter of the pupil. When each contract, what is the effect on the pupil? What nerves control each? Are these muscles voluntary or involuntary? Made of smooth muscle or skeletal muscle tissue?

TOOLBOX

Extra-ocular Muscles
(in Axial Musculature Chapter)

eye model

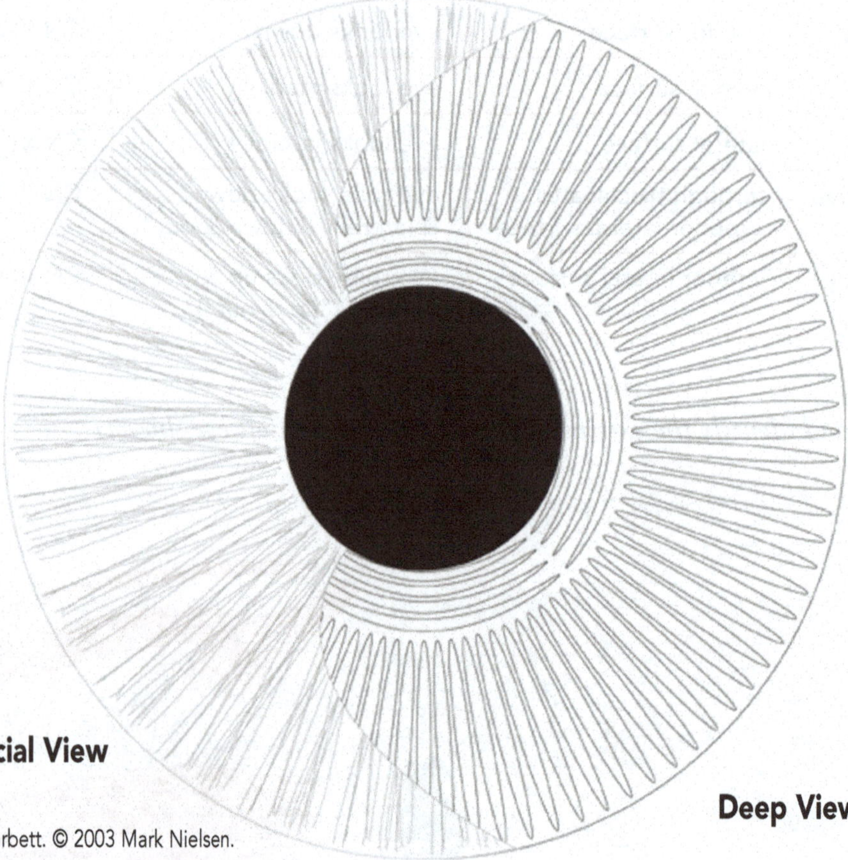

Illustration by Jamey Garbett. © 2003 Mark Nielsen.

MUSCLE	ACTION	EFFECT ON PUPIL	INNERVATION
pupillary sphincter (circular m.)	contract		CN III Oculomotor n. (PSNS)
pupillary dilator (radial m.)	contract		post ganglionic nerve from the superior cervical ganglion (SNS)

OUTCOME 5: Define accommodation. Describe the action of the ciliary muscle on the suspensory ligaments during accommodation. When the ciliary muscle contracts, what is the resulting shape of the lens? Is that action suited for near or far vision? When the ciliary muscle relaxes, what is the resulting shape of the lens? Is that action suited for near or far vision? Which cranial nerve controls the ciliary muscle?

TOOLBOX

The Lens and the Chambers of the Eye

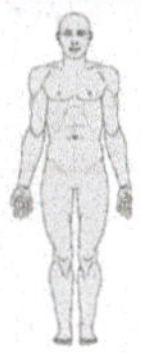
eye model

Fill in the blanks to demonstrate the relationship of the action of the ciliary muscle on the suspensory ligaments and lens.

IMAGE DEMONSTRATING *FAR VISION*

ciliary muscle = ______________

suspensory ligaments = ______________

lens shape = ______________

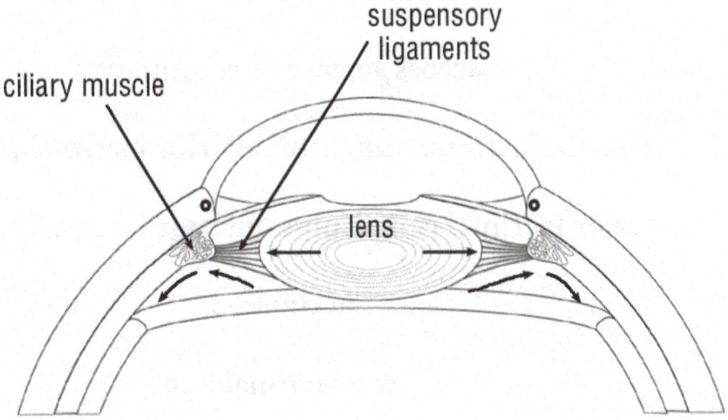

IMAGE DEMONSTRATING *NEAR VISION*

ciliary muscle = ______________

suspensory ligaments = ______________

lens shape = ______________

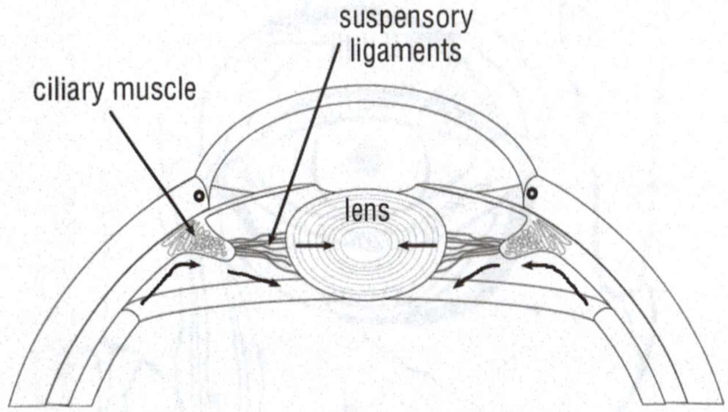

- ***Which cranial nerve controls the ciliary muscle?***
- ***Define accommodation:***

OUTCOME 6: Describe the pathway of lacrimal (extrinsic) fluid by listing the structures it passes through/over in the correct order. Be able to determine where the fluid was immediately before a specific location and where it will be immediately after a specific location.

TOOLBOX

Accessory Structures of the Eye

eye model

lacrimal gland

lacrimal gland ducts

across surface of conjunctiva

superior lacrimal punctum   inferior lacrimal punctum

superior lacrimal canaliculus   inferior lacrimal canaliculus

lacrimal sac

nasolacrimal duct

nasal cavity

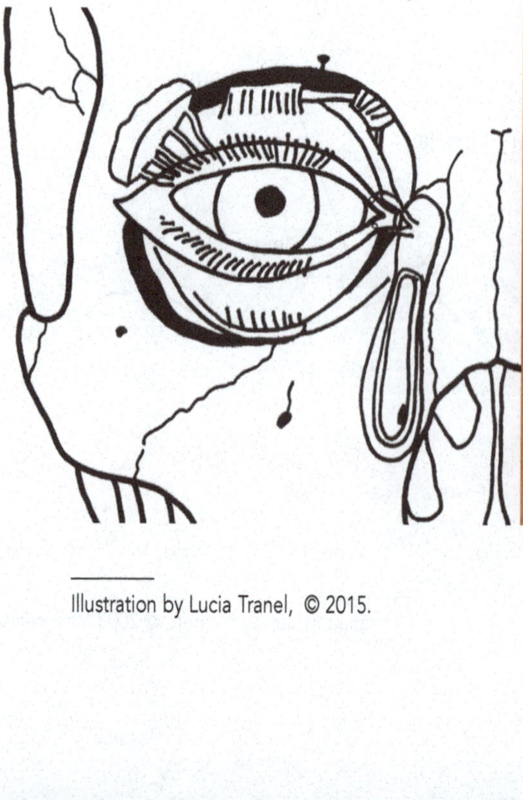

Illustration by Lucia Tranel, © 2015.

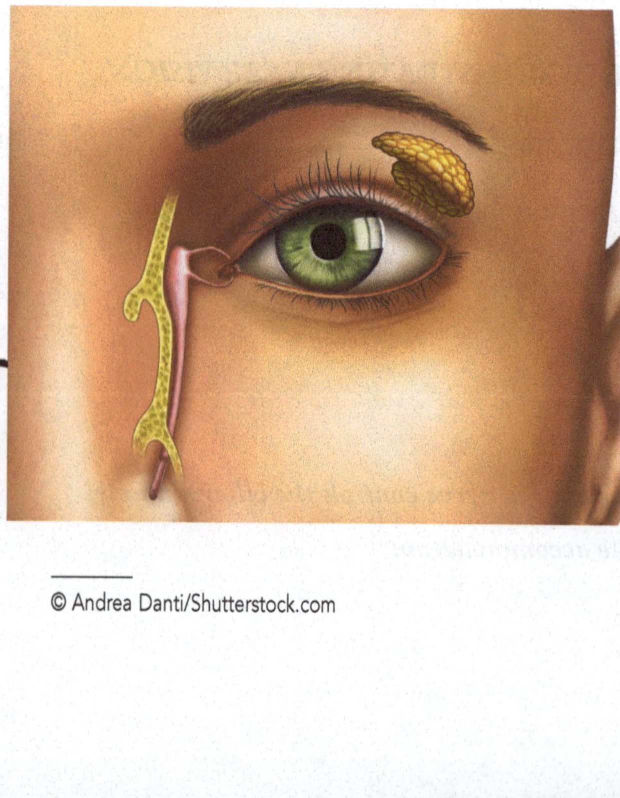

© Andrea Danti/Shutterstock.com

OUTCOME 7: Describe the pathway of aqueous (intrinsic) fluid by listing the structures it passes through in the correct order. What structure produces the aqueous fluid? Be able to determine where the fluid was immediately before a specific location and where it will be immediately after a specific location.

TOOLBOX

The Circulation of the Aqueous Humor

eye model

■ ***Label the structures. Draw arrows to indicate the flow pattern:***

ciliary body

(epithelial cells of)

posterior chamber of the anterior cavity

posterior cavity

pupil

anterior chamber of the anterior cavity

canal of Schlemm (scleral venous sinus)

cornea

iris

lens

suspensory ligaments

ciliary body

retina

choroid

sclera

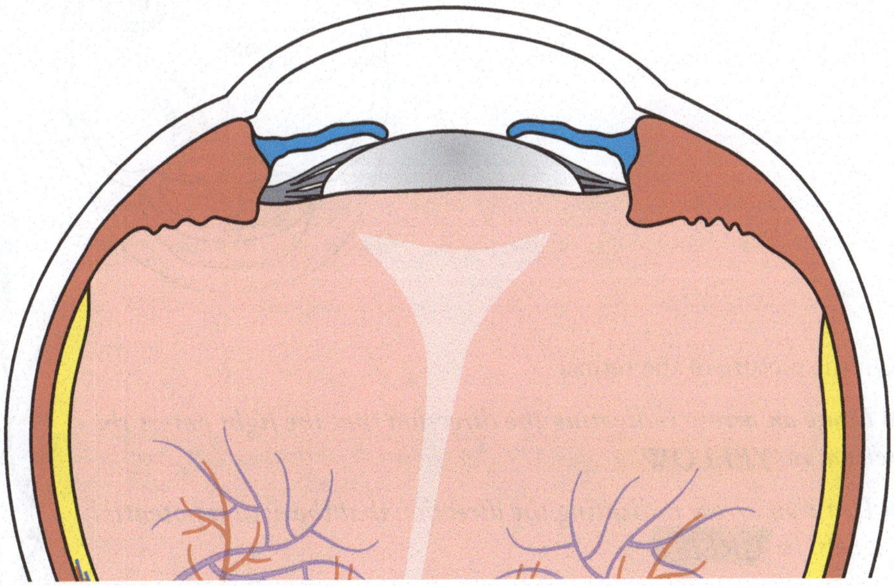

OUTCOME 8: Name the three layers of cells that compose the retina. Which cells are more internal (inner)? Which cells are more external (outer)? Describe the pathway of light as it travels through the cells of the retina. Describe the pathway of the action potential.

TOOLBOX

Retinal Organization

On the picture of the retina:

■ ***Label the details of the retina:***

pigmented cell layer
Three cell layers:
- **photoreceptors (rods and cones)**
- **bipolar cells**
- **ganglion cells**

dendritic ends of each cell layer
cell bodies of each cell layer
axonal ends of each cell layer
outer synaptic layer
inner synaptic layer
outer nuclear layer
inner nuclear layer

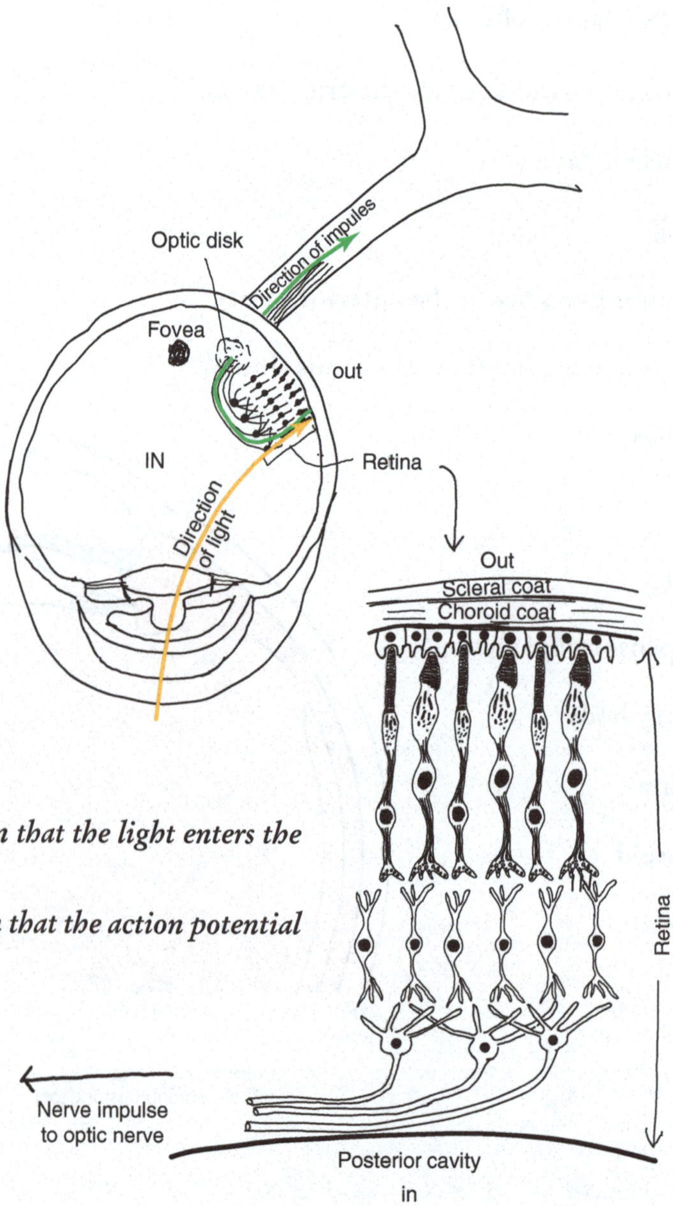

On the picture of the retina:

■ ***Draw an arrow indicating the direction that the light enters the retina, in YELLOW***

■ ***Draw an arrow indicating the direction that the action potential travels, in GREEN***

Illustration by Lucia Tranel, © 2015.

E. *Ear*

OUTCOME 9: Locate, label, and identify the external and internal structures of the ear.

TOOLBOX

Anatomy of the Ear

ear model

external ear
- external acoustic meatus
- tympanic membrane
- auricle

middle ear
- auditory ossicles (malleus, incus, stapes)
- tympanic cavity
- internal auditory (Eustachian) tube
- oval window
- round window

inner ear
- semicircular canal
- vestibule
- cochlea
- CN VIII Vestibulocochlear n.

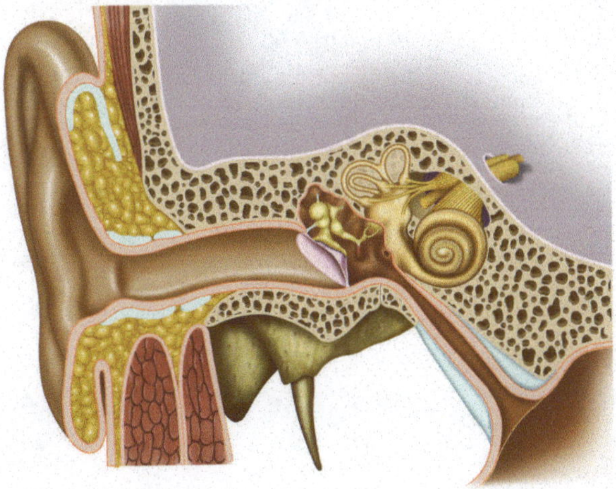

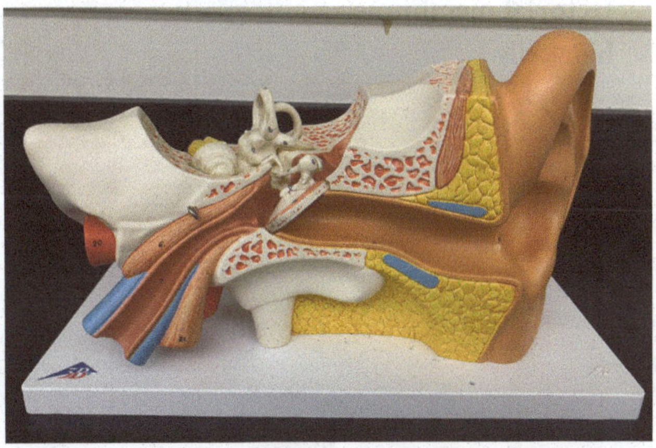

TOOLBOX

The Cochlea and the Organ of Corti

ear model

vestibular nerve

between the terms cochlear nerve and scala vestibuli

cochlear nerve

scala vestibuli (vestibular duct)
- contains perilymph
- vibrations from oval window

scala media (cochlear duct)
- contains endolymph
- organ of Corti

scala tympani (tympanic duct)
- contains perilymph
- vibrations to round window

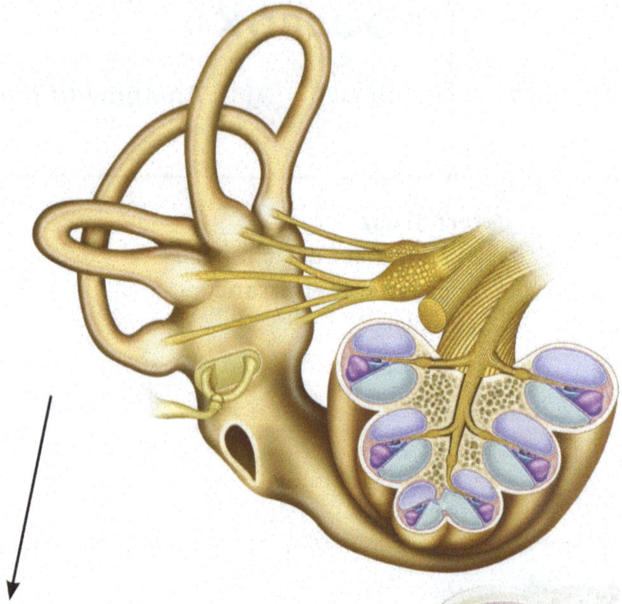

scala vestibuli

scala media

scala tympani

organ of Corti
- basilar membrane
- tectorial membrane
- hair cells

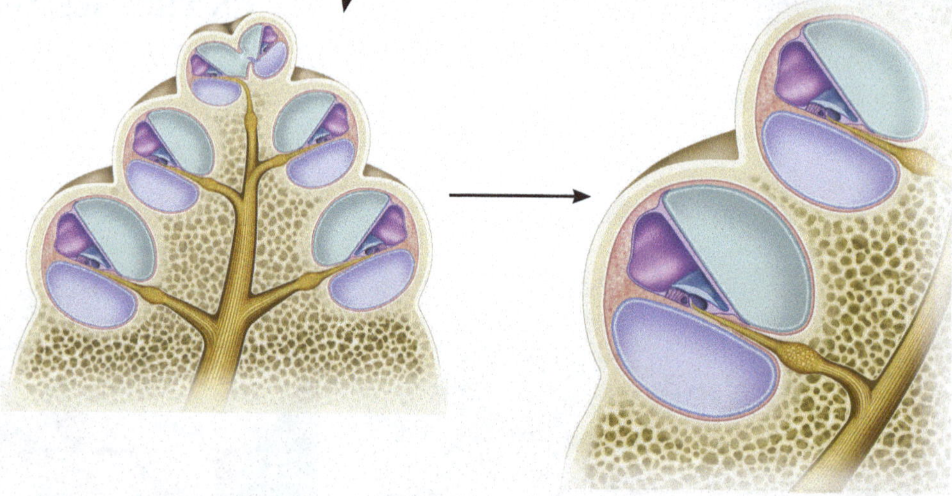

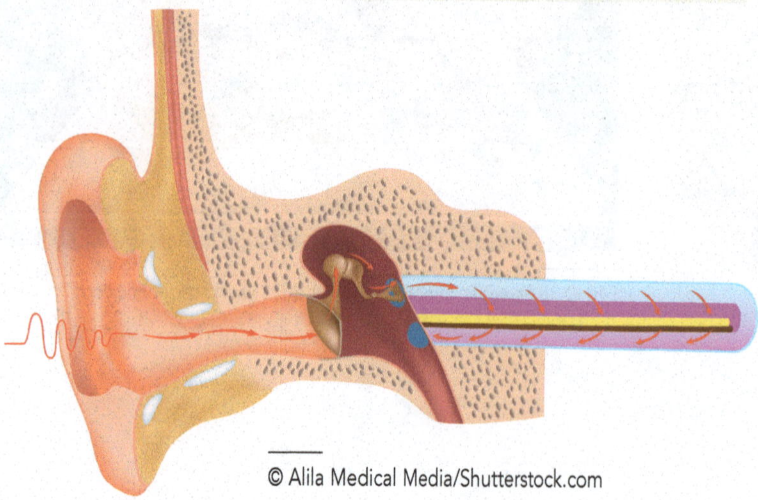

F. *Histological Examination*

OUTCOME 10: Examine the retina and cochlea under the microscope. Identify the structures listed in the tables. Draw and label what you see.

TOOLBOX

Retinal Organization

■ ***Locate the following structures within the image of the retina.***

STRUCTURE	YOUR NOTES	IMAGE	HISTOLOGY
pigmented layer			
rods and cones (outer nuclear layer)		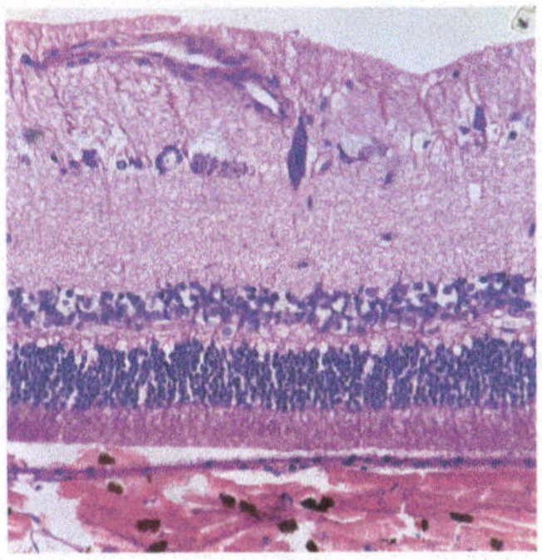	
outer synaptic layer			
bipolar cells (inner nuclear layer)			
inner synaptic layer			
ganglion cells (nuclei)			

TOOLBOX

The Cochlea and Organ of Corti

■ ***Locate the following structures within the image of the cochlea.***

STRUCTURE	YOUR NOTES	IMAGE	HISTOLOGY
scala vestibuli filled with perilymph			
scala media filled with endolymph		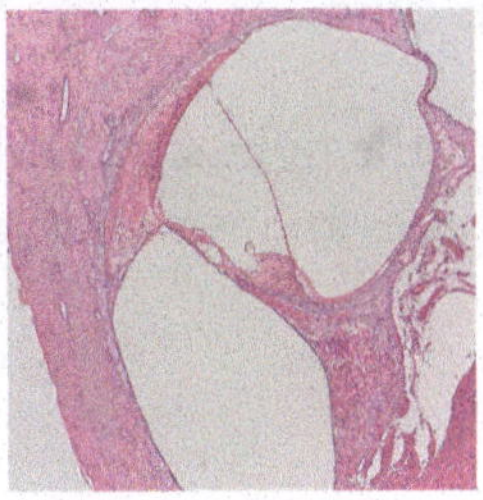	
scala tympani filled with perilymph			
organ of Corti			

G. Comparative Anatomy—Dissection

OUTCOME 11: Dissect a beef eye. Locate and identify the structures listed below.

TOOLBOX

✂

Examine the external anatomy of the eye. Locate:

- **adipose**
- **optic nerve**
- **extrinsic muscles**
- **sclera**
- **cornea**

Make a coronal section, cutting the anterior portion of the eye from the posterior portion. Start by inserting the tip of a scalpel or scissors through the sclera at a location approximately 1 cm from the margin of the cornea. Locate:

- **sclera**
- **choroid**—the iridescent area is the tapetum lucidum. This area is not in human eyes.
- **retina**—notice that it is not attached in any place except at the optic nerve
- **optic disc**—location of optic nerve exiting the eye
- **ciliary body**—the ciliary muscle is within the ciliary body
- **lens**—remove slowly so you can see the suspensory ligaments
- **pupil**
- **iris**—the intrinsic muscles can not be seen; they are within
- **cornea**
- **anterior cavity**—with aqueous humor (fluid)
- **anterior chamber of the anterior cavity**
- **posterior chamber of the anterior cavity**
- **posterior cavity**—with vitreous humor (body)

Review the flow of the aqueous fluid.

Hold the anterior portion up to the light, and look through the cornea and pupil. The cornea and lens become opaque once the tissue dies and is preserved.

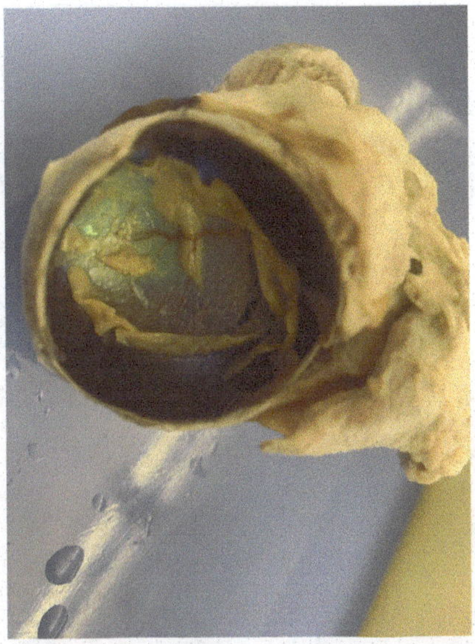

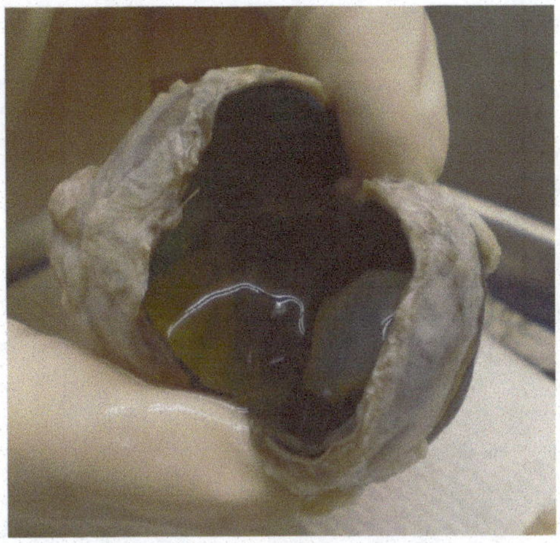

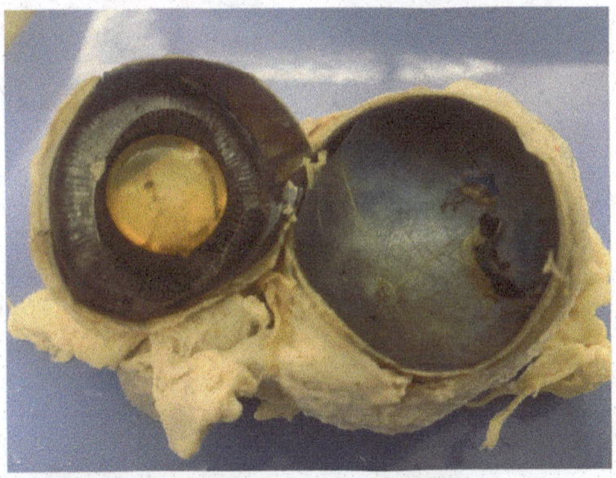

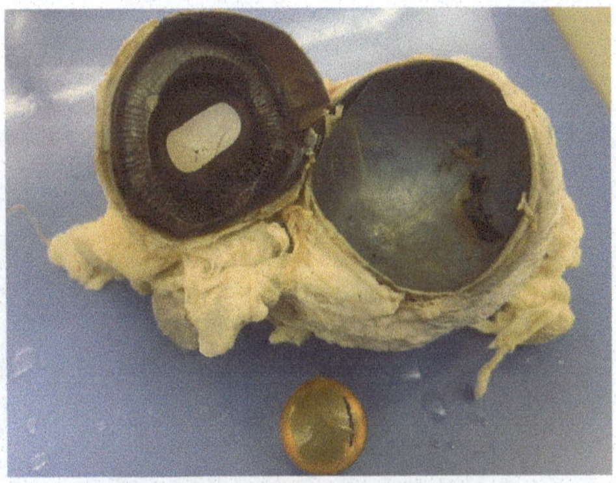

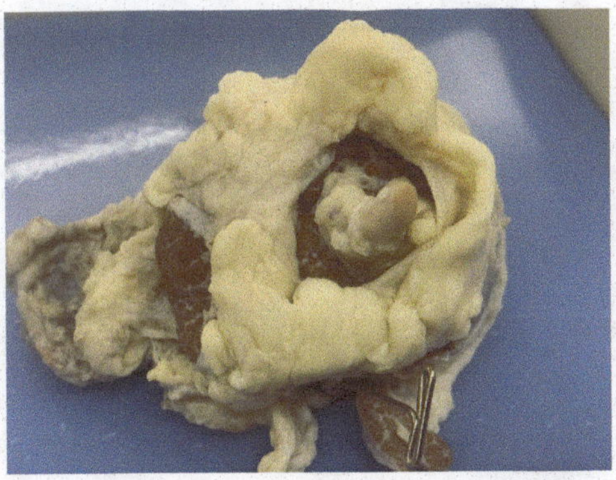

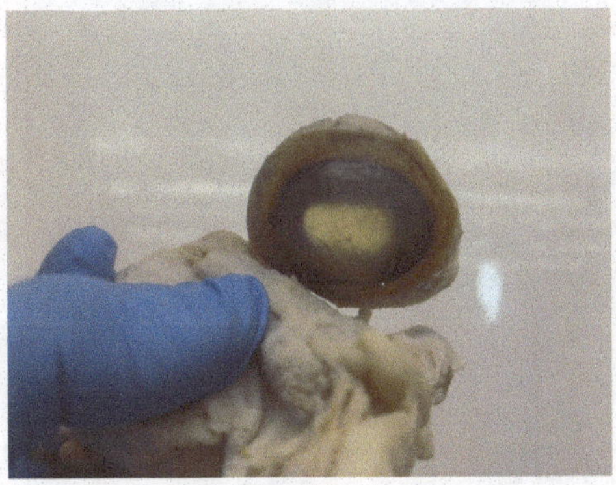

Source: Ben Jellen, 2021.

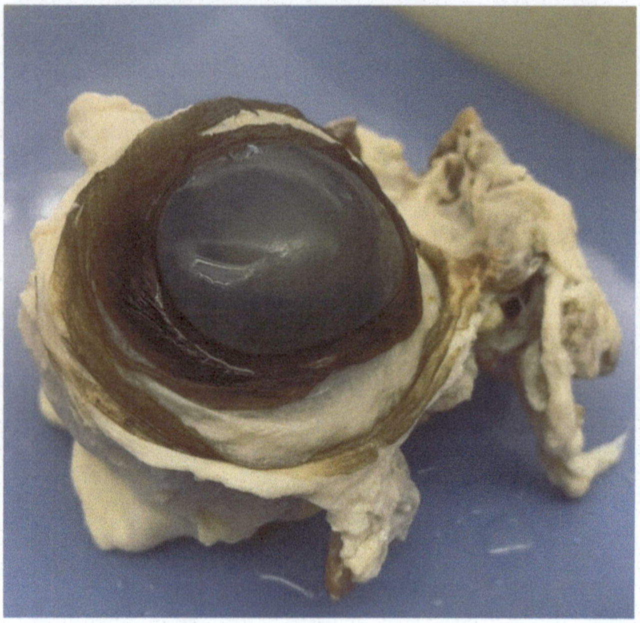

Source: Ben Jellen, 2021.

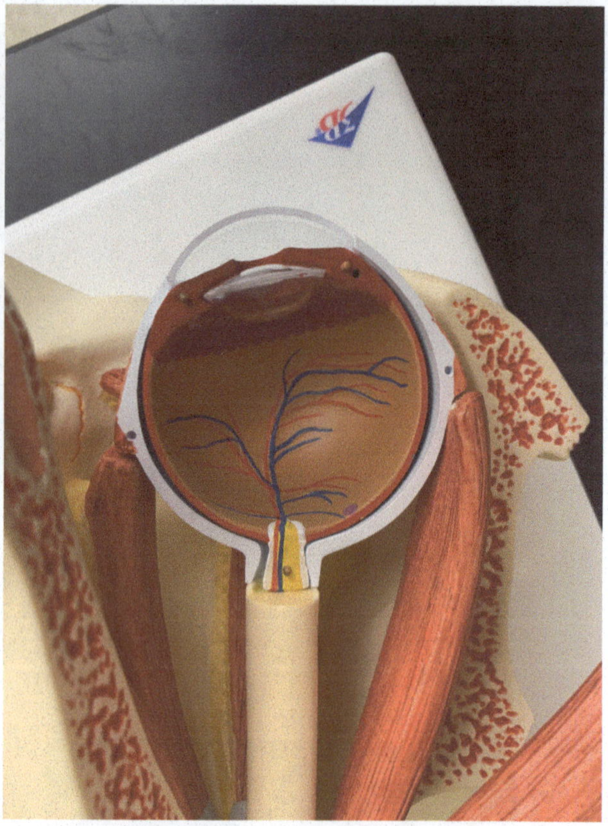

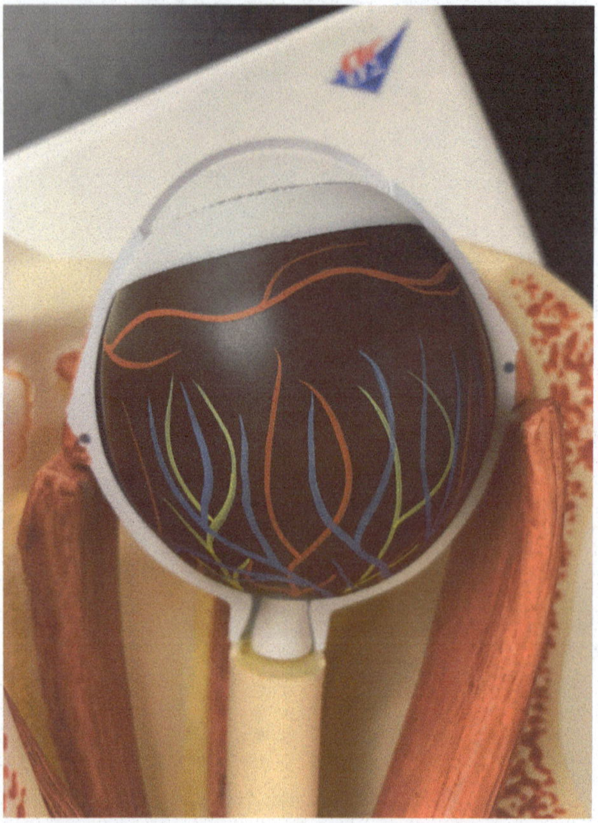

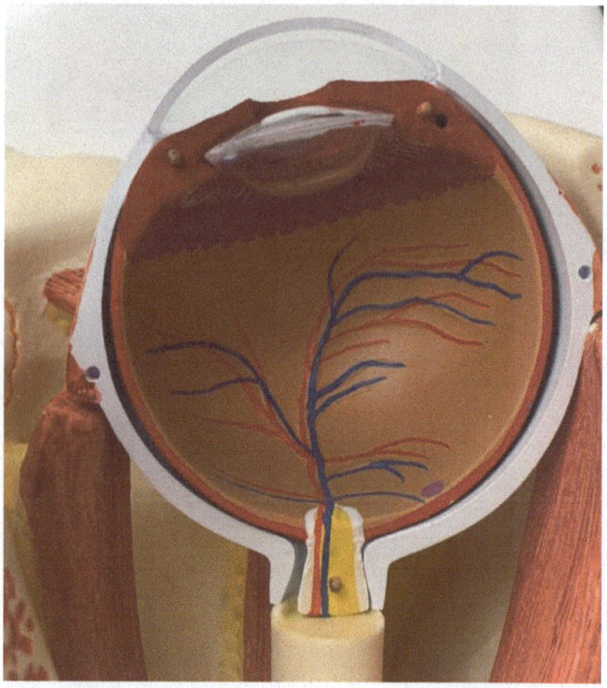

CHECKLIST: Summary Checklist Nervous II

STRUCTURE	YOUR NOTES
Mulipolar/Motor Neuron	
Pseudounipolar/Sensory Neuron	
Cell Body (Soma) of Neuron	
Axon of Neuron	
Dendrite of Neuron	
Nucleus of Neuron	
Gray Matter	
White Matter	
Axon Hillock	
Myelin Sheath	
Nodes of Ranvier	
Anterior/Ventral Column	
Posterior/Dorsal Column	
Lateral Column	
Anterior White Commissure	
Posterior White Commissure	
Anterior Gray Commissure	

STRUCTURE	YOUR NOTES
Anterior/Ventral Horns	
Posterior Gray Commissure	
Lateral Horns	
Posterior/Dorsal Horns	
Central Canal of Spinal Cord	
Anterior Median Fissure	
Posterior Median Sulcus	
Ventral Roots	
Dorsal Roots	
Dorsal Root Ganglion	
Spinal Nerves	
White Ramus	
Gray Ramus	
Lens	
Sclera	
Cornea	
Choroid	
Vitreous Humor/Body	

STRUCTURE	YOUR NOTES
Aqueous Humor	
Optic Disc	
Optic Nerve	
Circular Muscle of Iris (Pupillary Sphincter)	
Radial Muscle of Iris (Pupillary Dilator)	
Iris	
Pupil	
Suspensory Ligaments	
Posterior Chamber of Anterior Cavity	
Posterior Cavity	
Anterior Chamber of Anterior Cavity	
Retina	
Pigmented Cell Layer of Retina	
Photoreceptors (Rods and Cones) of Retina	
Bipolar Cells of Retina	
Ganglion Cells of Retina	
Outer and Inner Synaptic Layer	
Outer and Inner Nuclear Layer	

STRUCTURE	YOUR NOTES
Dendritic and Axonal Ends of Each Layer of Retina	
Ciliary Muscle	
Ciliary Body	
Lacrimal Gland	
Lacrimal Gland Ducts	
Conjunctiva Surface/ Ocular Conjunctiva	
Superior/Inferior Lacrimal Punctum	
Superior/Inferior Lacrimal Canaliculus	
Lacrimal Sac	
Nasolacrimal Duct	
Nasal Cavity	
Auricle of Ear	
External Acoustic Meatus	
Tympanic Membrane	
Temporal Bone	
Middle Ear	
Auditory Ossicles	
Maleus, Incus, and Stapes	

STRUCTURE	YOUR NOTES
Internal Auditory (Eustachian) Tube	
Tympanic Cavity	
Oval Window	
Round Window	
Inner Ear	
External Ear	
Semicircular Canal	
Vestibule	
Cochlea	
CN VIII (Vestibulocochlear Nerve)	
Scala Vestibuli (Vestibular Duct)	
Scala Media (Cochlear Duct)	
Scala Tympani (Tympanic Duct)	
Endolymph	
Perilymph	

EXERCISE 8

Cardiovascular System: Blood Vessel Architecture, Blood, and Heart

A. Introduction

The focus of this chapter is on the organs and tissues of the **cardiovascular system.** Exercise 9 will cover dissection of the actual blood vessels. The **organs** and **tissues** of this system include the **heart (cardio),** the **vessels (vascular),** and the **blood.** The lymph system is a one-way series of vessels considered to be related to the cardiovascular system and the immune system. The lymph fluid carried by the lymphatic vessels indirectly derives from the blood and returns directly to the blood and it functions to provide immune protection.

B. *Outcomes for Exercise 8*

1. Examine circulatory patterns: pulmonary circuit, systemic circuit, hepatic portal system, and lymph flow.

2. Compare the histological characteristics of arteries, veins, and capillaries. Describe the similarities and differences of each. Locate, label and identify the tunicae (layers) and the tissues that compose each tunica. Examine the changing features of blood vessels in the progression from an elastic artery to a vein.

3. Examine several arteries and veins under the microscope. Locate and identify the tunicae of each. Look for characteristics that will enable you to distinguish one from the other. Identify the tissues in each of the tunicae. Draw and label what you see.

4. Examine a blood smear under the microscope. Use the dichotomous key to aid in the location and identification all the components: plasma, erythrocyte, neutrophil, eosinophil, basophil, lymphocyte, monocyte, and thrombocyte. Sketch each cell as you locate it. Know clinical values and specific functions for each element. Draw and label what you see.

5. Examine the blood smears of acute inflammation (leukocytosis) and sickle cell anemia. Distinguish between each of the slides, describe the characteristics of each. Make notes so that you can identify which condition is evident.

6. Locate, label, and identify the external and internal structures of the heart.

7. Trace a drop of blood through the heart by listing the structures the blood travels through in the correct order. Start with the venae cavae and end with the aorta. Include all details.

8. Locate and identify the vessels of coronary circulation. Trace a drop of blood through the coronary circulation pathway by listing the structures the blood passes through in the correct order. Include all details.

9. Locate and identify the structures on a dissected mammalian heart.

C. *Patterns of Circulation*

OUTCOME 1: Examine the patterns of circulation: pulmonary circuit, systemic circuit, hepatic portal system, and lymph system.

Humans have a **closed** and **complete** pattern of cardiovascular circulation; that is, blood is contained within a closed circuit of vessels, and oxygenated blood is completely separated from de-oxygenated blood. Examine the image. Notice that there are two blood circulatory patterns: 1) the **pulmonary circuit,** from the heart to the lungs and back; and 2) the **systemic circuit,** from the heart to the body and back. **Arteries** are the vessels that carry blood away from the heart, and veins are vessels that carry blood back to the heart. The name designation of these vessels is based on **direction of flow** and NOT on the amount of oxygen in the blood. Thus, the pulmonary arteries typically carry blood that is low in oxygen, while the systemic arteries typically carry blood that is high in oxygen levels. The pulmonary veins carry blood that is oxygen rich, while the systemic veins carry low oxygenated blood. Often the color RED is used to designate a vessel carrying oxygenated blood, and BLUE is used to designate a vessel carrying de-oxygenated blood. The color purple typically denotes a site of gas exchange. Examine the image.

■ ***Fill in the table using the words* artery *and* vein. *Match the structure to its correct amount of oxygen levels.***

	OXYGENATED BLOOD	DE-OXYGENATED BLOOD
Pulmonary Circuit		
Systemic Circuit		

Occasionally, there are detours in the systemic circulation. Such is the case with the hepatic portal system. Examine the image. Notice that blood coming from the digestive organs goes through the liver before flowing into the general circulation.

In addition, there is a flow of lymph fluid that involves vessels known as lymphatics. Unlike blood, lymph flow is not a circuit; it is a one-way flow. The lymph fluid is picked up by lymph capillaries that lay alongside the blood capillaries, and the lymph fluid is transported back to the blood circulation. During the transport process, the lymph system removes cellular debris and pathogens. It provides an alternate route for some constituents of the fluid compartments. Lymph flow is designated GREEN in the image.

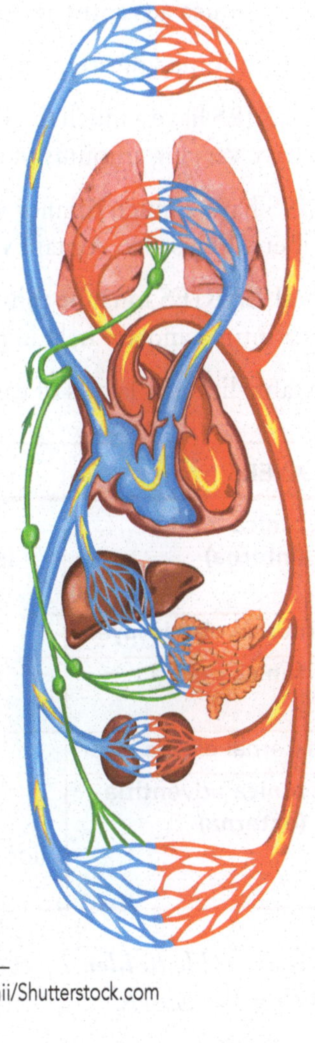

D. *General Anatomy of Blood Vessels*

OUTCOME 2: Compare the histological characteristics of arteries, veins, and capillaries. Describe the similarities and differences of each. Locate, label and identify the tunicae (layers) and the tissues that compose each tunica. Examine the changing features of blood vessels in the progression from an elastic artery to a vein.

TOOLBOX

Histological Organization of Blood Vessels

Structural Differences between Arteries and Veins

In general, the walls of arteries and veins have three distinct layers, or tunicae:

1. tunica intima or interna – the layer that lines the vessel
2. tunica media – the layer that is in the middle of the vessel wall
3. tunica adventitia or externa – the layer that lies on the outside of the vessel

There are some important structural differences among the three vessel types.

ARTERIES have a much thicker wall due to the more developed smooth muscle in the tunica media. They also have varying amounts of elastic fibers and layers so they can spring back to their original shape.

VEINS have a much thinner wall. They have very little smooth muscle in the tunica media, and no elastic fibers. Veins also contain valves.

CAPILLARIES must be thin enough for exchange of gases and other substances, so they have only the tunica intima and are held in place by the tunica adventitia.

The table lists the layers of each vessel and the specific histological tissue that composes that layer.

LAYERS	ARTERY	VEIN	CAPILLARY
1. tunica intima (interna)	endothelium (simple squamous epithelial tissue)	endothelium (simple squamous epithelial tissue)	endothelium (simple squamous epithelial tissue)
internal elastic layer	elastic fibers	X	X
2. tunica media	smooth muscle and elastic fibers	smooth muscle	X
external elastic layer	elastic fibers	X	X
3. tunica adventitia (externa)	connective tissue proper (loose, dense, adipose)	connective tissue proper (loose, dense, adipose)	connective tissue proper (loose, dense, adipose)

■ ***Use the table to identify the vessels in the diagram on the next page. Make sure to label each layer within the vessels.***

■ ***Label the details of the artery, vein, and capillary using the decriptions on the previous page.***

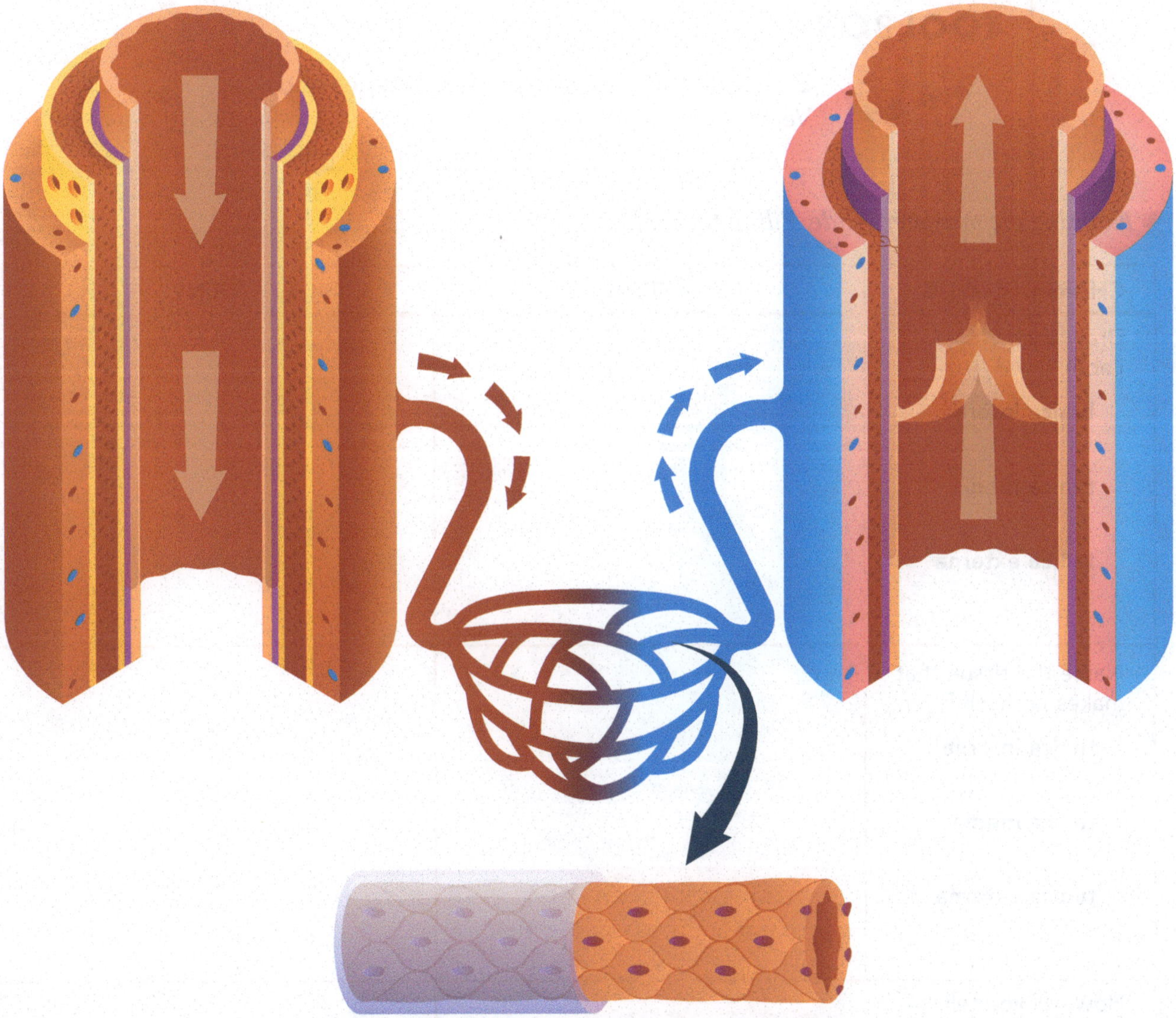

E. *Microscopic Examination of an Artery and a Vein*

OUTCOME 3: Examine several arteries and veins under the microscope. Locate and identify the tunicae of each. Look for characteristics that will enable you to distinguish one from the other. Identify the tissues in each of the tunicae. Draw and label what you see.

TOOLBOX

Structural Differences between Arteries and Veins

■ ***Examine the mesentery slide. Fill in the table.***

CHARACTERISTIC	ARTERY	VEIN
Draw each vessel. Label: **tunica interna** **tunica media** **tunica externa**		
Name the tissue that makes up each layer. **tunica interna** **tunica media** **tunica externa**		
How will you tell the vessels apart in the microscope?		
What is the function of each vessel?		

F. *Examination of Blood*

OUTCOME 4: Examine a blood smear under the microscope. Use the dichotomous key to aid in the location and identification all the components: plasma, erythrocyte, neutrophil, eosinophil, basophil, lymphocyte, monocyte, and thrombocyte. Sketch each cell as you locate it. Know clinical values and specific functions for each element. Draw and label what you see.

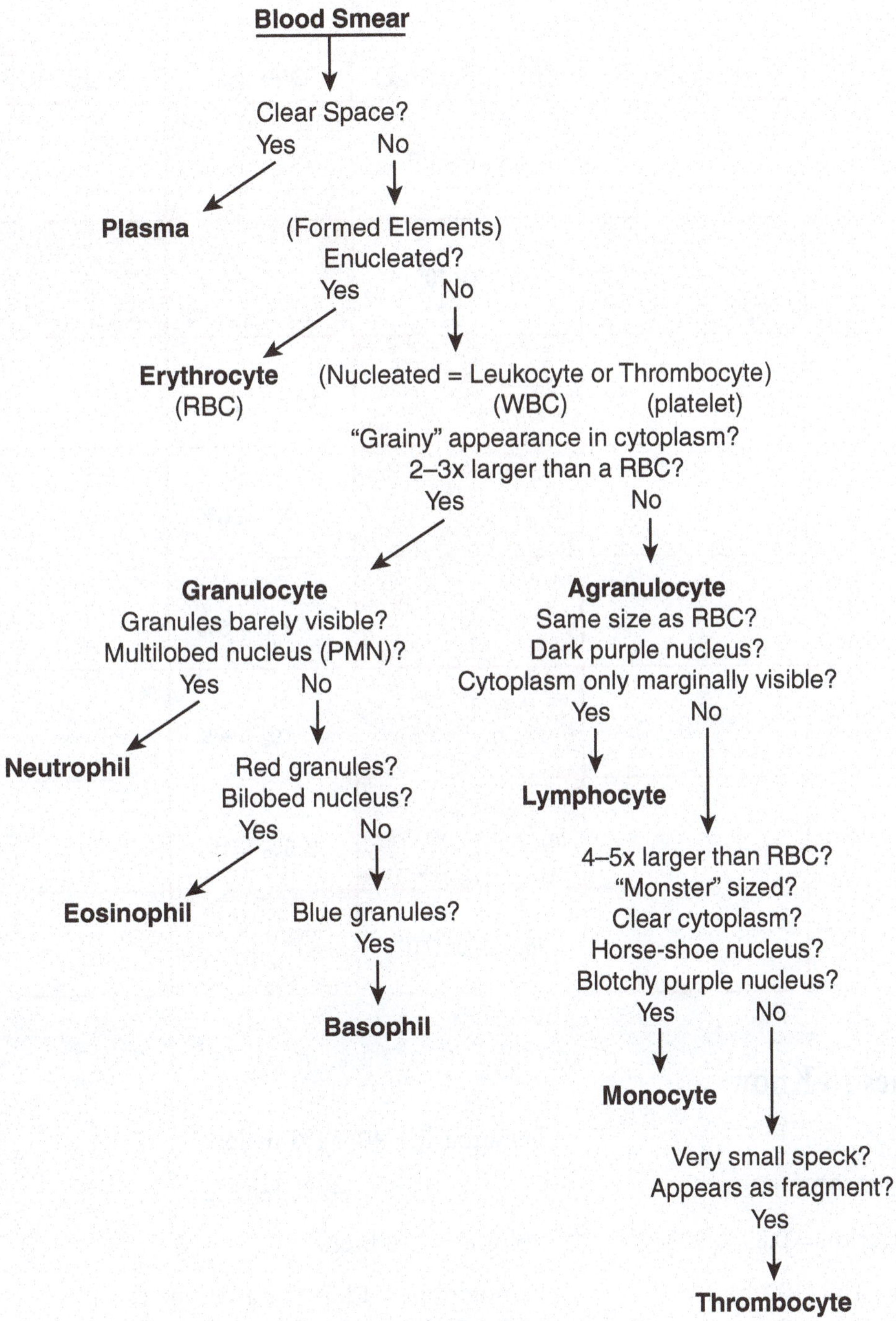

TOOLBOX

Composition of Whole Blood

A Review of the Formed Elements of the Blood

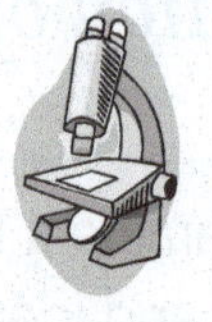

- ***Before coming to lab, use your textbook to fill in the specific function of each formed element.***
- ***In lab, examine the blood slide to locate, identify, and draw each formed element.***

NAME	DRAWING	CBC (CELLS/μL)	DIFF (%)	SPECIFIC FUNCTION
erythrocyte		$4.5 - 6 \times 10^6$		
leukocyte		$6 - 9 \times 10^3$		
Never neutrophil			50% – 70%	
Let lymphocyte			20% – 30%	
Mom monocyte			2% – 6%	
Eat eosinophil			1% – 5%	
Bacon basophil			0.5% – 1%	
thrombocyte		$150 - 500 \times 10^3$		

Other Values to Know

pH = 7.35–7.45

osmolarity = 280–300 mOsm

saline concentration = 0.85–0.90%

temperature = 38°C (100°F)

hematocrit = 40–54% male;
= 37–47% female

volume = 5–6 liters

hemoglobin = 13.5–17.5 g/dL male;
12.0 – 15.5 g/dL female

OUTCOME 5: Examine the blood smears of acute inflammation (leukocytosis) and sickle cell anemia. Distinguish between each of the slides, describe the characteristics of each. Make notes so that you can identify which condition is evident.

- ***Examine the demonstration slides of each pathological condition.***
- ***Make notes so that you can identify which condition is evident on each slide.***

sickle-cell anemia

acute inflammation (leukocytosis)

G. *Anatomy of the Heart*

OUTCOME 6: Locate, label, and identify the external and internal structures of the heart.

TOOLBOX

Superficial Anatomy of the Heart

heart model

■ ***Locate and label the listed structures on the diagrams.***

right atrium
right ventricle
left atrium
left ventricle
superior vena cava
inferior vena cava
aorta
brachiocephalic trunk
left common carotid artery
left subclavian artery
pulmonary trunk
ligamentum arteriosum
left pulmonary artery
right pulmonary artery
left coronary artery
bifurcation of left coronary artery
anterior interventricular sulcus
anterior interventricular artery
great cardiac vein
circumflex artery
atrioventricular sulcus
right coronary artery
left pulmonary veins
right pulmonary veins
posterior interventricular sulcus
posterior interventricular artery
middle cardiac vein
coronary sinus
apex
base

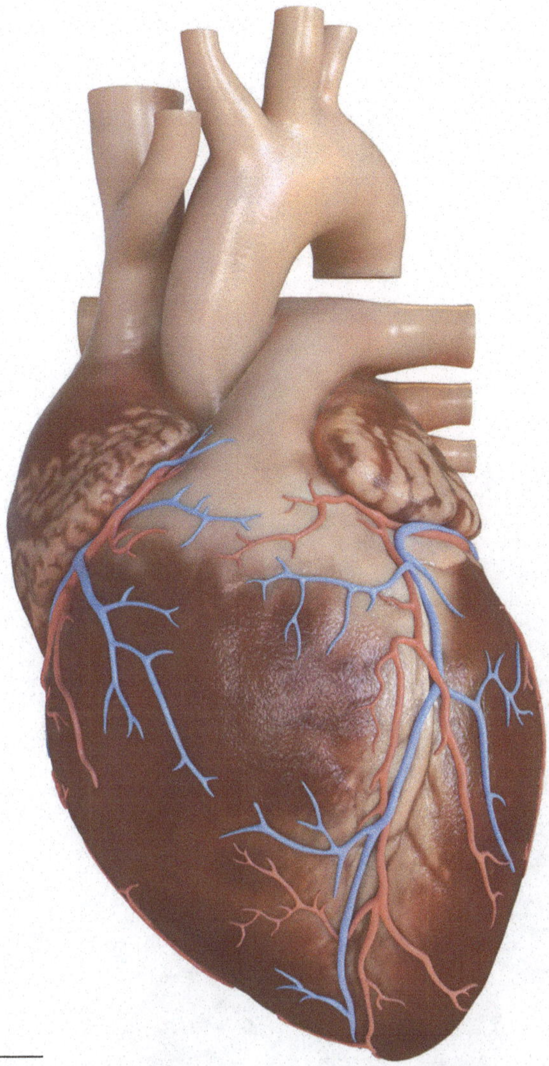

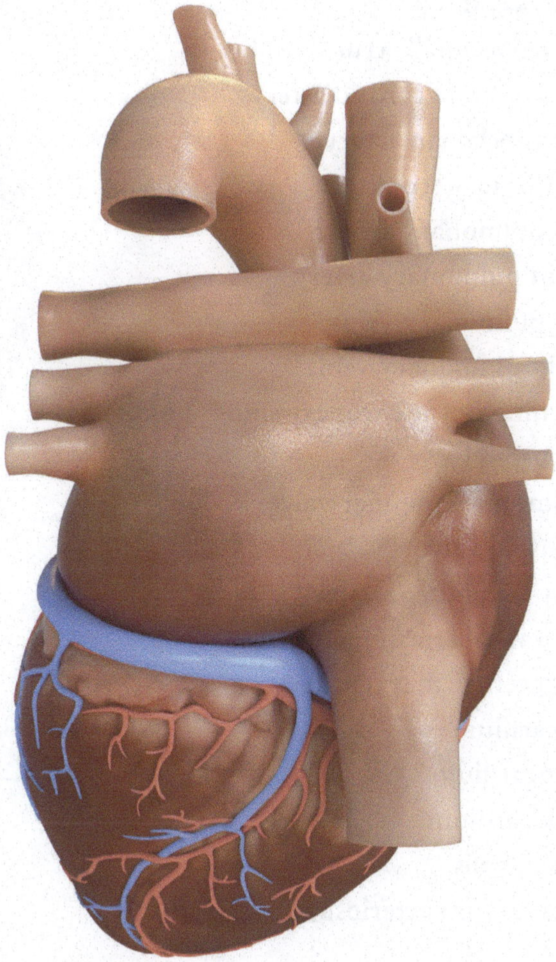

TOOLBOX

Sectional Anatomy of the Heart

heart model

right atrium
right ventricle
left atrium
left ventricle
superior vena cava
inferior vena cava
aorta
  ascending
  arch
  descending
brachiocephalic trunk
left common carotid artery
left subclavian artery
pulmonary trunk
left pulmonary artery
right pulmonary artery
left pulmonary veins
right pulmonary veins
opening of the coronary sinus
tricuspid valve
pulmonic semilunar valve
bicuspid valve
aortic semilunar valve
interatrial septum
interventricular septum
epicardium
myocardium
endocardium
fossa ovalis
ligamentum arteriosum

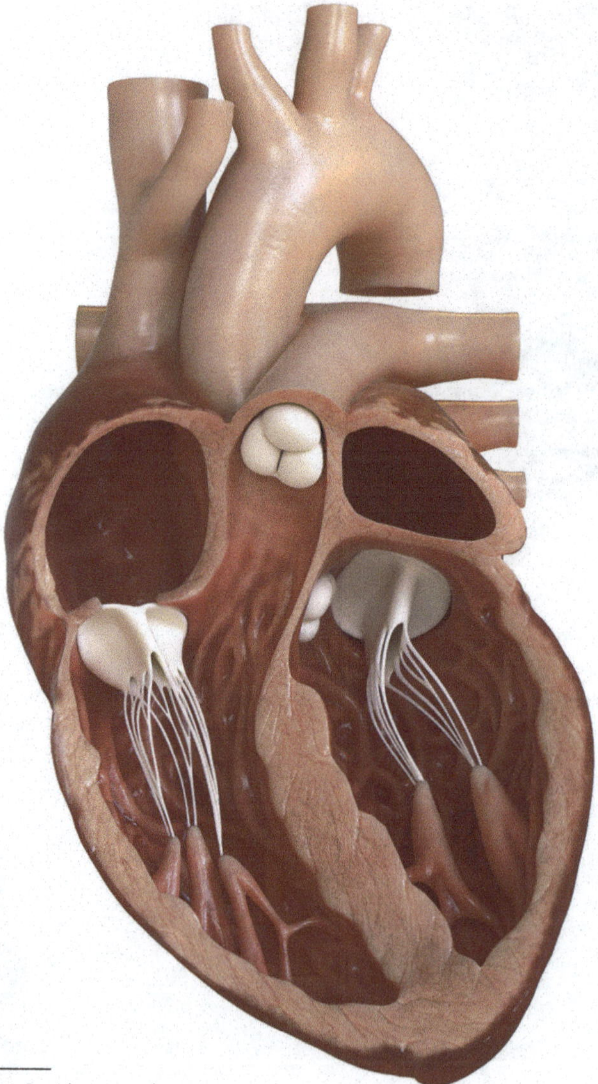

TOOLBOX

Sectional Anatomy of the Heart

heart model

left atrium

left ventricle

pectinate muscles

cusps of the bicuspid valve

papillary muscles

chordae tendineae

trabeculae carneae

aortic arch

brachiocephalic trunk

left common carotid artery

left subclavian artery

pulmonary trunk

right pulmonary artery

left pulmonary artery

inferior vena cava

right pulmonary veins

left pulmonary veins

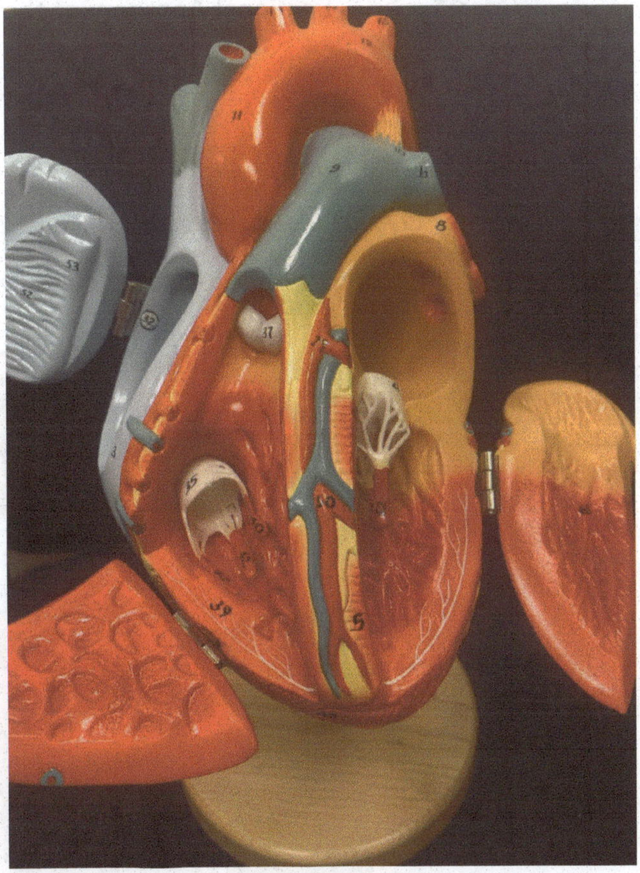

OUTCOME 7: Trace a drop of blood through the heart by listing the structures the blood travels through in the correct order. Start with the venae cavae and end with the aorta. Include all details.

TOOLBOX

Sectional Anatomy of the Heart

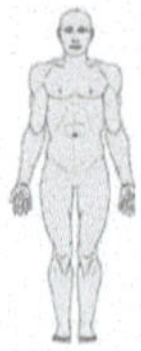

heart model

- ***Use arrows to indicate the pathway through the heart.***
- ***Use RED arrows for O_2-rich blood (oxygenated).***
- ***Use BLUE arrows for O_2-poor blood (deoxygenated).***

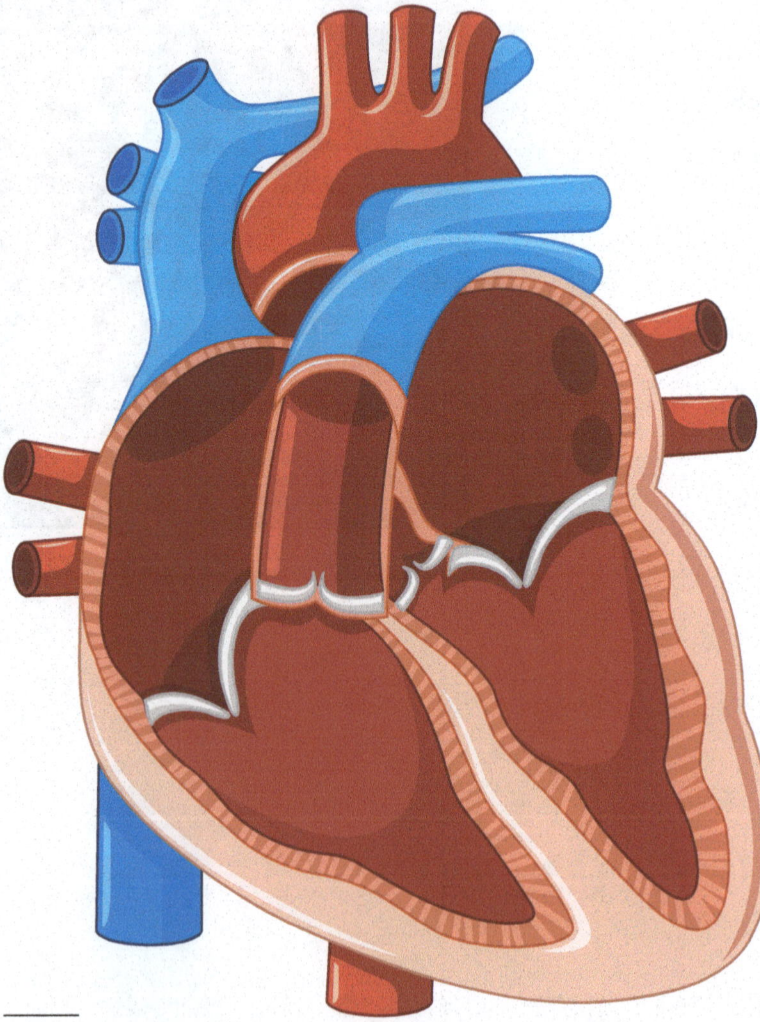

OUTCOME 8: Locate and identify the vessels of the coronary circulation. Trace a drop of blood through the coronary circulation by listing the structures the blood passes through in the correct order. Include all details.

TOOLBOX

Coronary Circulation

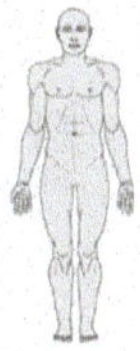

heart model

START: right coronary artery artery	**START:** left coronary artery	**START:** left coronary
↓	↓	↓
	anterior interventricular a.	circumflex artery
	↓	↓
END: right atrium	**END:** right atrium	**END:** right atrium

H. *Comparative Anatomy – Dissection of a Mammalian Heart*

OUTCOME 9: Locate and identify the structures of the mammalian heart.

TOOLBOX

■ ***External Anatomy of the Mammal Heart, pericardial sac intact***

Examine a heart with the **pericardial sac** intact (see photo below). Notice that the membranous sac, although thin, is relatively tough. *Why do you think this might be?* You should also notice that there is quite a bit of adipose tissue attached to the outer surface of the sac. Most of it has been removed from this one. *What is the purpose of the adipose tissue?*

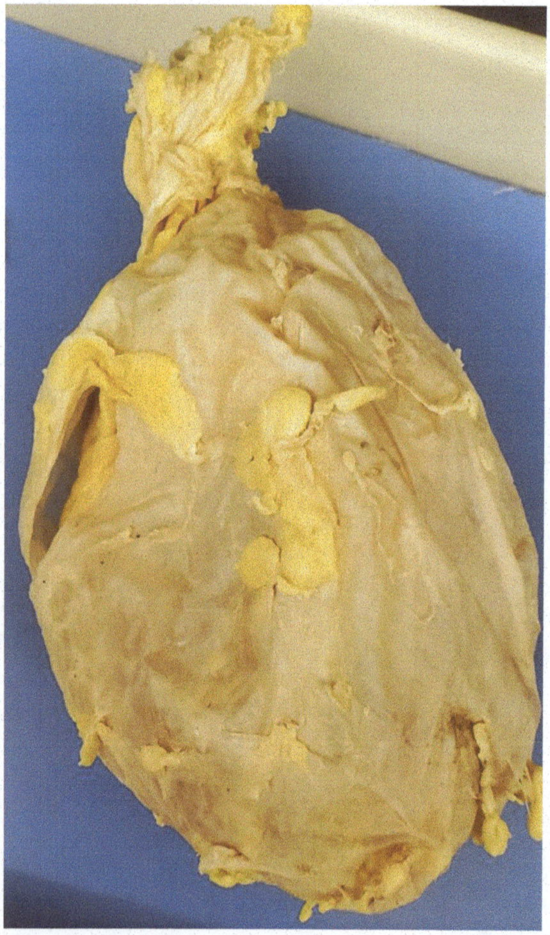

© Sarah Schlereth.

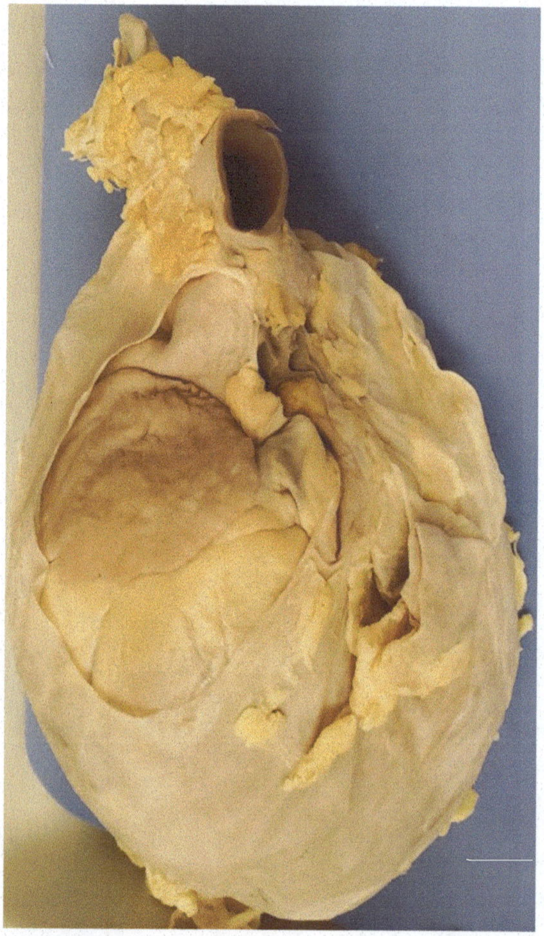

© Sarah Schlereth.

1. Identify the **parietal pericardium** – the serous membrane that adheres to the inner wall of thesac, composed of simple squamous epithelium
2. Identify the **visceral pericardium** – the serous membrane that adheres to surface of the heart, composed of simple squamous epithelium
3. Identify the **pericardial cavity** – the space between the parietal pericardium and the visceral pericardium

The simple squamous epithelial cells of these two serous membranes secrete a watery fluid to provide lubrication in order to reduce friction between the two membranes.

■ *External Anatomy of the Mammal Heart (Pericardial Sac Removed)*

4. The first thing to notice is that the heart comes to a point at one end, called the **apex,** and is broad with vessels entering and leaving at the opposite end, the **base.**
5. Place the heart in the dissecting tray with the base away from you and the apex pointing towards you. You must determine the difference between the **ventral/anterior surface** and the **dorsal/posterior surface.** The ventral surface will be evident because the **pulmonary trunk** is easily viewed. This large artery is visible as a thick, white tube-shaped vessel curving over the upper surface of the heart towards the left. *Remember that left and right on dissection specimens always refer to that of the specimen, so as you are looking at it, it will curve to YOUR right.* Immediately lateral to this vessel, you will see the flap-like **auricle of the left atrium** of the heart. On the right side of the aorta, locate the flap-like **auricle of the right atrium** of the heart.

(Auricle = the external flap; Atrium = the internal chamber).

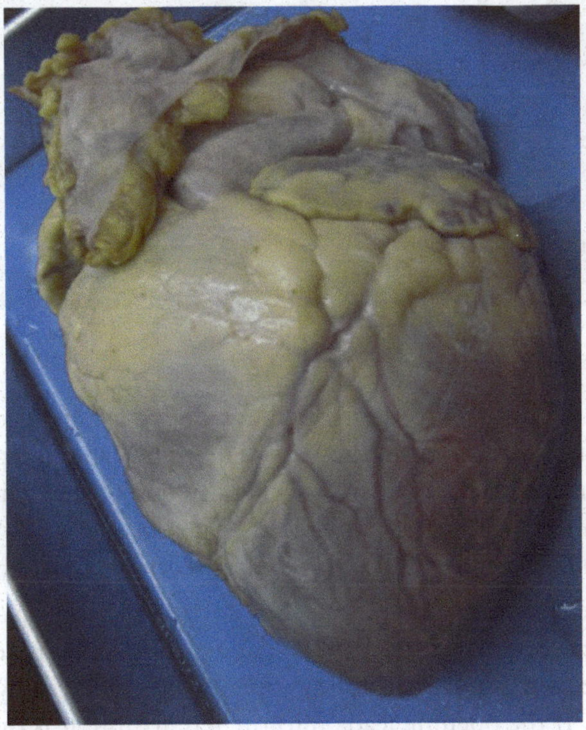

6. Note the surface of the heart is smooth and shiny. This is evidence of the **visceral pericardium.**
7. When viewing the ventral surface of the heart, locate a large groove, or **sulcus,** that travels from the left atrium to the right ventricle. This groove corresponds to the separation of the two ventricles; hence, it is the **anterior interventricular sulcus.** The grooves that run between the atria and the ventricles are called the **atrioventricular sulci.** Notice large amounts of adipose tissue on the surface of the heart, particularly within the sulci and along the surface of the ventricles.
8. Identify the **right ventricle and left ventricle.** Flip the heart over to its dorsal surface and locate the groove between the right and left ventricles on the dorsal surface; the **posterior interventricular sulcus.**
9. Label the structures you identified on the accompanying image of the ventral surface of the heart.
10. Flip the heart over to view the dorsal surface. As you view this surface of the heart, you should immediately notice a number of openings and vessels near the base. These are usually obscured with adipose tissue, so you might need to carefully remove some of the adipose tissue to see them more clearly. View the dorsal surface of the dissected hearts, as well as the following images as you attempt to identify the remaining structures listed below. The image on the left below shows the dorsal surface of the heart before any of the adipose tissue was removed.

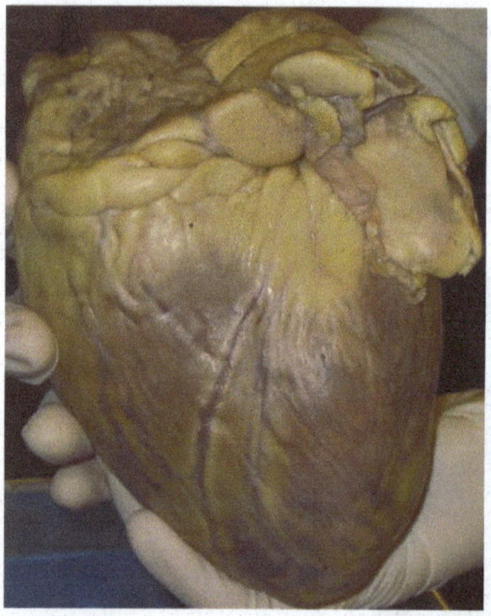

© Sarah Schlereth.

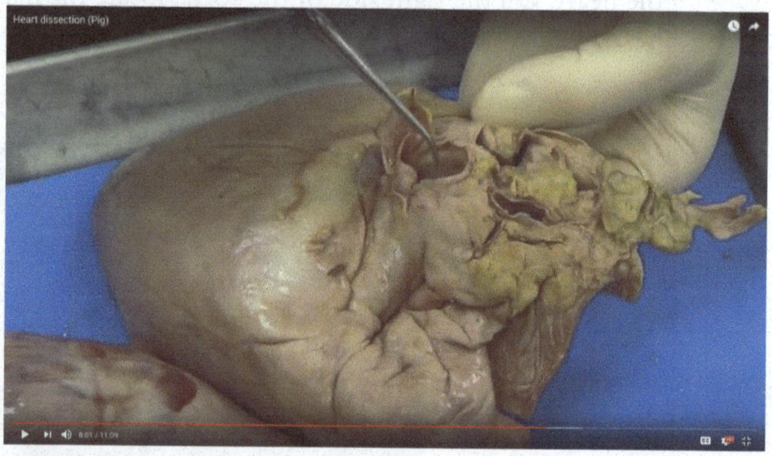

© Sarah Schlereth.

11. Very gently insert the probe into these openings one at a time to determine their connection, and therefore what structure/vessel they are. The thin-walled, flap-like **posterior vena cava,** located near the atrioventricular sulcus, is generally one of the easiest to locate on this surface (indicated by the probe in the image above). If you gently insert your probe into the posterior vena cava, you should be able to pass into the right atrium of the heart.
12. Two slightly smaller openings are located superior to the vena cava. Using your probe, gently (so that you do not create an artificial opening) pass the probe into each of the remaining openings along the base of the heart and follow them to see where they lead to (if a vein) or come from (if an artery). If the opening leads into the left atrium, it is one of the **pulmonary veins.** If it comes from the right atrium, it is likely one of the **pulmonary arteries,** branching off of the pulmonary trunk.

■ *Internal Anatomy of the Mammal Heart*

You will now examine the internal anatomy of the heart. Create a coronal / frontal section similar to the one pictured in the following image by inserting a scalpel in the lateral edge of the apex and cutting superiorly to the right atrium. Repeat this procedure on the left side of the heart and ending at the left atrium. As you look at the dissected heart, you will notice blood vessels, valves, and other structures visible in the heart.

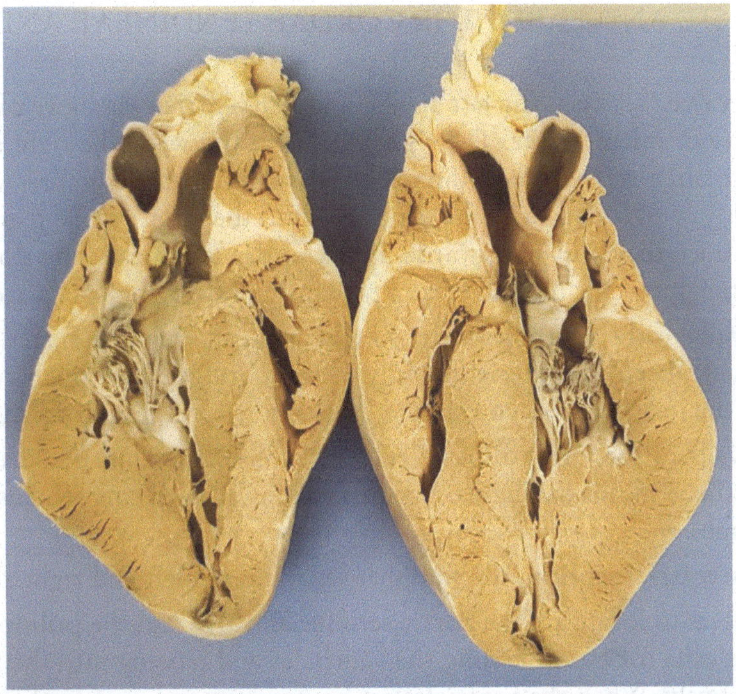

13. Notice that the muscle in the walls of the heart (called the **myocardium**) varies in thickness among the different chambers. In this view, one of the first things you should do is attempt to orient yourself, so that you can determine the difference between the left side of the heart and the right side. The easiest way to do this is to look at the thickness of the wall of the lower chambers/ventricles of the heart. Remember that the right ventricle functions to pump blood to the lungs (**pulmonic circulation**), whereas the left ventricle functions to pump blood to the entire body (**systemic circulation**). Thus, the left ventricle has a thicker myocardial layer than that of the right ventricle. In the image above, you can see spaces between the muscular walls. Those spaces are the ventricular chambers. The left and right ventricles of the heart are also separated from one another by a thick section of muscle, called the **interventricular septum**. LOCATE AND LABEL THE LEFT VENTRICLE, RIGHT VENTRICLE, INTERVENTRICULAR SEPTUM, MYOCARDIAL LAYER.
14. Locate the left and right atria. They are more difficult to determine, as the chambers are rather small.
15. Within each atrial and ventricular chamber is a thin inner lining, covering the walls of the chambers. This lining is the **endothelium,** a layer made of **simple squamous epithelial tissue**. The endothelium forms a continuous sheet of simple squamous epithelium from the chambers all the way through the blood vessels into the capillary walls.

16. Within each ventricle, locate string-like structures, attached on one end to a finger-like muscle in the wall of the ventricle, and on the other end, to a flap-like structure. These string-like structures are called the **chordae tendineae.** The finger-like muscles they are attached to are called **papillary muscles,** and the flap-like structures are the **atrioventricular (AV) valves,** which separate the atria from the ventricles of the heart. Think of these valves as EXITS from the atria to the ventricles. LOCATE AND LABEL THE **RIGHT AV VALVE** AND THE **LEFT AV VALVE.** (Use the thickness of the walls of the ventricles to determine left and right). LOCATE AND LABEL THE **CORDAE TENDINEAE** AND THE **PAPILLARY MUSCLES.**
17. The **left AV valve** (between the left atrium and left ventricle) is more often referred to as the **bicuspid valve** or **mitral valve.** The **right AV valve** (between the right atrium and right ventricle) is more often referred to as the **tricuspid valve.** These structures work together to prevent the backflow of blood from the ventricles into the atria. As the papillary muscles in each ventricle contract, they pull downward on the chordae tendineae. This does NOT close the AV valve, but rather, prevents the valve from **prolapsing** (swinging backwards into the atrium).
18. Spread the walls of the ventricles apart so that you can view the ridges of muscle within the chambers, the **trabeculae carneae** (literally, "meaty bridges"). Though you will not be able to label this on the image above, you should be able to clearly identify it in the dissected hearts.
19. Spread the walls of the atria to locate ridges of muscles within the chambers, the **pectinate muscles.** Pectinate means having narrow projections or divisions set closely in a row (like the teeth of a comb).
20. Locate the **interatrial septum,** the wall that separates the left and right atria.
21. Two more valves to locate are the **aortic semilunar valve** and the **pulmonic semilunar valve.** These are the exits for blood leaving the ventricles and passing into the aorta and pulmonary trunk, respectively. When the ventricles are contracting and pushing blood out of the ventricles, these pockets/cusps are pushed flat up against the walls of the vessels to allow the blood to leave the ventricles. Their construction prevents backflow of blood into the ventricles. When the ventricles relax, the cusps fill with blood and close valve. Gently pass the probe through each of these vessels and valves to verify which one is which.

CHECKLIST: Summary Checklist Cardiovascular I

TERM/STRUCTURE	YOUR NOTES
Cardiac Muscle	
Intercalated Disc	
Muscle Fibers/Cells	
Tunica Intima/Interna	
Endothelium	
Subendothelial Layer (Composed of Areolar Connective Tissue)	
Internal Elastic Layer	
Tunica Media	
Tunica Externa/Advantitia	
Artery	
Vein	
Capillary	
Erythrocyte	
Thrombocyte	
Leukocyte	
Neutrophil	
Lymphocyte	
Monocyte	
Eosinophil	
Basophil	

(continued)

TERM/STRUCTURE	YOUR NOTES
Pericardial Sac	
Fibrous Pericardium	
Parietal Pericardium	
Pericardial Cavity	
Visceral Pericardium	
Myocardium	
Endocardium	
Apex of Heart	
Base of Heart	
Anterior Interventricular Sulcus	
Posterior Interventricular Sulcus	
Atrioventricular/Coronary Sulcus	
Right and Left Atrium	
Right and Left Ventricle	
Pulmonary Trunk	
Pulmonary Arteries	
Ascending Aorta	
Aortic Arch	
Superior Vena Cava	
Inferior Vena Cava	

TERM/STRUCTURE	YOUR NOTES
Pulmonary Veins	
Right Coronary Artery	
Anterior Interventricular Artery	
Posterior Interventricular Artery	
Circumflex Artery	
Great Cardiac Vein	
Middle Cardiac Vein	
Coronary Sinus	
Tricuspid Valve/Right Atrioventricular Valve	
Pulmonary Semilunar Valve	
Bicuspid/Mitral/Left Atrioventricular Valve	
Aortic Semilunar Valve	
Brachiocephalic Trunk	
Left Subclavian Artery	
Pectinate Muscle	
Papillary Muscle	
Chordae Tendineae	
Trabeculae Carneae	
Interventricular Septum	
Interatrial Septum	

(continued)

TERM/STRUCTURE	YOUR NOTES
Fossa Ovalis	
Opening of the Coronary Sinus	
Ligamentum Arteriosum	
Left Common Carotid Artery	

EXERCISE 9

Cardiovascular System: Dissection of Arteries and Veins

A. *Introduction*

Important things to remember when learning the circulatory patterns of the body:

1. Think of the vessels as roadways. Imagine that you are driving along a road and picture the pattern of vessels much the way you might view a map.
2. The more you observe a map, the more familiar you become with the major thoroughfares and their branches. Likewise, the more you examine the dissection and diagrams, the more familiar you will become with the blood vessels and their branching patterns.
3. Practice "traveling" through the vessels by selecting a starting place and a destination. At first, try traveling along the roadway by examining the diagrams of the vessels. Then try doing the same thing without looking.
4. It always helps to write it down and draw it out.
5. Getting to the heart and pulmonary circuit is half the battle. Once there, the pathway is always the same.
6. Remember, the only way to get from an artery to a vein is through a capillary bed or shunt. The only way to get from a vein to an artery is through the heart.
7. **You must always indicate LEFT or RIGHT.**

   You must always indicate ARTERY or VEIN.

B. *Outcomes for Exercise 9*

1. Locate, label, and identify the major arteries and veins of the body. Be able to determine where the blood was immediately before and after a specific vessel.
2. Locate, label, and identify the abdominal arteries and veins and their branches. Be able to determine where the blood was immediately before and after a specific vessel.
3. Locate, label, and identify the veins of the hepatic portal system. Be able to determine where the blood was immediately before and after a specific vessel.
4. Review human fetal circulatory patterns, and describe the changes in the fetal structures after birth.
5. Dissect the arteries and veins in the preserved mammal and locate the vessels described in the text. Be able to determine where the blood was immediately before and after a specific vessel.

C. *Major Arteries and Major Veins*

OUTCOME 1: Locate, label, and identify the major arteries and major veins of the body. Be able to determine where the blood was immediately before and after a specific vessel.

TOOLBOX

An Overview of the Systemic Arterial System

An Overview of the Systemic Venous System

Mr. Muscle
Mr. Flip Chart
arm, leg

pulmonary arteries
coronary arteries
aorta
brachiocephalic artery
common carotid artery
internal carotid artery
external carotid artery
subclavian artery
axillary artery
brachial artery
radial artery
ulnar artery
thoracic aorta
abdominal aorta
celiac trunk
superior mesenteric artery
renal artery
gonadal (spermatic or ovarian) artery
inferior mesenteric artery
common iliac artery
internal iliac artery
external iliac artery
femoral artery
popliteal artery
posterior tibial artery
anterior tibial artery

■ *Use a line to connect the artery to its name.* ■ *Color the arteries* ***RED***.

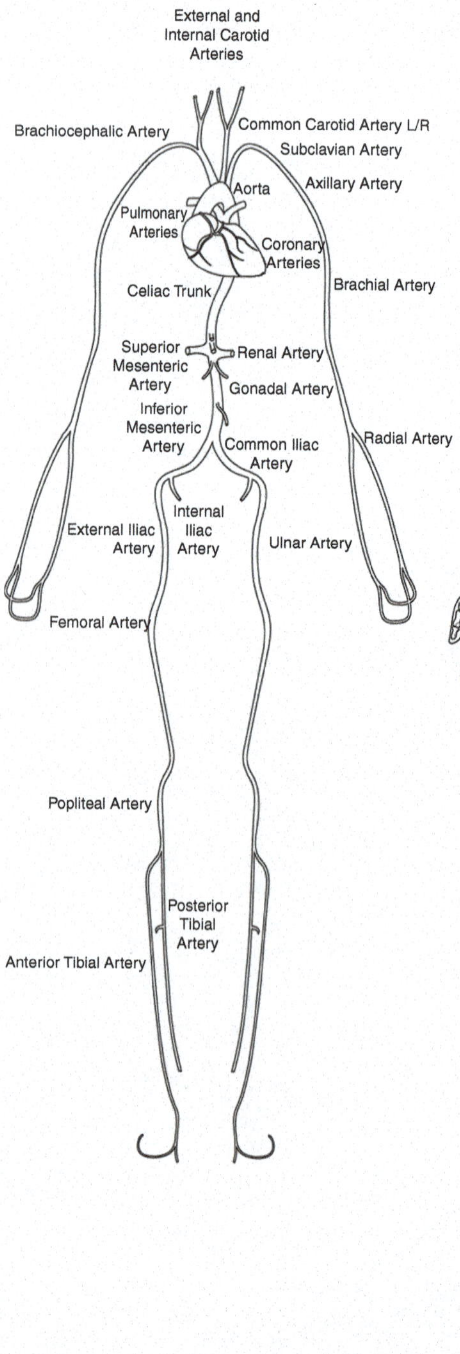

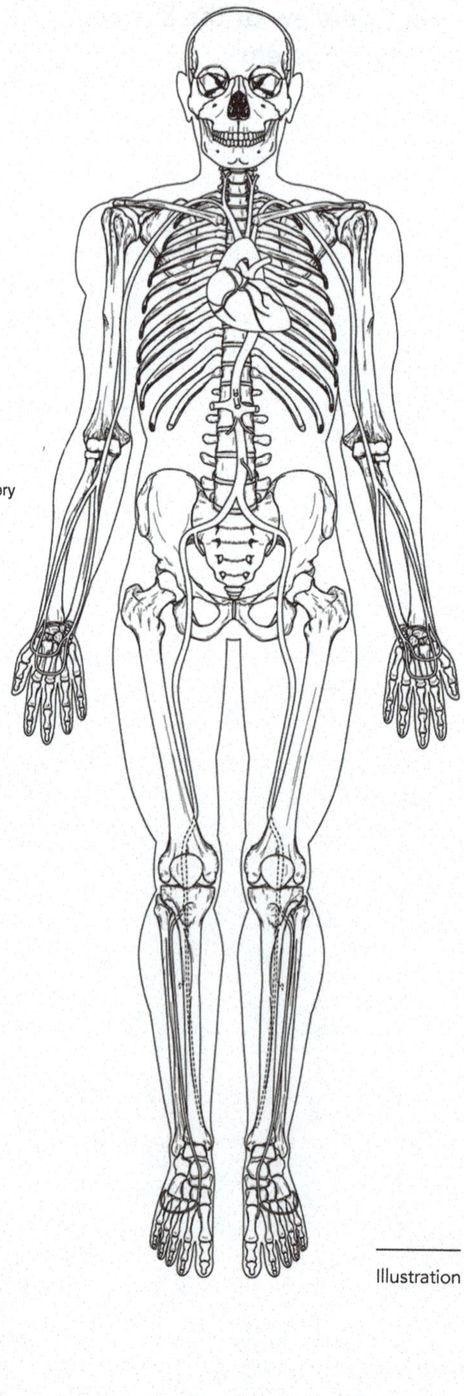

Illustration by Jamey Garbett. © 2003 Mark Nielsen.

cardiac veins
pulmonary veins
superior vena cava
brachiocephalic vein
internal jugular vein
external jugular vein
L/R subclavian vein
axillary vein

brachial vein
radial vein
ulnar vein
inferior vena cava
hepatic veins
renal vein
gonadal (spermatic or ovarian) vein

common iliac vein
internal iliac vein
external iliac vein
femoral vein
popliteal vein
posterior tibial vein
anterior tibial vein
great saphenous vein

■ *Use a line to connect the vein to its name.*

■ *Color the veins BLUE.*

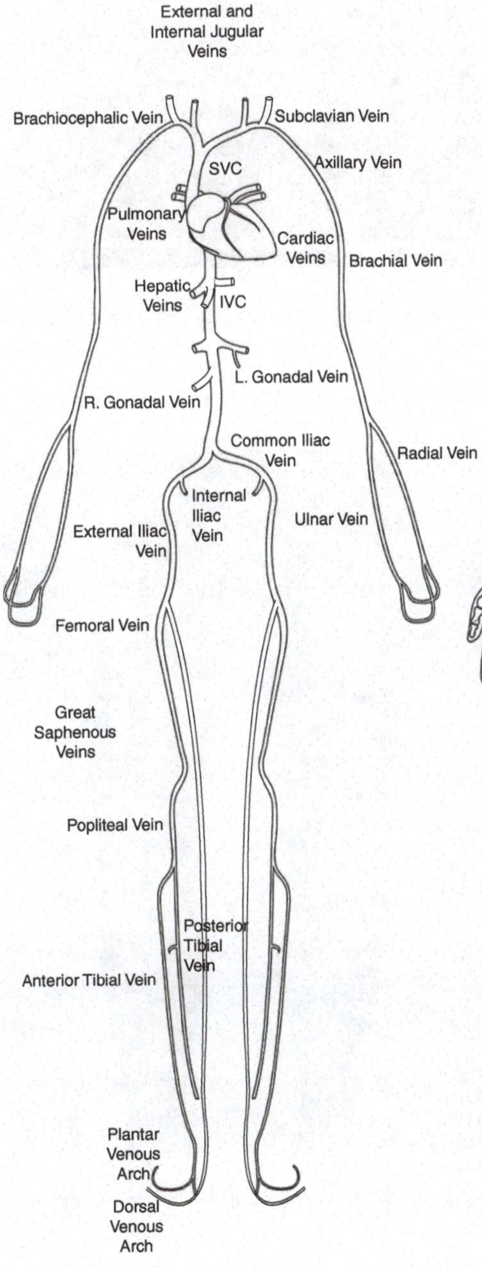

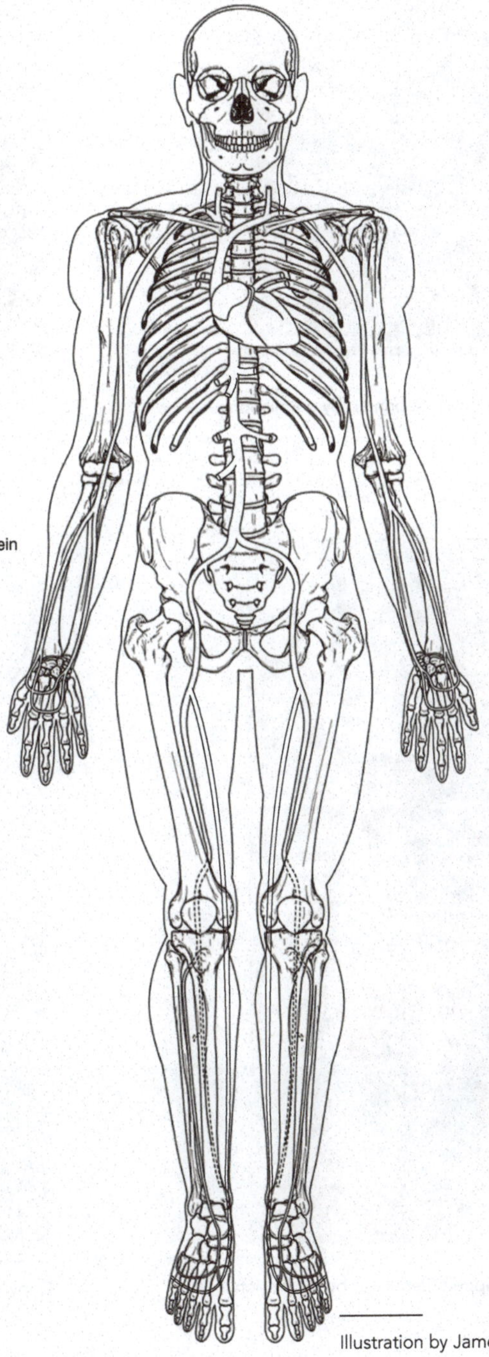

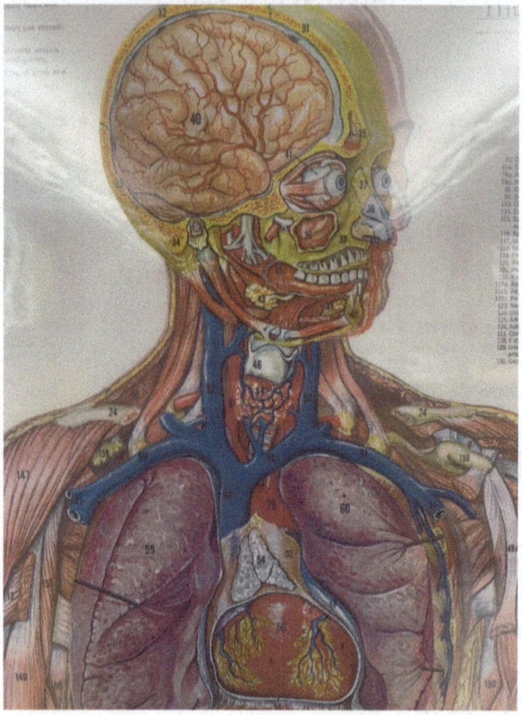

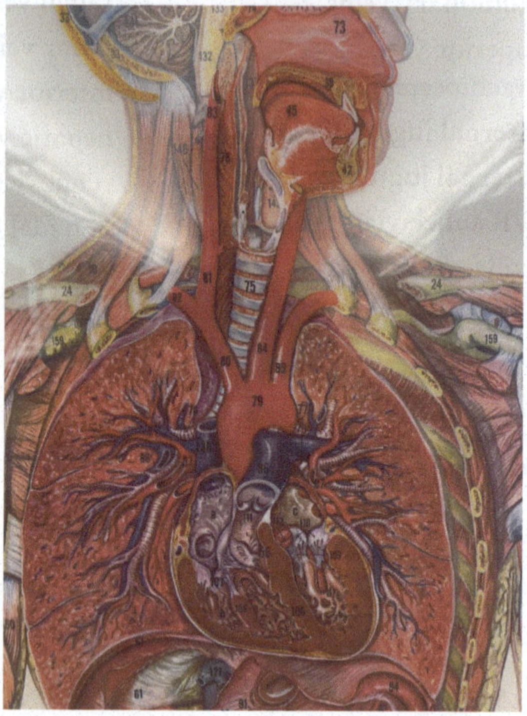

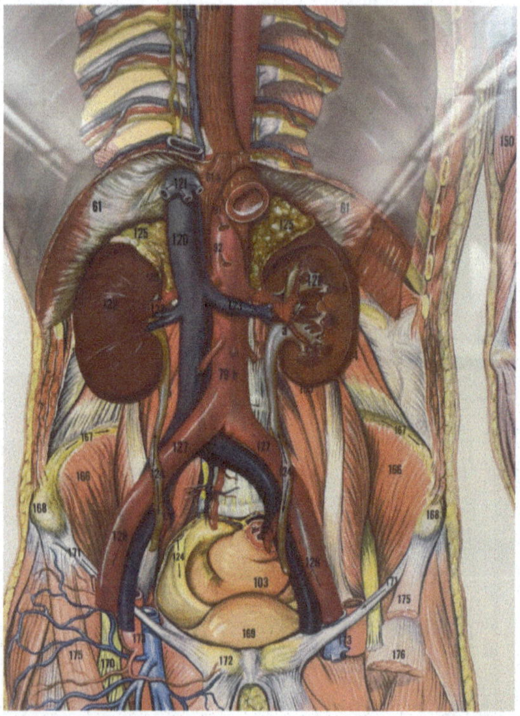

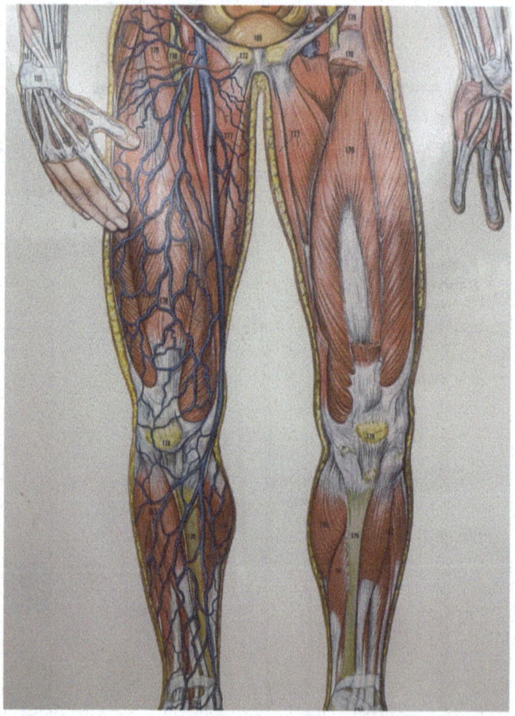

D. *Special Flow Patterns*

OUTCOME 2: Locate, label, and identify the abdominal arteries and veins and their branches. Be able to determine where the blood was immediately before and after a specific vessel.

TOOLBOX

Major Arteries of the Trunk
Arteries of the Abdomen
Venous Drainage of Trunk and Upper Limb

Mr. Muscle
Mr. Flip Chart

abdominal aorta

celiac trunk
see small image
for branches

superior mesenteric artery

L/R renal artery

L/R gonadal (spermatic or ovarian)

inferior mesenteric artery

L/R common iliac artery

L/R femoral artery

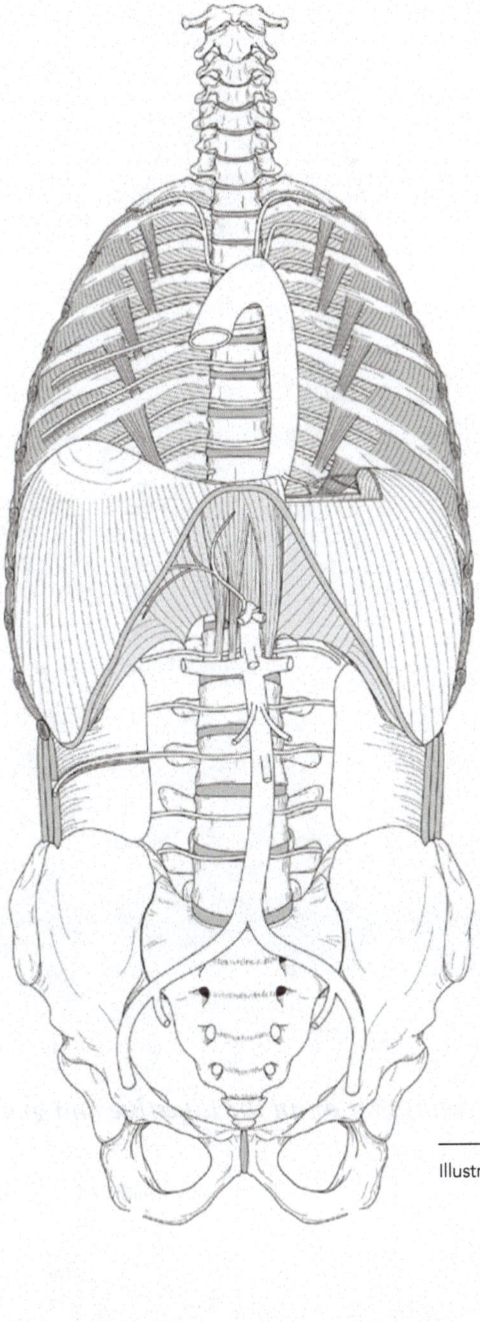

Branches of celiac trunk:
common hepatic artery
left gastric artery
splenic artery

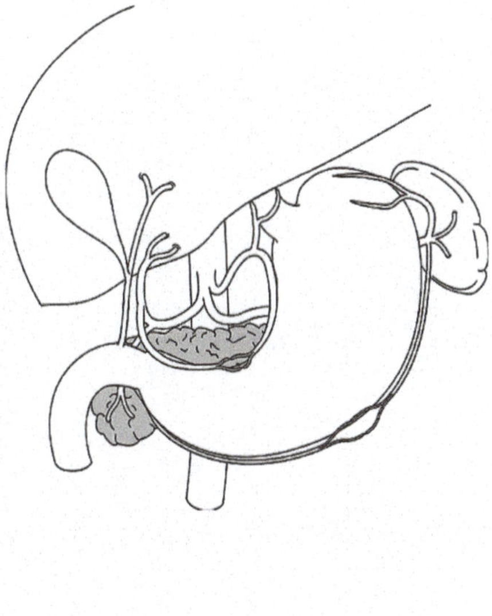

Illustration by Jamey Garbett. © 2003 Mark Nielsen.

inferior vena cava

hepatic veins

L/R renal veins

L/R spermatic or ovarian vein

L/R common iliac vein

L/R internal iliac vein

L/R external iliac vein

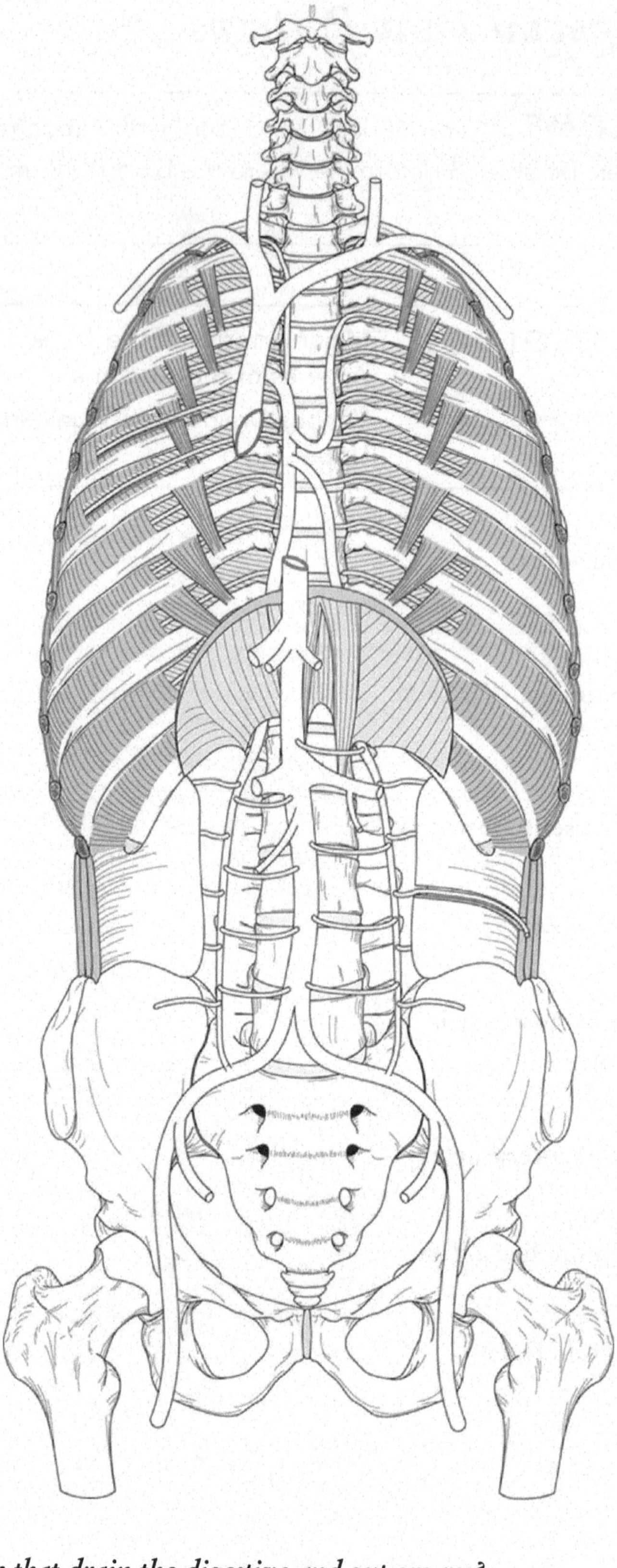

■ ***Where are the veins that drain the digestive and gut organs?***

OUTCOME 3: Locate, label, and identify the veins of the hepatic portal system. Be able to determine where the blood was immediately before and after a specific vessel.

TOOLBOX

The Hepatic Portal System

Mr. Muscle,
Mr. Flip Chart

L/R gastric vein

splenic vein

inferior mesenteric vein

superior mesenteric vein

hepatic portal vein

into liver

(microscopic structures within the liver)

sinusoid capillaries of liver lobules

central veins

out of liver

hepatic veins

inferior vena cava

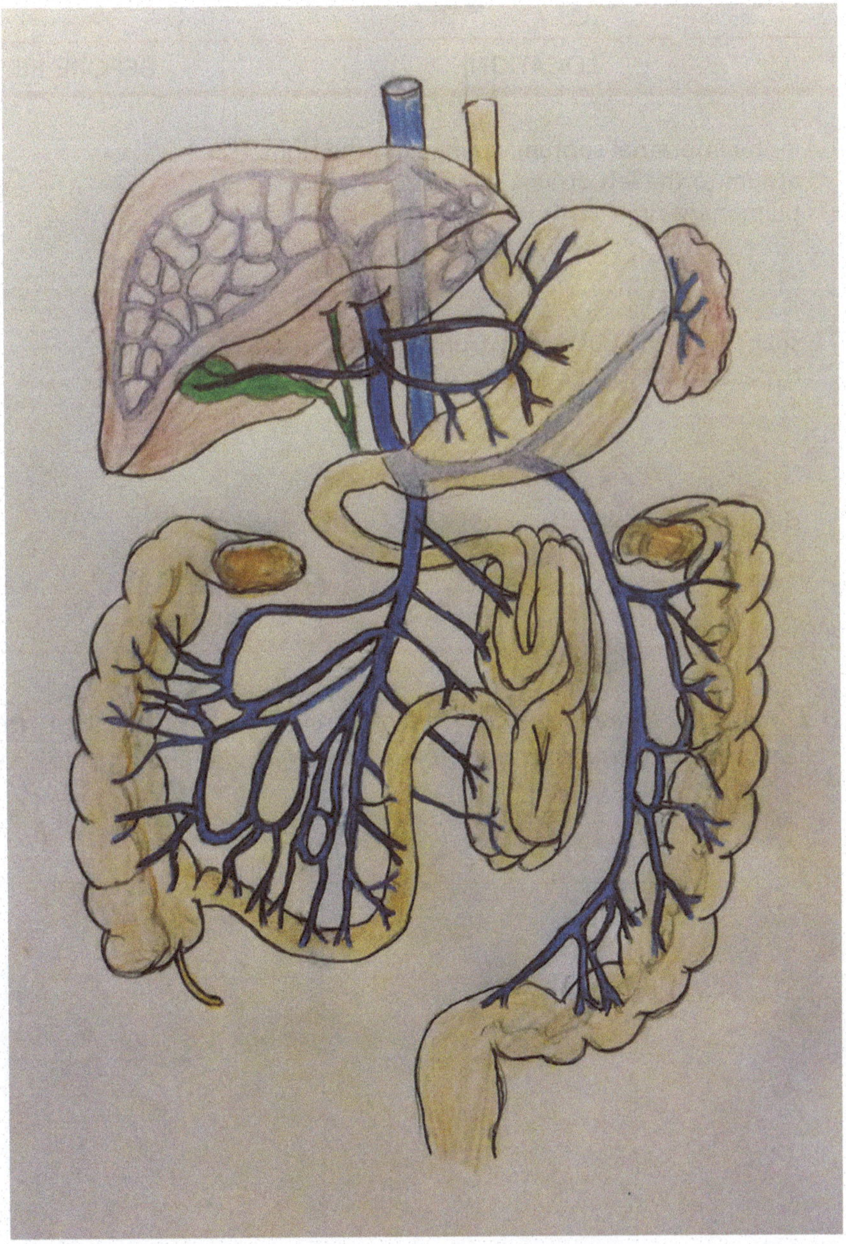

OUTCOME 4: Review human fetal circulatory patterns, and describe the changes in the fetal structures after birth.

TOOLBOX

Fetal Circulatory Patterns

■ ***Name the Fetal Circulatory Structures based on the described location:***

LOCATION	BEFORE BIRTH	AFTER BIRTH
in the interatrial septum, connecting the right atrium to the left atrium, bypassing the pulmonary circuit		
shunt from the pulmonary trunk to the aorta, bypassing the pulmonary circuit		
shunt from umbilical vein to inferior vena cava, bypassing the fetal liver		

■ ***Trace blood through the human fetal circulation using the two alternate pathways through the heart. In each case, start with the venae cavae and end with the aorta.***

E. *Comparative Anatomy—Dissection*

OUTCOME 5: Dissect and identify the arteries and veins in the preserved specimen. Be able to determine where blood was immediately before a specific vessel and where it will be immediately after a specific vessel.

TOOLBOX

✂

EXTERNAL ANATOMY

Examine the animal's external anatomy before beginning the dissection. Determine whether the animal is a male or a female. To do this, examine the ventral and posterior regions of the animal, looking for specific structures of the urogenital system.

1. Lay the animal on its dorsum so that you view the ventral surface.
2. On approximately the midline of the abdomen, find the umbilical cord. Examine the region immediately surrounding the umbilical cord. Notice that it has been tied off with a string.
   a) Male: there will be a small opening just posterior to the umbilical cord. Note: look closely, as it may have accidently been tied up with the umbilical cord. This opening is the opening of the penis, or the **male urogenital orifice**, an opening shared by the urinary and reproductive systems in the male. The remainder of the penis is located beneath the skin on the ventral surface, between this opening and the opening for the anus. You will dissect out the penis later when you do the urogenital dissection.
   b) Female: there will NOT be an opening here. Instead, the **female urogenital orifice** is located near the tail, on the ventral surface, just anterior to the opening to the anus.
3. Find the anal opening, or **anal orifice**, in both the male and female. It is located just ventral to the tail.
4. Additional external genitalia may be used to distinguish between male and female. Both are located near the base of the tail, near the anal orifice.
   a) Male: Notice an enlarged, sac-like region, just ventral and anterior to the tail. This is the **scrotum** or **scrotal sac**. The male **gonads**, called **testes,** are located within this sac. You will open the scrotum and examine one testis when you complete the urogenital dissection.
   b) Female: Notice a small finger-like structure just anterior to the female urogenital orifice. This is the **urogenital papilla** and will later develop into the **clitoris** of the adult female. The raised tissue lateral to the urogenital orifice are the **labia**. Together, the labia and the urogenital orifice make up the **vulva or external genitalia** of the female.
5. On both male and female, find the two parallel rows of **mammary papillae (nipples)** located on each side of the midline on the surface.

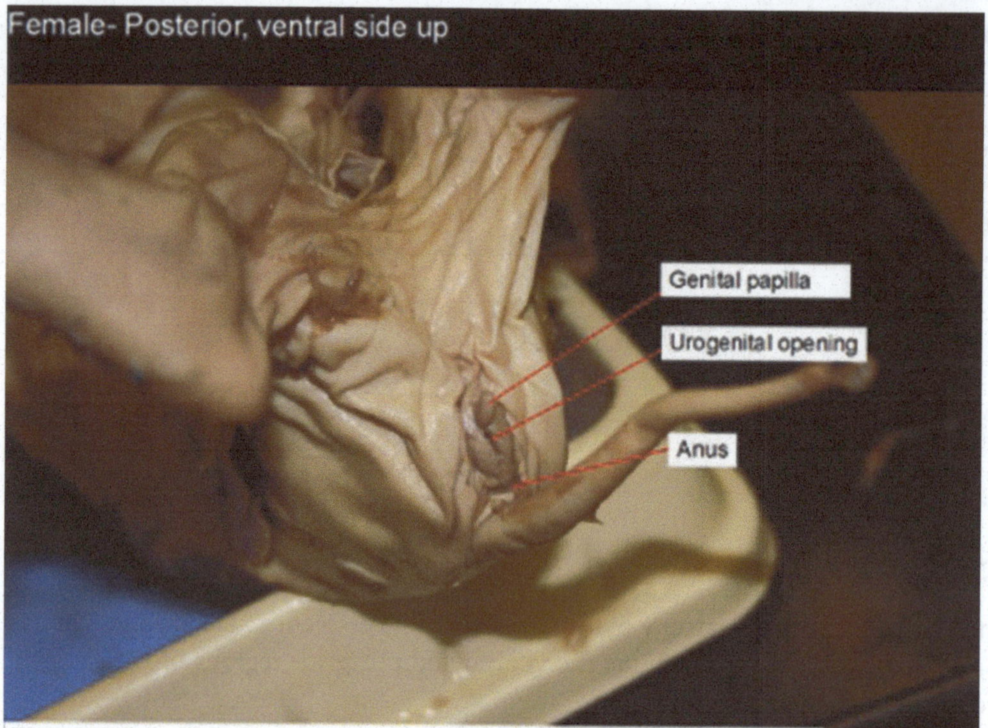

The urogenital opening is an opening to both the urinary and reproductive system.

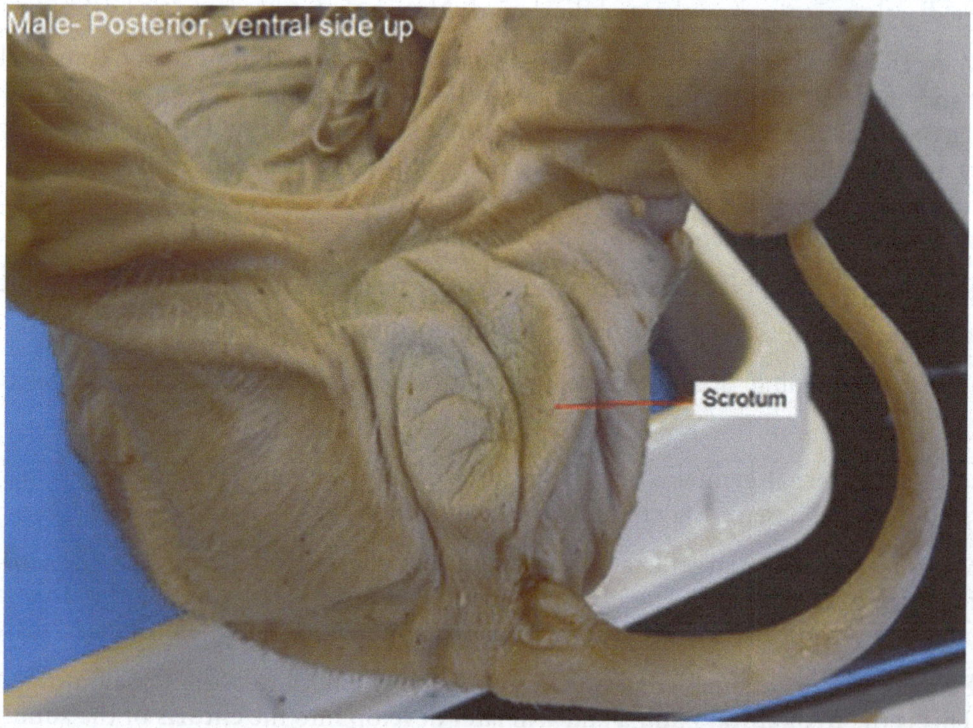

Michael J. Gregory, Ph.D., Clinton Community College, CC BY-SA-NC 3.0

OPENING CUTS

1. Lay the animal on its dorsal surface in the dissecting tray.

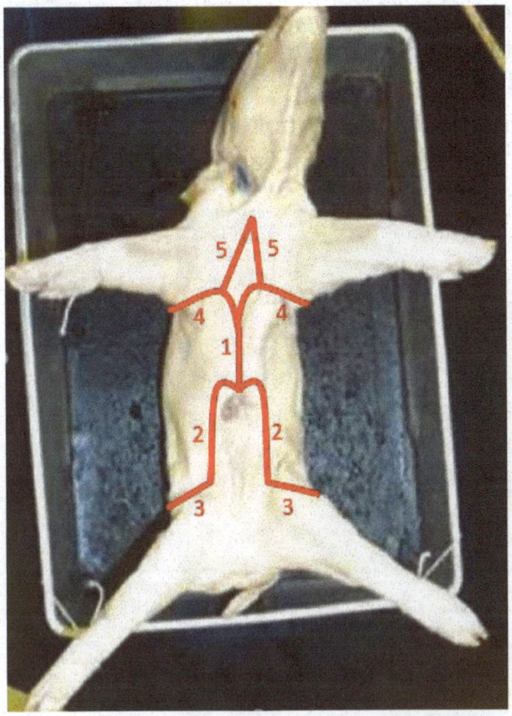

Contributed by Sarah Schlereth. © Kendall Hunt Publishing Company.

2. Cut along the midline of the body. **See Cut #1** in the image. Start just anterior to the umbilical cord and extend the cut anteriorly until you reach the diaphragm. You will not be able to see the diaphragm because it is internal, but it will be just deep to the point where you begin to meet resistance while making your cut. It is approximately even with or just posterior to where the forelimbs meet the body. ***Be careful not to cut too deeply***. You should cut through a few layers of muscle into the abdominopelvic cavity. Some of the internal organs will become visible.
3. Next, extend the posterior end of this cut laterally and down past the umbilical cord. **See Cuts #2.** Remember, in males the penis is just deep to the surface of the skin in this region; so, DO NOT make a midsagittal cut. Instead, extend your cut around the umbilical cord and about a finger's width on either side, straight down to where the hind limbs meet the body. There will be a flap between the two **Cuts #2**. The flap will be attached internally to the umbilical vein, which will disappear into the liver. Clip the vein, leaving a stub at the umbilicus and fold it toward the tail. Thus, this part of the anatomy is moved out of the way. If you are unsure, examine the image and ask the lab instructor for clarification.
4. Once you have made the cuts along both sides of the umbilical cord, extend the cuts laterally. **See Cuts #3.** At the anterior end of cut #1, make two more lateral incisions. **See Cuts #4.**
5. Once the abdominopelvic cavity is open, locate the muscular diaphragm at the anterior end of the cavity. It should be hugging the anterior-most surface of the liver. The diaphragm separates the thoracic and abdominopelvic cavities. Extend Cut #1 just slightly anteriorly and observe a small opening into the thoracic cavity. Gently insert your finger in that opening and separate the diaphragm from its attachment to the ventral wall. It will become draped over the liver, like a blanket. Pull down on the diaphragm and liver. You are able to see into the thoracic cavity and see part

of the lungs in their pleural cavities and the apex of the heart within its pericardial sac. Run your finger over the ventral surface of the pericardial sac to separate it from the sternum.

6. Before continuing, carefully examine **Cuts #5** in the image. Once the pericardiac sac has been detached from the sternum, make diagonal cuts from Cut #4. Start approximately 1-2 finger widths lateral to the midline. Have the cuts meet just anterior to the forelimbs. This will allow the removal of the sternum to expose the thoracic cavity and its contents.
7. When all of the above cuts are completed, your animal should look similar to the animal in the image. The next step will be to expose the veins and other structures anterior to the heart.

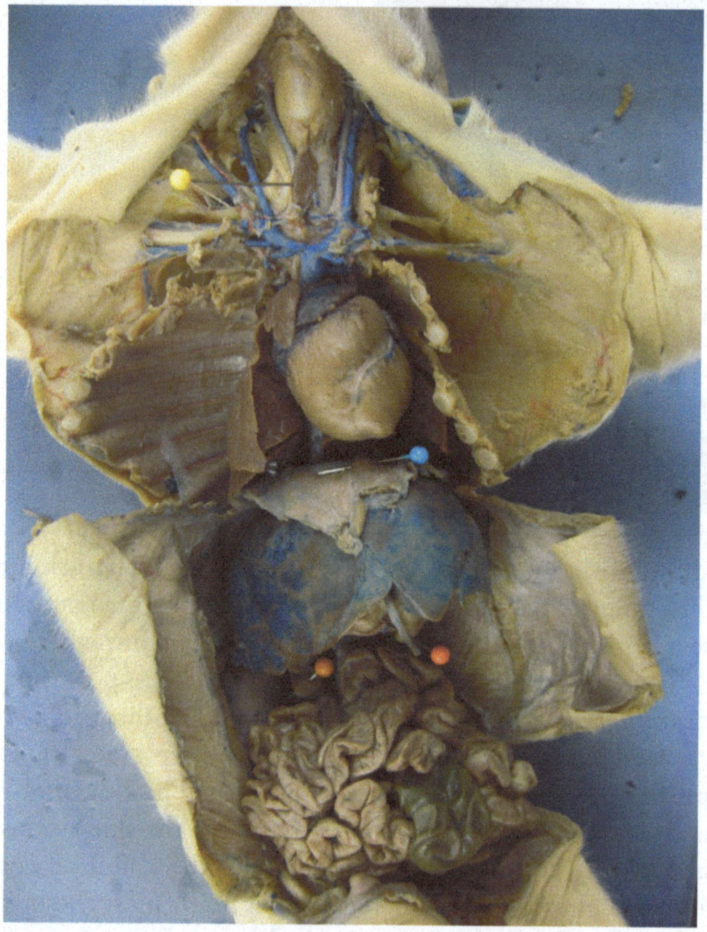

The vessels of these specimens are injected with colored latex for ease of identification. Find the two ***injection incisions*** on either side of the neck. These were used to access the vessels to make the injections. The arteries are filled with red latex and the veins with blue latex. Sometimes, the injection process does not work as expected. You may discover vessels injected with the wrong colors or a mixture of red and blue or without any latex at all. Regardless of whether the vessels are properly injected or not, you can often distinguish veins and arteries by the thickness of their walls—veins have thinner walls than arteries. You can also use the locations of other vessels and structures to help you identify the vessels.

Thoracic Veins

1. To expose the veins in the thoracic cavity, remove the overlying tissues without destroying the vessels. The easiest way to do this is to start with the ***injection incisions***. The blue jugular veins can be observed just inside these incisions. Carefully follow this vessel as it travels medially toward the heart. As you slowly and carefully expose this vessel, you will begin to see more of the vessels. When you have exposed them all, you will see vessels that look like those pictured on the previous page.

2. Locate and identify the following, by first locating some non-vessel landmarks. The bulb-shaped structure at the anterior of the exposed neck is called the **larynx**, and the ringed tube extending posteriorly from it is the **trachea**, both structures of the respiratory system.

3. The two, large blue vessels lateral to the trachea are
   - **internal jugular veins, return blood from the brain.**

4. Immediately medial to each of these jugular veins are
   - **common carotid arteries, sending blood to the face and brain.** They should be injected with red latex, but they may also simply appear white. This happens because the walls of these arteries are so thick that you cannot see the latex through the walls. Look between the common carotid arteries and the internal jugular veins. Find thin, white string-like structures. These are the **vagus nerves.**

5. Follow the internal jugular veins posteriorly to find more veins. Sometimes, there will be second jugular veins, lateral to the internal jugular veins. These are
   - **external jugular veins, bring blood from the facial region.**

6. The veins coming from the forelimb are the
   - **subclavian veins, return blood from the forelimb.**

7. The internal jugular veins, the external jugular veins and the subclavian veins from the left and right flow into the
   - **left and right brachiocephalic veins.**

8. The two brachiocephalic veins carry blood into
   - **anterior vena cava,** which carries the blood into the right atrium of the heart.

After locating the vessels in the animal, label the vessels on the diagram:

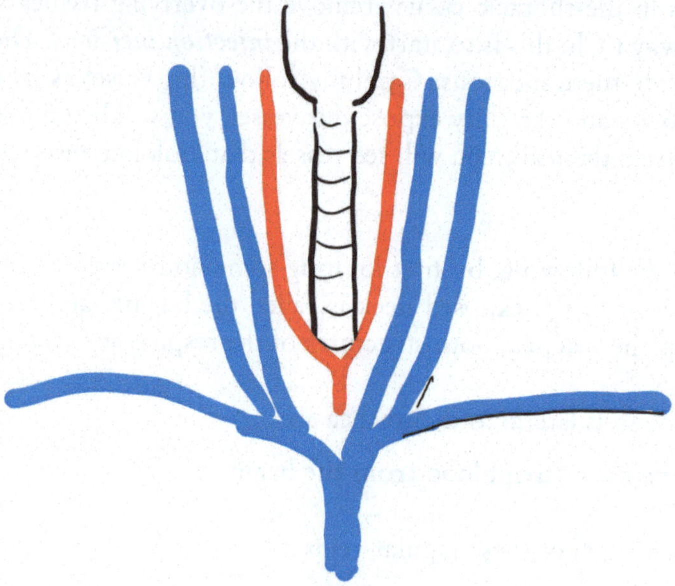

Contributed by Sarah Schlereth. © Kendall Hunt Publishing Company.

Thoracic Arteries

1. The thoracic veins will have to cut out of the way in order to see the thoracic arteries more clearly. ***In order to still be able to review all the veins later, it is suggested that you leave the veins intact at one animal at your table*** and ***remove them from a second animal to view the arteries in that animal.*** With the thoracic veins removed, you will see several arteries extending from the base of the heart and from the aorta. Blood leaving the left ventricle of the heart passes through the **aortic semilunar valve** before entering the first part of the aorta, referred to as the **ascending aorta**. The part of the aorta arching from that point is referred to as the **aortic arch**.

2. The branches off of the **aortic arch** in most mammals are different from the branches in humans. While there are three branches off of the aortic arch in humans, there are only two branches in this animal. As you view the aortic arch and imagine the blood flowing from the heart, the first branch off of the arch is the brachiocephalic **artery.** This artery will branch into three others:
   - **right subclavian artery carrying blood to the right forelimb,**
   - **right common carotid artery carrying blood to the right side of the face and brain,**
   - **left common carotid artery carrying blood to the left side of the face and brain.**

3. You might observe a short branch that appears to extend from the brachiocephalic before the two common carotid arteries emerge. This as the **bicarotid trunk.** It is often not present.

4. The second branch off of the arch is the
   - **left subclavian artery carrying blood to the left forelimb.**

Locate the vessels in the animal. This labeled diagram will help you be aware of the differences between humans and other mammals.

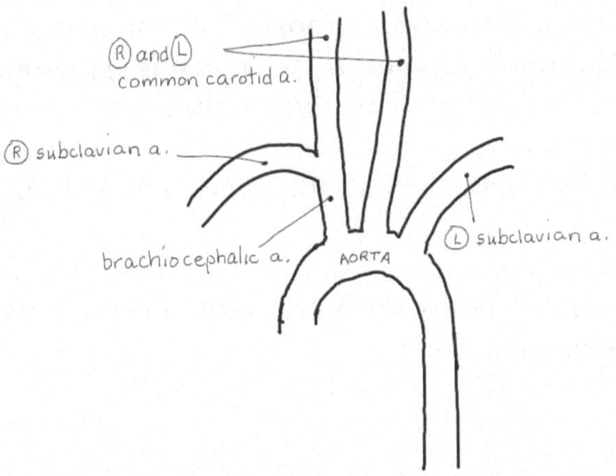

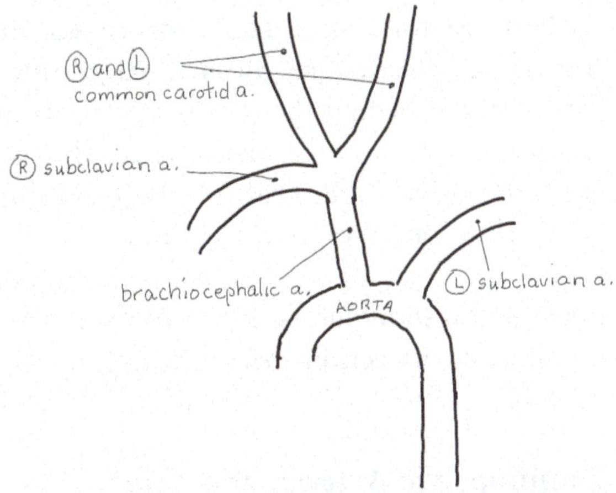

Source: Illustrations by Lucia Tranel, 2021.

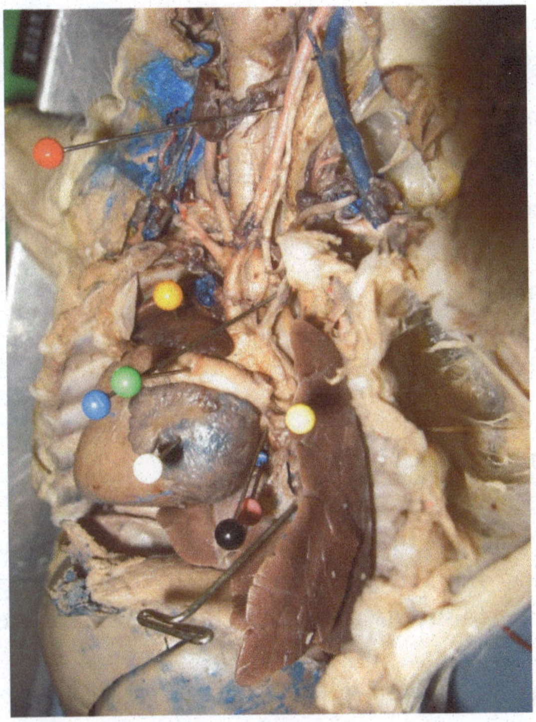

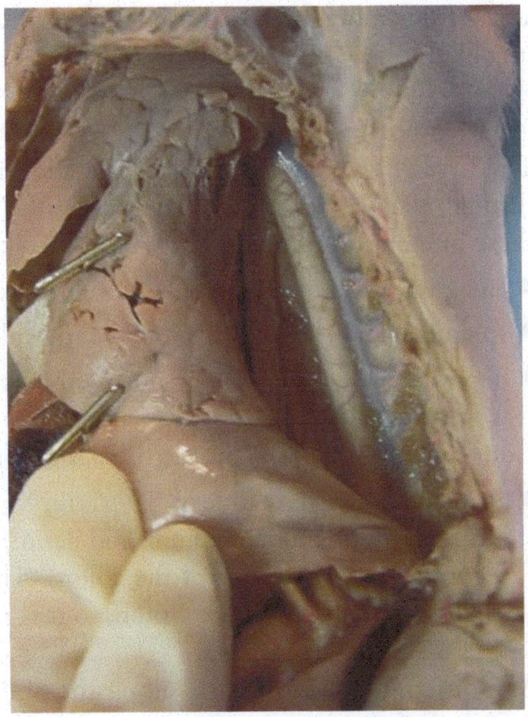

Contributed by Sarah Schlereth. © Kendall Hunt Publishing Company.

1. Attempt to identify the four chambers of the heart: the **right atrium, right ventricle, left atrium,** and **left ventricle**. The thick, white artery located on the ventral surface of the heart, between the right and left atria, is the **pulmonary trunk**. Blood from the right ventricle of the heart enters this artery after passing through the **pulmonary semilunar valve**. The pulmonary trunk branches to the **left** and **right** to form the **pulmonary arteries,** which carry blood to the **lungs**.

2. Follow the pulmonary trunk as it travels behind the heart. It appears to flow directly into the aorta. This is a fetal extension called the **ductus arteriosus.** It connects the blood from the pulmonary

trunk to the aorta, bypassing the lungs. Beyond the ductus, try to locate the **left pulmonary artery**. If you lift the heart at its apex and peak underneath, you might also see the **right pulmonary artery.**

3. Follow the aortic arch until it straightens out to become the **thoracic aorta**. You can see the thoracic aorta more clearly by ***carefully pulling the lungs and heart over to the right side of the animal*** and looking behind them. When you do this, you will see a white, thick-walled tube extending posteriorly toward the **diaphragm**. This is the thoracic aorta. Just ventral to that, you will see a similarly-sized tube. That tube is the **esophagus**, which you will identify when you perform the dissection for the digestive system.

Your dissection should look similar to the images provided. **Identify the bolded structures and vessels above on your dissection specimen, then label the structures on:**

Abdominopelvic Arteries and Veins

1. Just posterior to the diaphragm, along the dorsal body wall, you should now expose the anterior-most portion of the **abdominal aorta**, and should easily be able to see the first few branches off the aorta. Remember, because of the thickness of the walls of arteries in comparison to those of veins, it is likely that you may not be able to see the red latex that has been injected into the arteries, so these vessels may appear white in color.

2. The first *unpaired* branch off of the abdominal aorta, almost immediately after the aorta passes through the diaphragm, is the **celiac trunk**. The celiac trunk has three *unpaired* branches. Locating these branches is difficult, but you MUST know them by name and know what organs they feed.
   - **left gastric artery**, supplying blood to the stomach
   - **splenic artery**, supplying blood to the spleen
   - **hepatic artery**, supplying blood to the liver.

3. Continuing posteriorly, the next *unpaired* branch off of the abdominal aorta is
   - **anterior mesenteric artery**. This artery has branches that supply blood to all parts of the small intestine (duodenum, jejunum, and ileum), as well as the first two parts of the large intestine (the cecum and the first part of the colon). The celiac and anterior mesenteric arteries are unpaired (that is, there is only one of each, so it is incorrect to specify a left or right for either of them).

4. As you continue posteriorly along the abdominal aorta, you will next come to the first *paired* branches of arteries
   - **left** and **right renal arteries**, which carry blood to the **kidneys**.

5. The location where renal arteries branch off of the abdominal aorta is partially hidden by the blue
   - **renal veins**, carrying blood from the kidneys to the
   - **posterior vena cava** (the large blue vessel covering the ventral surface of the abdominal aorta from the diaphragm down to the renal veins). Beyond this point, the posterior vena cava begins to be dorsal to the abdominal aorta.

6. Approximately an inch or less posterior to the renal arteries, you should be able to see two thread-like arteries
   - **left and right gonadal (spermatic or ovarian) arteries.** NOTE: The name of the arteries depends upon the sex of the animal. In a female, these arteries extend to the left and right ovaries; hence, they are called the **left and right ovarian arteries.** In a male, these arteries will appear to leave the abdominal cavity. They extend to the left and right testes; hence, they are the **left and right spermatic arteries.**
7. Almost immediately medial to where the gonadal arteries branch off of the abdominal aorta is another unpaired artery
   - **posterior mesenteric artery**, which has branches that supply blood to the remainder of the colon as well as the rectum of the large intestine. If you are having difficulty finding it, look for tiny, thread-like vessels on the dorsal surface of the rectum and follow them back toward the abdominal aorta.
8. Continuing posteriorly, near the distal end of the abdominal aorta, you will find two sets of paired branches: a thinner, outer pair
   - **left and right external iliac arteries,** thinner, outer pairs, supplying blood to the hind limbs and a thicker, inner pair, called the
   - **left and right internal iliac arteries,** thicker outer pair, which supply the left and right umbilical arteries in a fetus and supply blood to pelvic organs like the urinary bladder after birth. *Note: Unlike humans, these animals do NOT have common iliac arteries!*
9. Your dissections should look similar to the images provided. **Identify the bolded structures and vessels above on your dissection specimen, then label the structures on the image.**

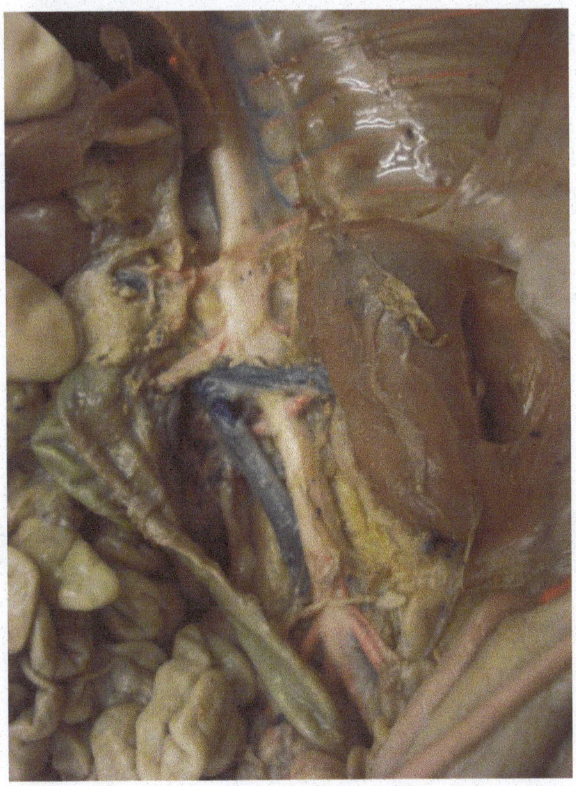

Contributed by Sarah Schlereth.

10. Located just dorsal to the iliac arteries are the **iliac veins**. In order to see those clearly, you will need to carefully cut and remove the iliac arteries, along with the distal end of the abdominal aorta. Just as when you had to remove the thoracic veins to see the thoracic arteries clearly, you will likely want to do this in a second animal at your table, so that you can still review the arteries later. *Since you have not yet dissected the reproductive and urinary system structures, you will need to use extra care exposing these vessels, so that you do not destroy those structures. Ask your lab instructor or TA for assistance.*

11. Once you have these arteries out of the way, you can follow the **posterior vena cava** down to its distal end, where it receives blood from
    - **left** and **right common iliac veins.** Flowing into these veins are two others.
    - **left** and **right internal iliac veins,** bringing blood from deep pelvic and gluteal organs to the common iliac veins
    - **left** and **right external iliac veins,** bringing blood from
    - **femoral veins** carry blood from the legs.

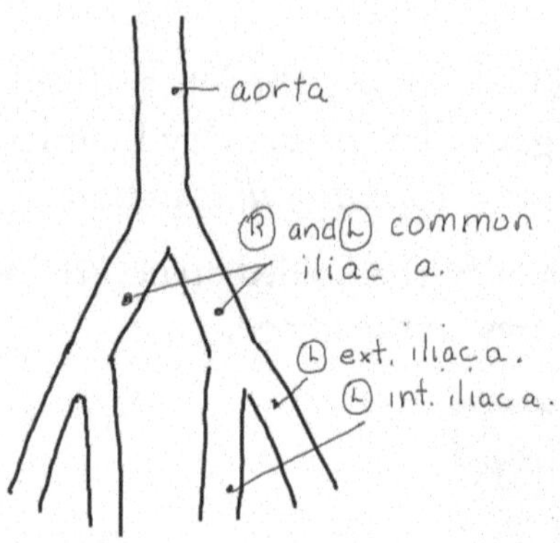

HUMAN

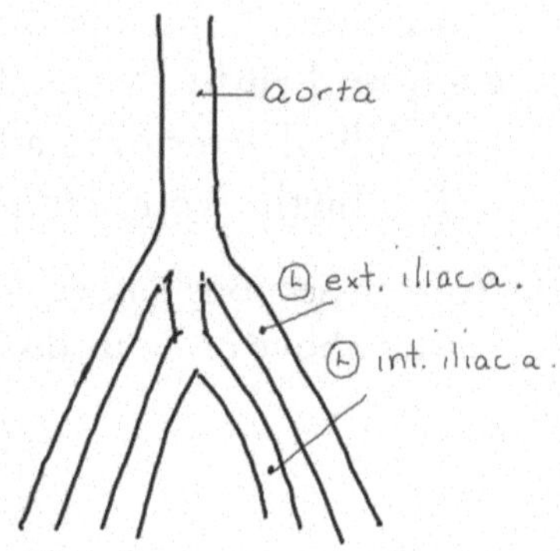

PRESERVED ANIMAL

Illustrations by Lucia Tranel, 2021.

CHECKLIST: Summary Checklist Cardiovascular II

TERM/STRUCTURE	YOUR NOTES
Pulmonary Arteries	
Pulmonary Veins	
Superior Vena Cava	
Brachiocephalic Veins	
Internal Jugular Veins	
Ascending Aorta	
Aortic Arch	
Descending Aorta	
Thoracic Aorta	
Abdominal Aorta	
Brachiocephalic Trunk	
Left Common Carotid Artery	
Left Subclavian Artery	
Right Subclavian Artery	
Right Common Carotid Artery	
Axillary Arteries	
Brachial Arteries	
Ulnar Arteries	

TERM/STRUCTURE	YOUR NOTES
Radial Arteries	
Internal Carotid Arteries	
External Carotid Arteries	
Axillary Veins	
Brachial Veins	
Ulnar Veins	
Radial Veins	
Cephalic Veins	
Common Iliac Arteries	
Common Iliac Veins	
External Iliac Arteries	
External Iliac Veins	
Internal Iliac Arteries	
Internal Iliac Veins	
Femoral Arteries	
Femoral Veins	
Great Saphenous Veins	
Popliteal Arteries	
Popliteal Veins	

TERM/STRUCTURE	YOUR NOTES
Anterior Tibial Arteries	
Anterior Tibial Veins	
Posterior Tibial Arteries	
Posterior Tibial Veins	
Celiac Trunk	
Left Gastric Artery	
Common Hepatic Artery	
Splenic Artery	
Superior Mesenteric Artery	
Renal Arteries	
Renal Veins	
Gonadal/Spermatic/Ovarian Arteries	
Gonadal/Spermatic/Ovarian Veins	
Inferior Mesenteric Artery	
Hepatic Veins	
Superior Mesenteric Vein	
Inferior Mesenteric Vein	
Splenic Vein	
Hepatic Portal Vein	

TERM/STRUCTURE	YOUR NOTES
Ductus Arteriosus	
Foramen Ovale	
Ductus Venosus	
Umbilical Arteries	
Umbilical Vein	
Subclavian Veins	
External Jugular Veins	
Vagus Nerves	
Liver Capillaries	

EXERCISE 10

Respiratory System

A. *Introduction*

The respiratory system consists of **upper** and **lower** respiratory systems. The major organs of the upper respiratory system include the **paranasal sinuses,** the **nasal cavity,** and the **pharynx.** Major organs of the lower respiratory system include the **larynx,** the **trachea and its branches,** the **lungs,** and the **diaphragm.** We examined the tissues of the respiratory system during our study of histology where we looked at **hyaline cartilage** and **ciliated pseudostratified columnar epithelium** in the trachea and **simple squamous epithelium** in the lung.

B. *Outcomes for Exercise 10*

1. Locate, label, and identify the basic anatomy of the respiratory system.

2. Locate, label, and identify the structures of the upper respiratory tract. Be able to determine where the air was immediately before a specific structure and where it will be immediately after a specific structure.

3. Locate, label, and identify the structures of the lower respiratory tract. Be able to determine where the air was immediately before a specific structure and where it will be immediately after a specific structure.

4. Locate, label, and identify the structures of the respiratory unit. Be able to determine where the air was immediately before a specific structure and where it will be immediately after a specific structure.

5. Dissect the respiratory system in the preserved animal and locate the structures described in the text. Be able to determine where the air was immediately before a specific structure and where it will be immediately after a specific structure.

6. Examine a slide of the trachea under the microscope. Identify the specific histological tissues that make each layer in the wall of the trachea. Draw and label what you see.

7. Examine a slide of the lung under the microscope. Identify the specific histological tissues listed. Compare the similarities and differences between a bronchus and a bronchiole. Locate each under the microscope and draw each to determine how you will distinguish them from each other. Draw and label what you see.

C. *Anatomy of Respiratory System Organs*

OUTCOME 1: Locate, label, and identify the basic anatomy of the respiratory system.

TOOLBOX

Structures of the Respiratory System

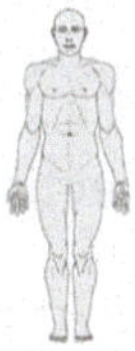

Thoracic model
Mr Flip Chart
Mr. Muscle

frontal sinus

sphenoid sinus

external nares

nasal cavity with nasal conchae

internal nares

pharynx

- nasopharynx
- oropharynx
- laryngopharynx

glottis

larynx

trachea

left and right primary (1°) bronchi

secondary (2°) bronchi

tertiary (3°) bronchi

bronchioles

left and right lungs

diaphragm

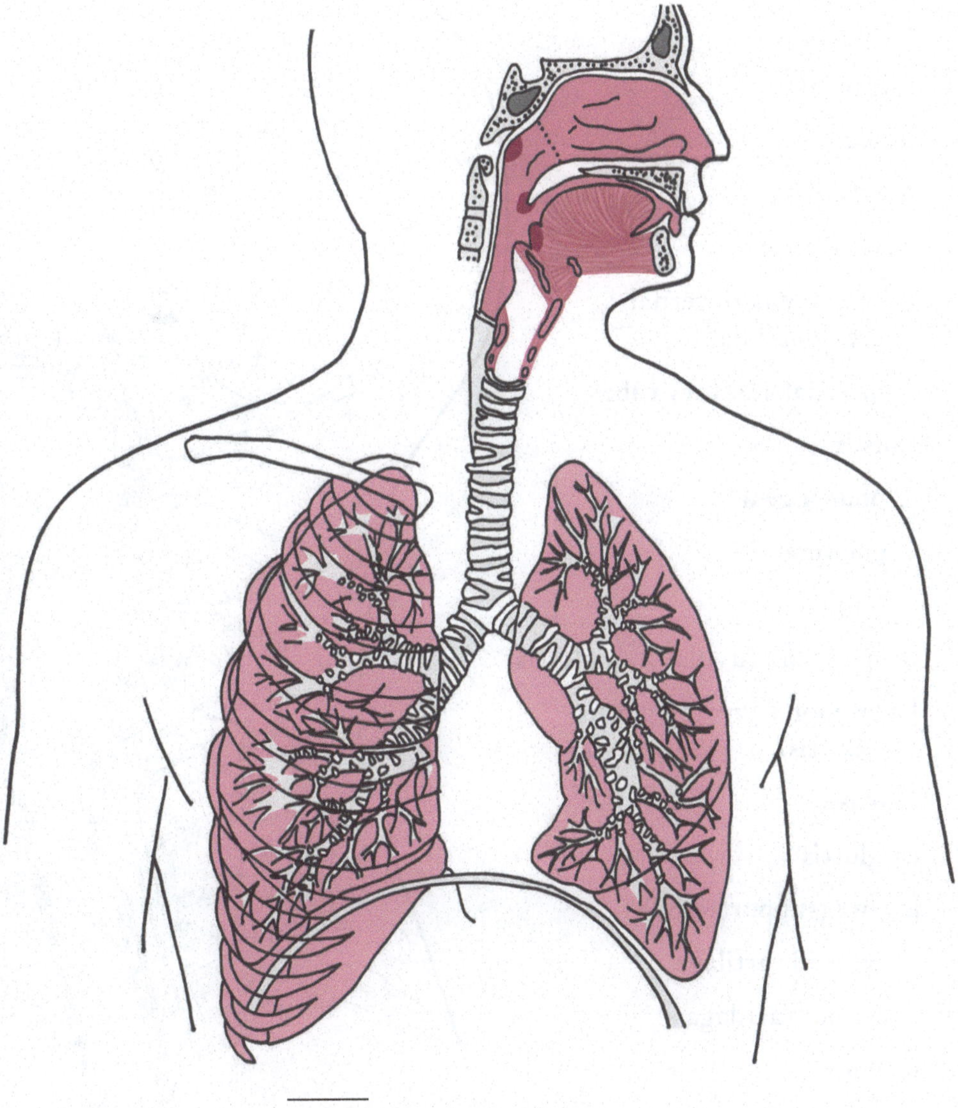

OUTCOME 2: Locate, label, and identify the structures of the upper respiratory tract. Be able to determine where the air was immediately before a specific structure and where it will be immediately after a specific structure.

TOOLBOX

Respiratory Structures of the Head

Sagittal section

Thoracic model

Mr Flip Chart

Mr. Muscle

right external naris

hard palate

soft palate

uvula

nasal cavity

nasal septum

nasopharynx (internal nares -> uvula)

- internal auditory tube

tonsils

- pharyngeal
- palatine
- lingual

oropharynx (uvula -> epiglottis)

laryngopharynx (epiglottis -> glottis)

glottis

epiglottis

larynx (supported by):

- thyroid cartilage
- cricoid cartilage

trachea

esophagus

tracheal cartilage

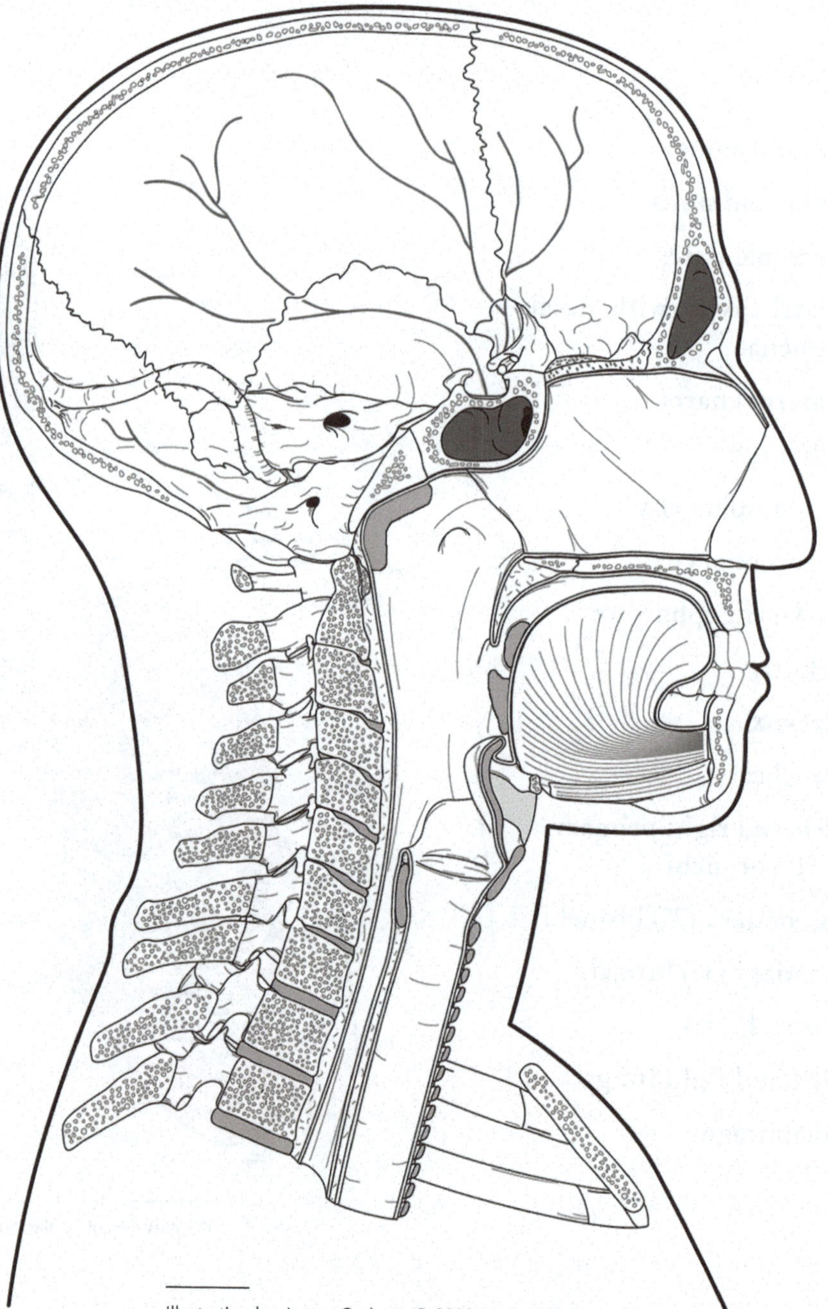

Illustration by Jamey Garbett. © 2003 Mark Nielsen.

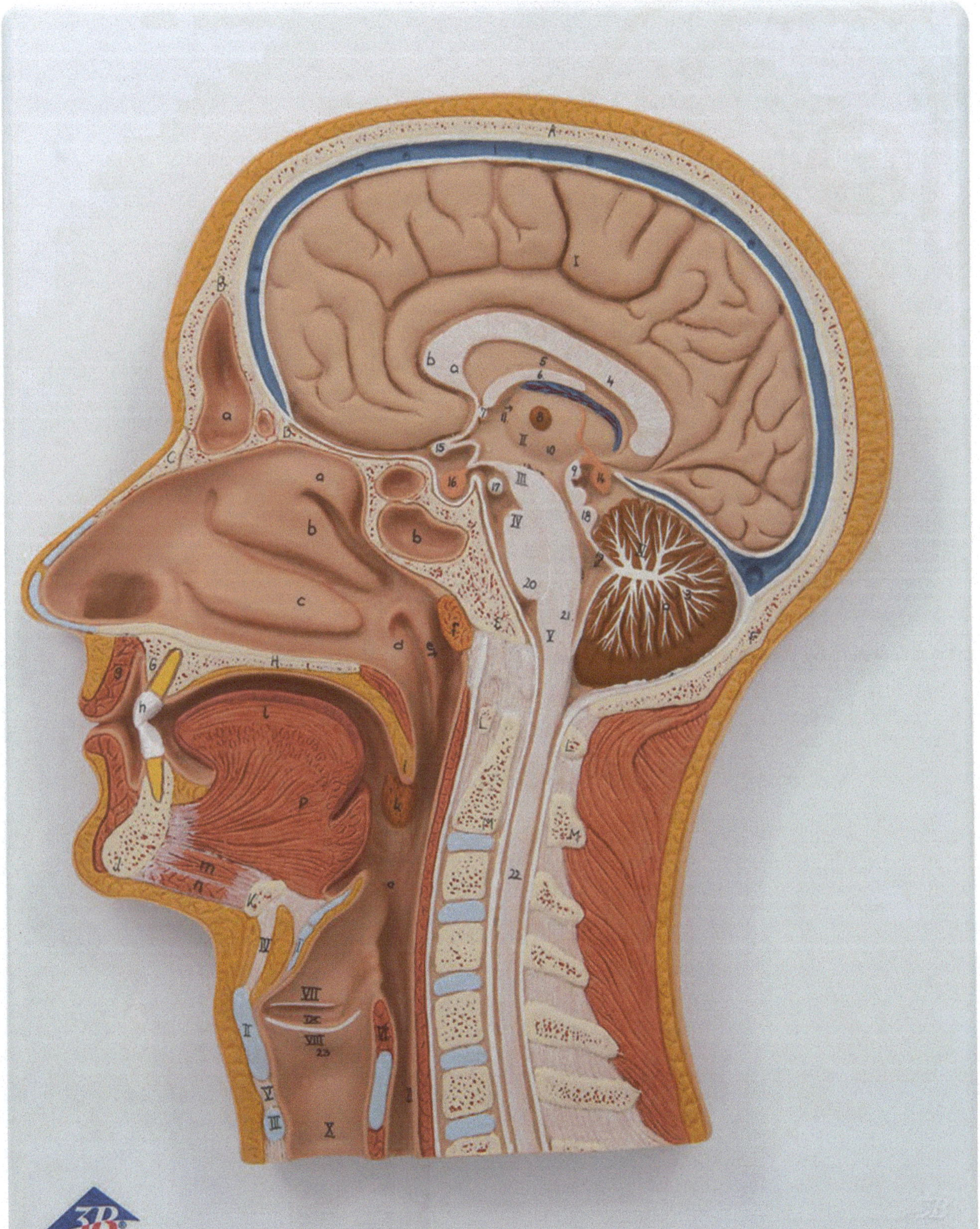

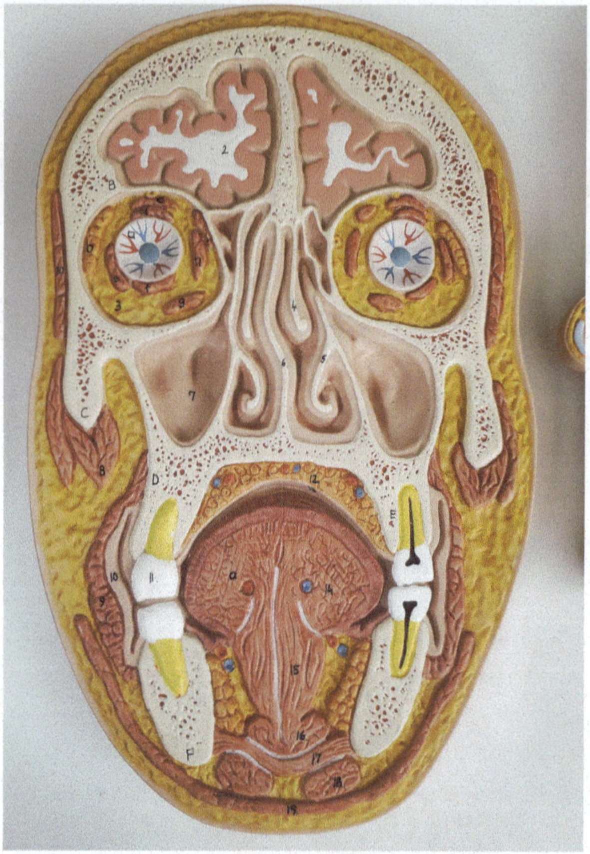

C12 © 3B Scientific GmbH, Germany, 2021. www.3bscientific.com

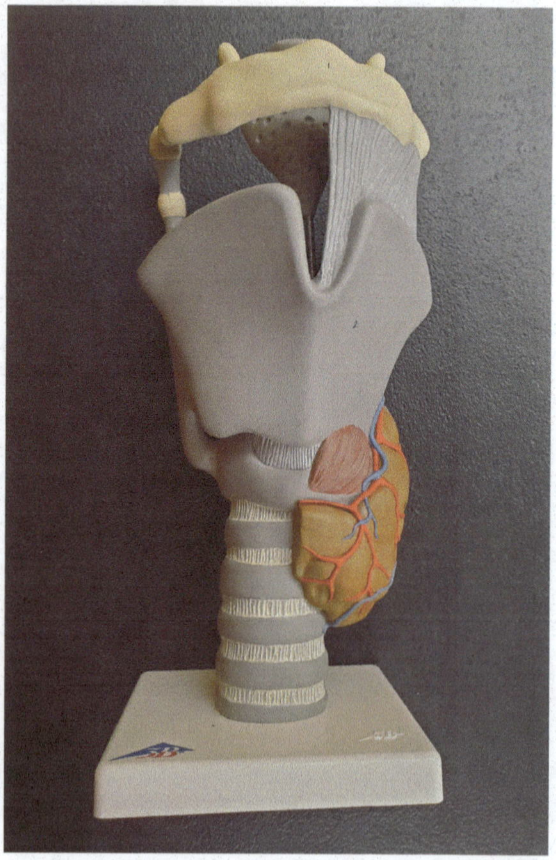

G20 © 3B Scientific GmbH, Germany, 2021. www.3bscientific.com

OUTCOME 3: Locate, label, and identify the structures of the lower respiratory tract. Be able to determine where the air was immediately before a specific structure and where it will be immediately after a specific structure.

TOOLBOX

Anatomy of the Trachea and Primary Bronchi

Thoracic model
Mr Flip Chart
Mr. Muscle

larynx (supported by):

- thyroid cartilage
- cricoid cartilage

trachea

tracheal cartilages

left and right primary (principle) bronchi

secondary (lobar) bronchi

tertiary (segmental) bronchi

left lung

- upper lobe
- lower lobe

right lung

- upper lobe
- middle lobe
- lower lobe
- visceral pleura

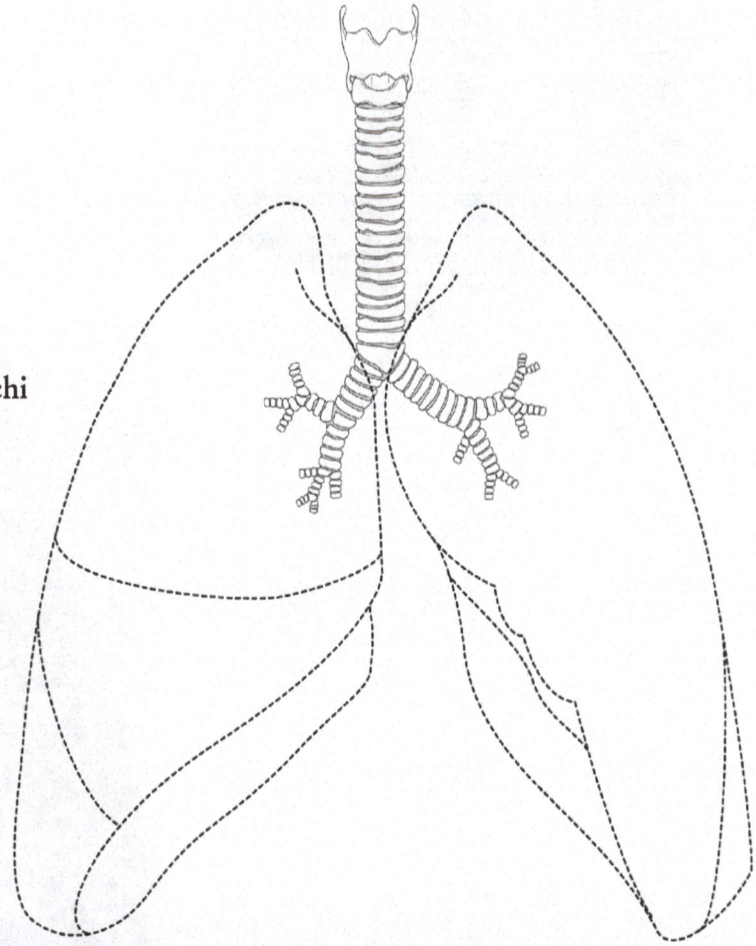

Illustration by Jamey Garbett. © 2003 Mark Nielsen.

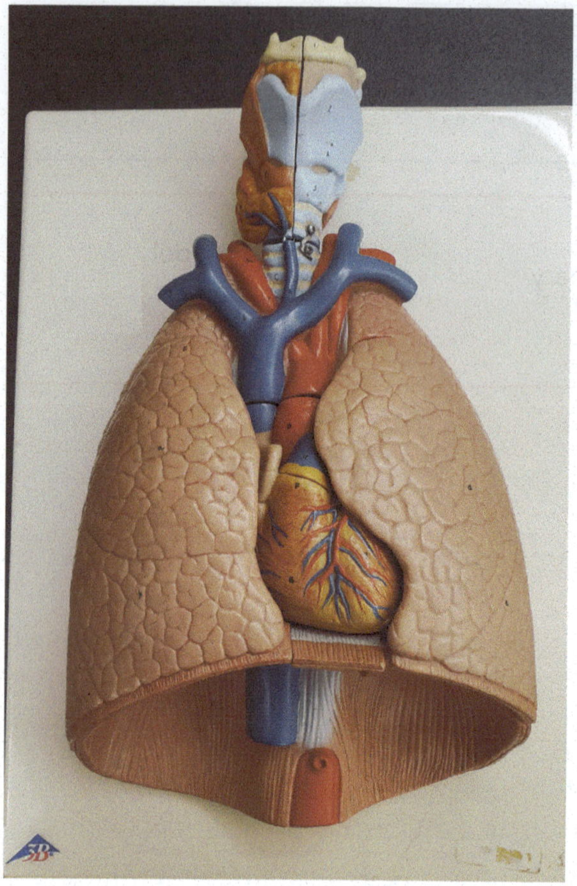

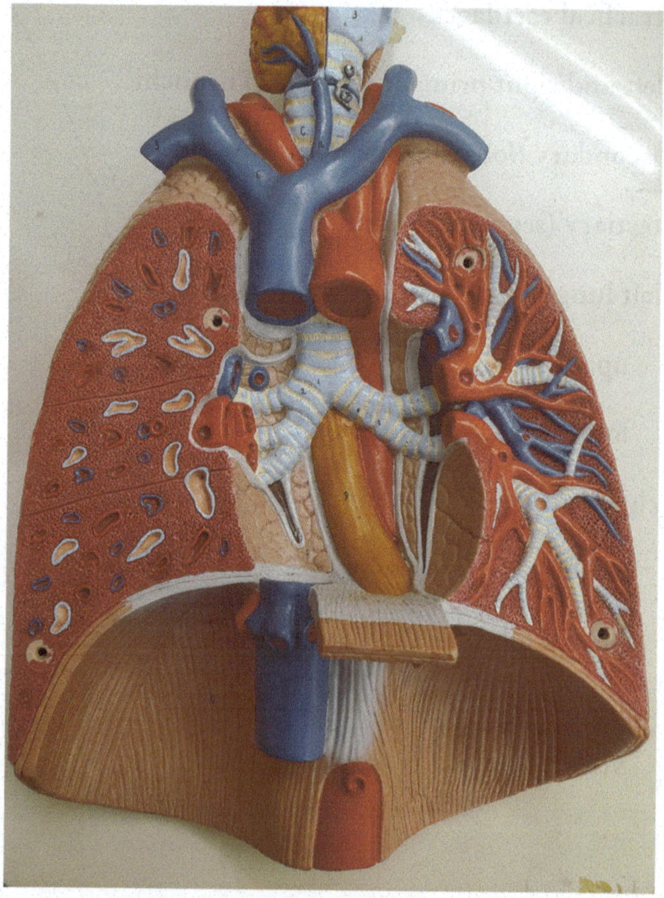

OUTCOME 4: Locate, label, and identify the structures of the respiratory unit. Be able to determine where the air was immediately before a specific structure and where it will be immediately after a specific structure.

TOOLBOX

Bronchi and Bronchioles
Alveolar Organization

primary bronchus
cartilage ring
secondary bronchus
tertiary bronchi
bronchioles
terminal bronchioles
respiratory bronchioles
lobules

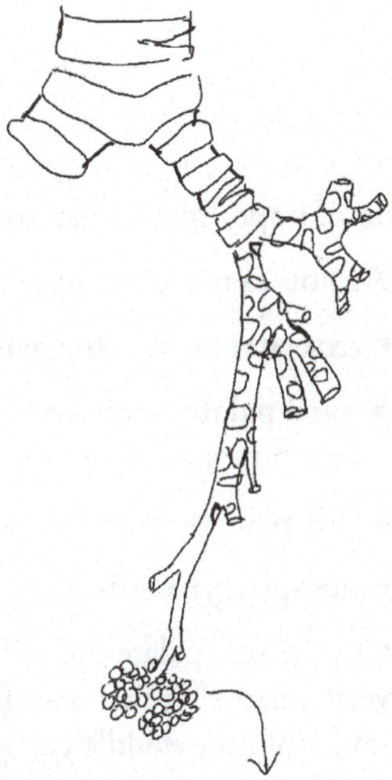

terminal bronchiole

respiratory bronchiole

alveolar sac

alveolar duct

alveolus

simple squamous epithelium

pulmonary arteriole

pulmonary capillary

pulmonary venule

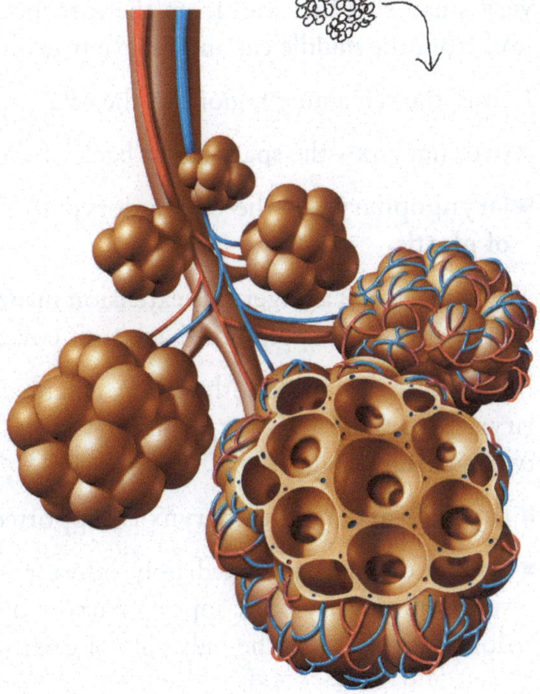

D. *Comparative Anatomy—Dissection*

OUTCOME 5: Dissect the respiratory system in the preserved animal and locate the structures described in the text. Be able to determine where the air was immediately before a specific structure and where it will be immediately after a specific structure.

TOOLBOX

Dissection of upper respiratory structures involves cracking the jaw of the preserved animal.

1. As you examine the upper respiratory tract, locate:
   - **external nares**—the paired nostrils (naris is singular)
   - **hard palate**—anterior surface on the roof of the mouth; roughened; the palatine and maxillary bones are superior
   - **soft palate**—posterior surface on the roof of the mouth; superior to this surface is the cavity
   - **nasopharynx**—the first of the three parts to the pharynx. **From internal nares to tip of uvula.**
2. Cut the soft palate opened to expose the nasopharynx. On the lateral surfaces of the cavity are very small slits that lead from the **internal auditory tubes (Eustachian tubes)**. These tubes lead from the middle ear and relieve pressure.
3. Locate the remaining regions of the pharynx:
   - **oropharynx**—the space in the back of the mouth. **From tip of uvula to tip of epiglottis.**
   - **laryngopharynx**—the cavity dorsal to the epiglottis. **From tip of epiglottis to opening of glottis.**
4. The **epiglottis** is a finger-like extension made of hyaline cartilage. When swallowing occurs, the tongue presses the epiglottis back so it covers the **glottis**, the opening to the larynx.
5. Now locate the **larynx**—the entire, enlarged superior end of the trachea (the "voice box"). The larynx separates the upper respiratory tract from the lower respiratory tract. It is supported by two cartilage bands:
   - **thyroid cartilage**—the larynx is supported by this broad band of hyaline cartilage.
   - **cricoid cartilage**—immediately inferior to the thyroid cartilage; this ring of cartilage also supports the larynx. It appears narrow on the ventral surface, but it becomes wider on the dorsal surface. It is the only ring of cartilage that completely encircles the trachea.

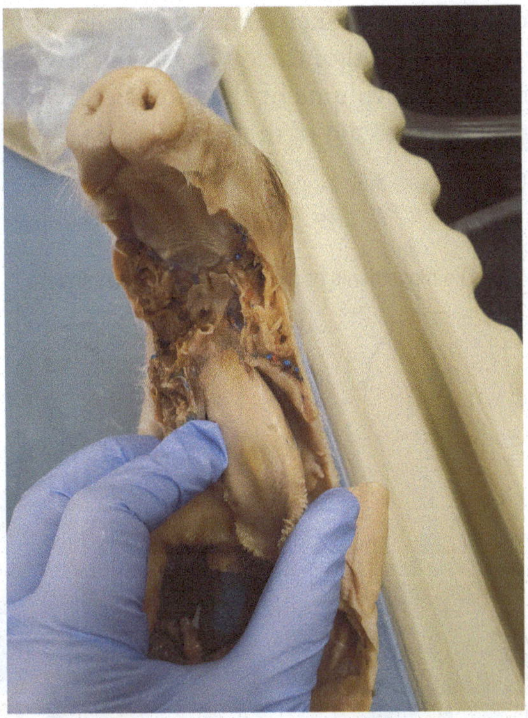

6. The **trachea** continues from the larynx. Notice that its cartilage rings do not completely encircle the tube. The dorsal surface of the trachea is soft and contains a smooth muscle known as the **trachealis muscle** (you will examine on the microscope). What structure lies immediately dorsal to the trachea?

7. The trachea branches left and right into the
   - **primary bronchi (bronchus is singular)**—These branches are sometimes referred to as principle bronchi.
   - **secondary bronchi**—branches of each primary bronchus into the lobes of the lungs (sometimes called lobar).
   - **tertiary bronchi**—These branches are difficult to see. If you scrape some of the lung tissue away, you might be able to locate them. These branches are sometimes called segmental, since they branch into the segments of the lungs.

8. Notice that the **lungs** are **lobed** to increase surface area three lobes on the left (upper, middle, lower) and four lobes on the right (upper, middle, lower, intermediate). This animal has one more lobe on each side compared to humans. Review:
   - **visceral pleura**
   - **parietal pleura**
   - **pleural cavity**

9. Locate the **diaphragm**. It is made of skeletal muscle tissue. When it contracts, it moves downward, increasing the volume of the thoracic cavity, so air moves into the lungs. When it relaxes, it moves upward decreasing the volume of the thoracic cavity so air moves out of the lungs. Muscles of the rib cage and thorax aid in the process.

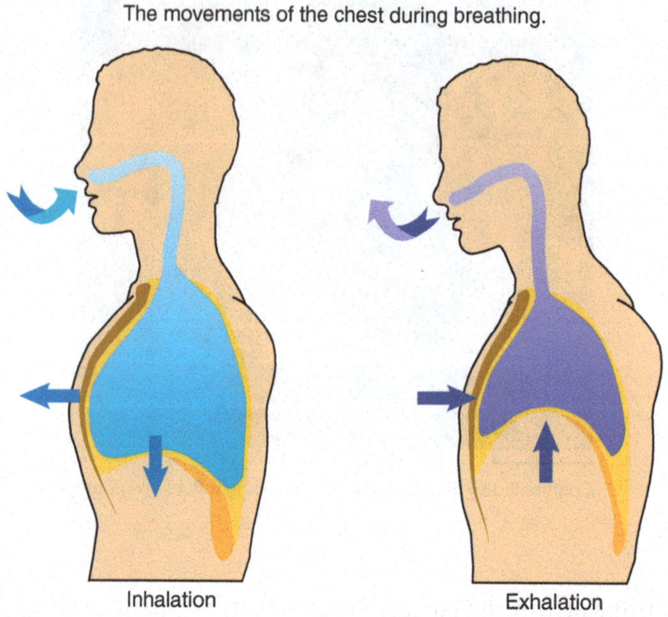

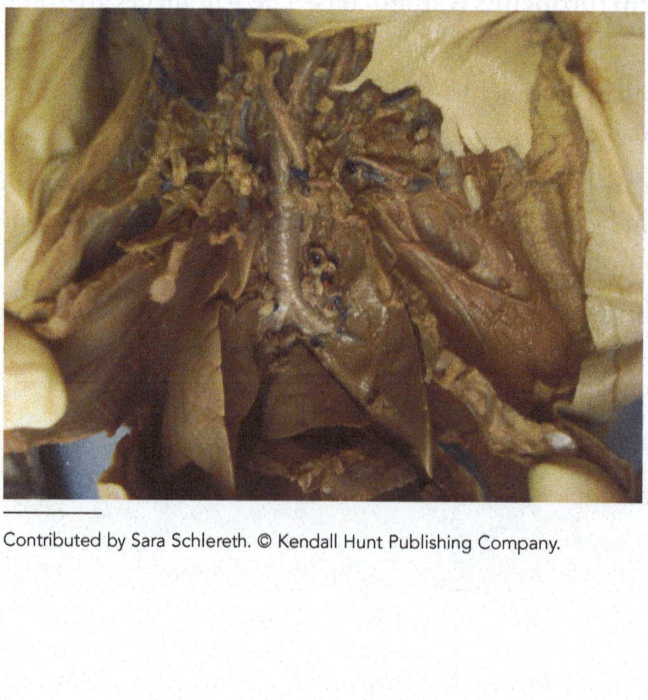

E. *Histological Examination*

OUTCOME 6: Examine a slide of the trachea under the microscope. Identify the specific histological tissues that make each layer in the wall of the trachea. Draw and label what you see.

TOOLBOX

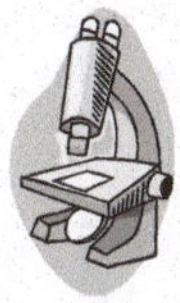

■ ***Locate the structures listed in the table. Draw and label what you see.***

LOCATE	TISSUE COMPOSITION	DRAWING
mucosa and lumen	ciliated pseudostratified columnar epithelium goblet cells	
cartilage ring	hyaline cartilage	
trachealis muscle	smooth muscle	

■ ***What structure lies immediately dorsal to the trachealis muscle?***

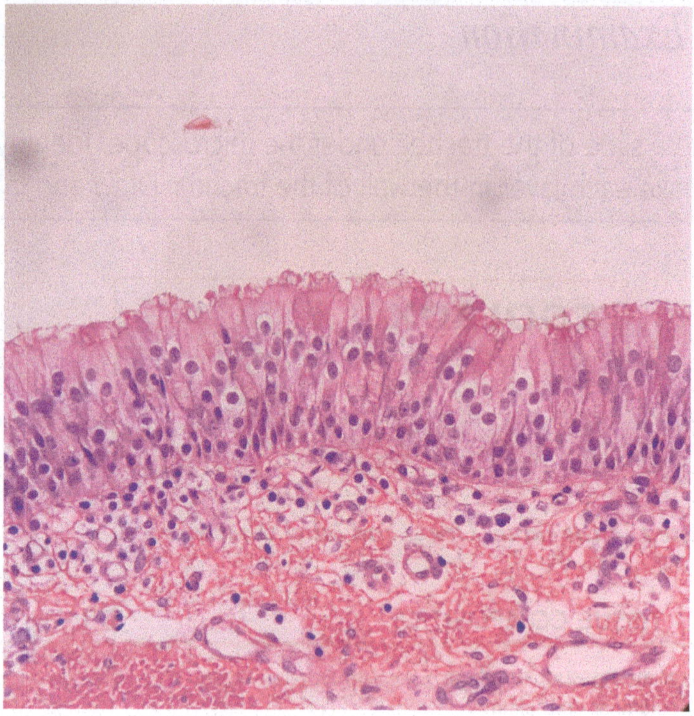

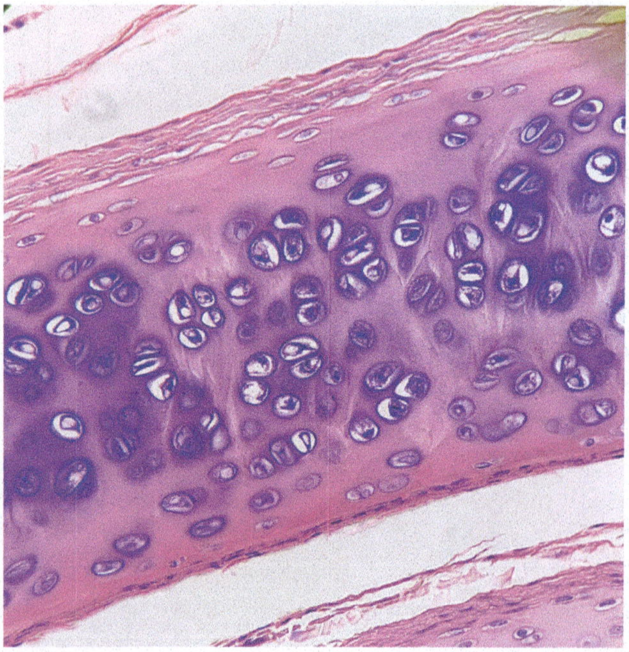

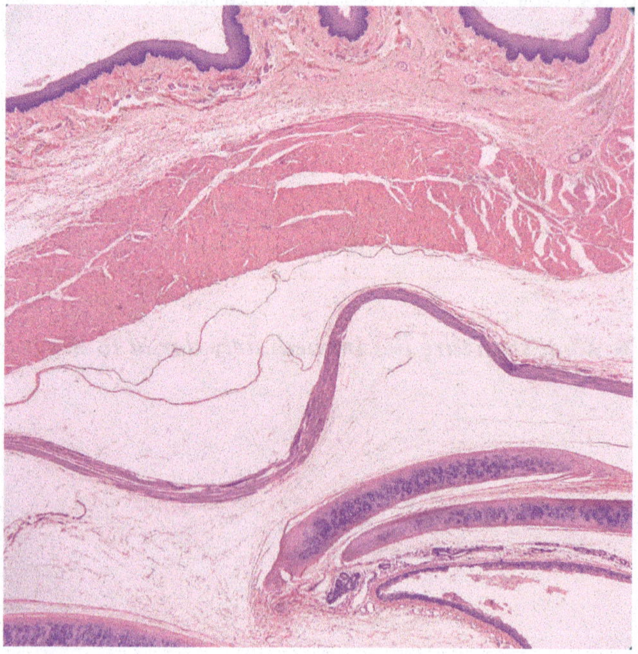

OUTCOME 7: Examine a slide of the lung under the microscope. Identify the specific histological tissues listed. Compare the similarities and differences between a bronchus and a bronchiole. Locate each under the microscope and draw each to determine how you will distinguish them from eachother. Draw and label what you see.

TOOLBOX

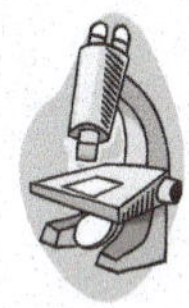

■ ***Locate each structure. Make sure you can determine the tissue composition and the characteristics of each structure listed in the table.***

LOCATE	TISSUE COMPOSITION	NOTES
alveolus and **alveolar lumen**	simple squamous epithelium	mucous membrane
bronchus	ciliated pseudostratified columnar epithelium	hyaline cartilage present smooth muscle present
bronchiole epithelial membrane thins as branching continues	largest bronchioles ↓ simple columnar epithelium simple cuboidal epithelium simple squamous epithelium smallest bronchioles (terminal)	no hyaline cartilage ↓ few cilia, but begin to disappear smooth muscle present throughout

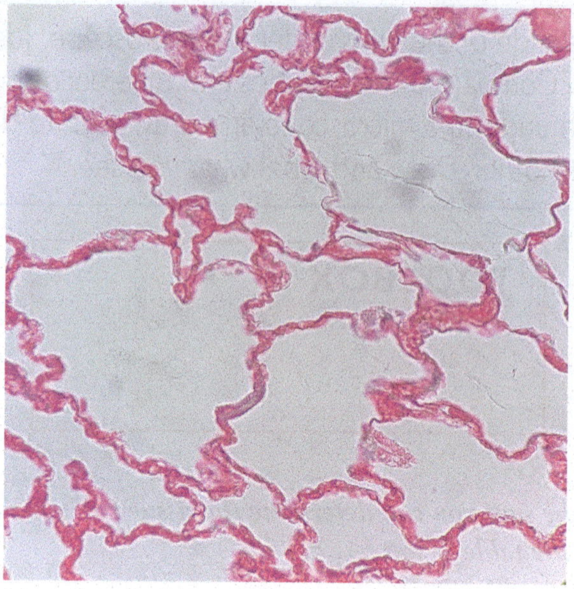

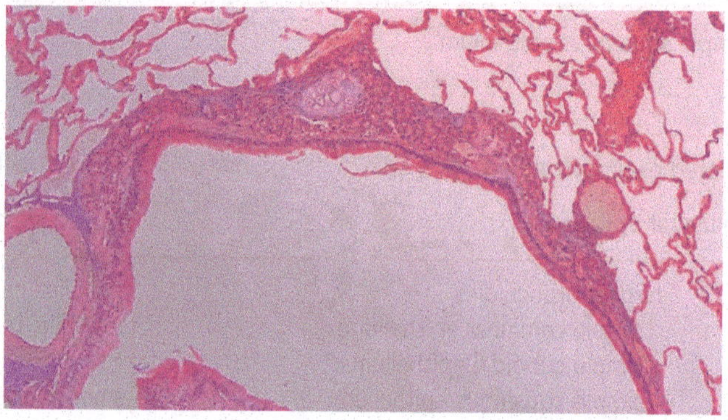

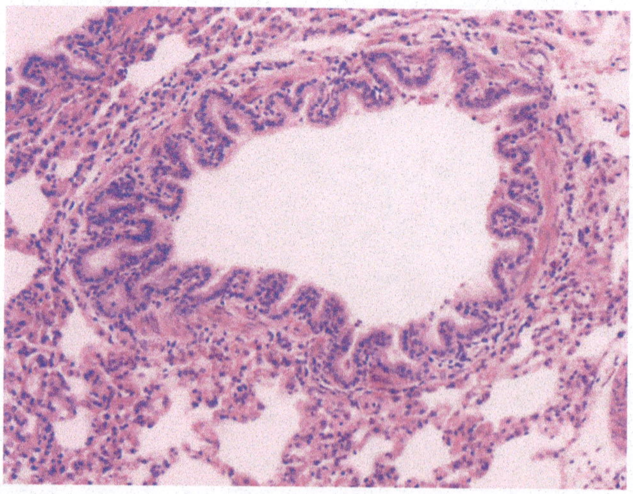

Contributed by Tony Wilmes. © Kendall Hunt Publishing Company.

CHECKLIST: Summary Checklist Respiratory

TERM/STRUCTURE	YOUR NOTES
Esophagus	
Trachea	
C-Shaped Tracheal Cartilage (Hyaline Cartilage)	
Trachealis Muscle (Smooth Muscle)	
Ciliated Pseudostratified Columnar Epithelium	
Goblet Cell	
Simple Squamous Epithelium	
Alveoli	
Bronchiole	
Bronchus	
External Nares	
Pharynx	
Nasopharynx	
Oropharynx	
Laryngopharynx	
Glottis	
Larynx	
Thyroid Cartilage	
Cricoid Cartilage	
Primary Bronchi	
Secondary Bronchi	

(continued)

TERM/STRUCTURE	YOUR NOTES
Tertiary Bronchi	
Pulmonary Arteries	
Pulmonary Veins	
Terminal Bronchioles	
Respiratory Bronchioles	
Alveolar Sac	
Alveolar Duct	
Mediastinum	
Pleural Cavity	
Parietal Pleura	
Visceral Pleura	
Left Lung (Upper and Lower Lobes)	
Right Lung (Upper, Middle, and Lower Lobes)	
Diaphragm	
Frontal Sinus	
Sphenoid Sinus	
Hyoid Bone	
Hard Palate	
Soft Palate	
Uvula	
Nasal Cavity	
Nasal Conchae (Superior, Middle, and Inferior)	

TERM/STRUCTURE	YOUR NOTES
Nasal Septum	
Internal Nares	
Internal Auditory Tube	
Pharyngeal Tonsils	
Palatine Tonsils	
Lingual Tonsils	
Epiglottis	
Laryngeal Prominence	
Pulmonary Arteriole	
Pulmonary Capillary	
Pulmonary Venule	

EXERCISE 11

Digestive System

A. Introduction

The focus of this chapter is on the organs and tissues of the **digestive system.** The usual reference to this system is the digestive system. However, it is important to remember that **digestion**—the mechanical and/or chemical process of converting food into absorbable chemical substances—is only one process in the entire system. The processes of **ingestion, movement, absorption, secretion**, and **excretion** also are part of it. The **digestive tract,** also known as the **alimentary canal** or **tubular gut,** consists of a tube that runs through the body from the **mouth** to the **anus.** The **gastrointestinal tract (GI tract)** is the portion that involves, more specifically, the stomach and the intestines. **Accessory organs,** such as the teeth, salivary glands, mucous glands, pancreas, liver, and gallbladder are organs necessary for the digestive process, but they are not part of the pathway for food. We will observe the **organs** and **tissues** of this system using images, models, dissection, and microscopic examination.

B. *Outcomes for Exercise 11*

1. Locate, label, and identify the structures of the digestive tract. Be able to determine where food was immediately before a specific structure and where it will be immediately after a specific structure.
2. Locate, label, and identify the details of the stomach, the small intestine, and the large intestine.
3. Locate, label, and identify the details of the accessory organs of the digestive system.
4. Dissect the digestive system in the preserved animal and locate the structures described in the text. Be able to determine where the food was immediately before a specific structure and where it will be immediately after a specific structure.
5. Review the histological organization of the digestive tract. Name the four layers. Identify the tissues that compose each layer. Describe the function of each layer.
6. Review the histological organization of a liver lobule. Determine direction of blood and bile flow.
7. Review the histological organization of a pancreatic lobule. Determine which cells are exocrine and which cells are endocrine.
8. Examine a slide of the jejunum under the microscope. Identify the four layers. Identify the tissues that compose each layer. Draw and label what you see.
9. Examine two slides of the liver under the microscope. The first slide is stained to enhance the view of a liver lobule. The second slide will allow observation of a portal triad. Identify and draw the details listed.
10. Examine a slide of the pancreas under the microscope. Locate the acinar cells. Locate the islets of Langerhans. Which is exocrine? Which is endocrine? Locate the pancreatic duct. Draw and label what you see.

C. *Anatomy of Digestive System*

OUTCOME 1: Locate, label, and identify the structures of the digestive tract. Be able to determine where food was immediately before a specific structure and where it will be immediately after a specific structure.

TOOLBOX

Components of the Digestive System
Oral Cavity

Large plaque of digestive system
Mr. Flip Chart
Mr. Muscle

oral cavity

tongue

hard palate

soft palate

uvula

oropharynx

laryngopharynx

epiglottis

esophagus

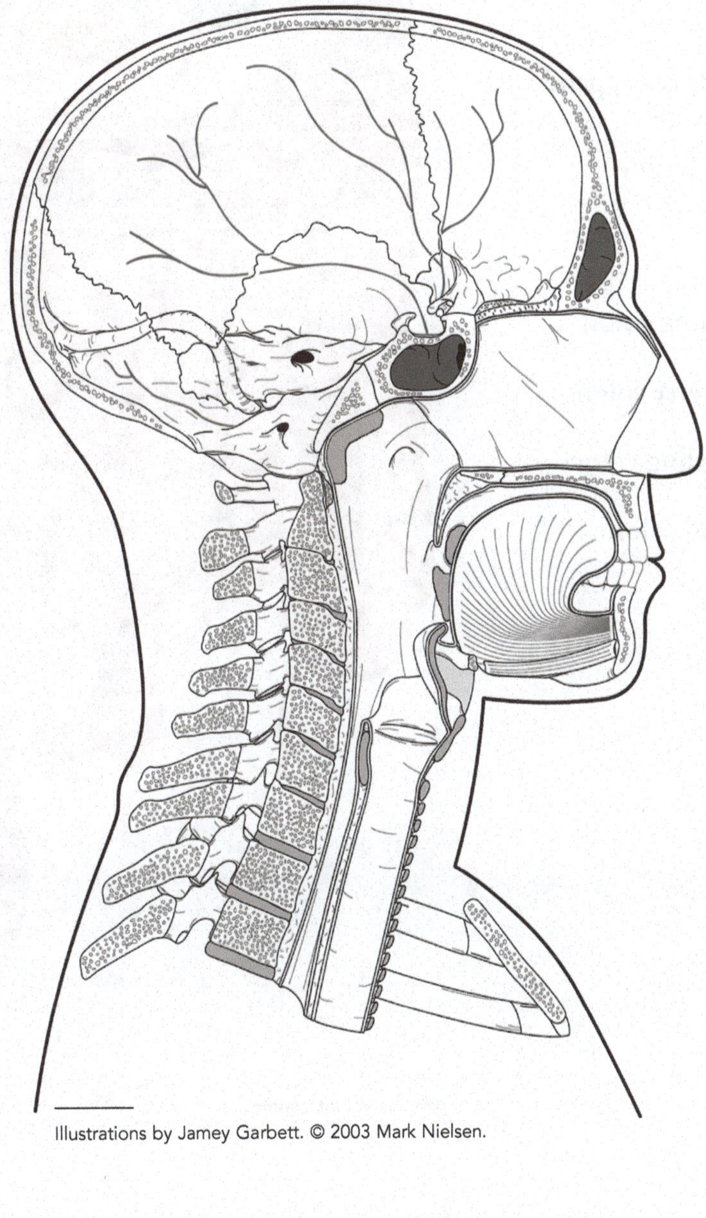

esophagus

stomach

spleen

liver

gall bladder

small intestine

duodenum

jejunum

ileum

vermiform appendix

large intestine

cecum

ascending colon

transverse colon

descending colon

sigmoid colon

rectum

anus

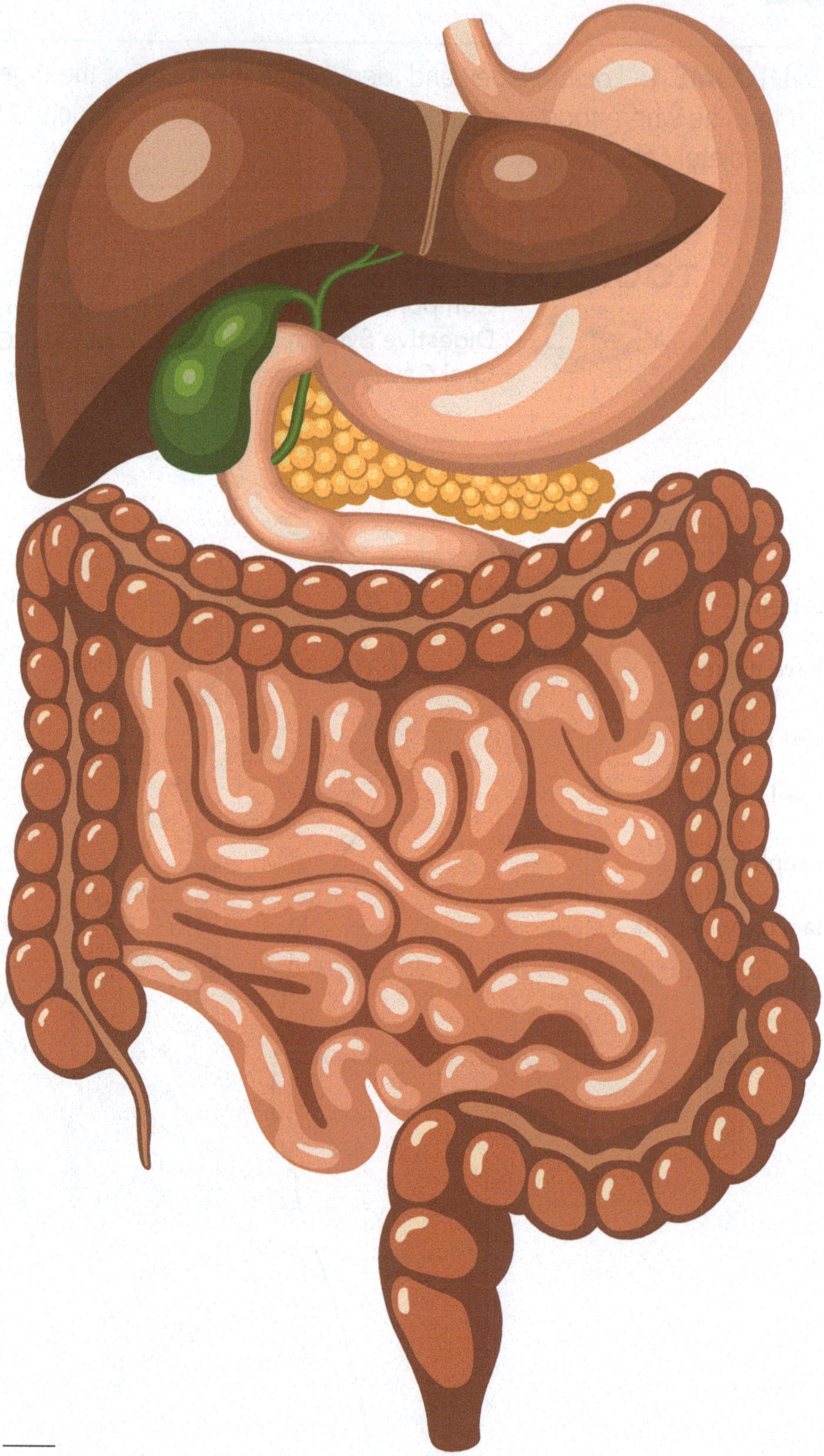

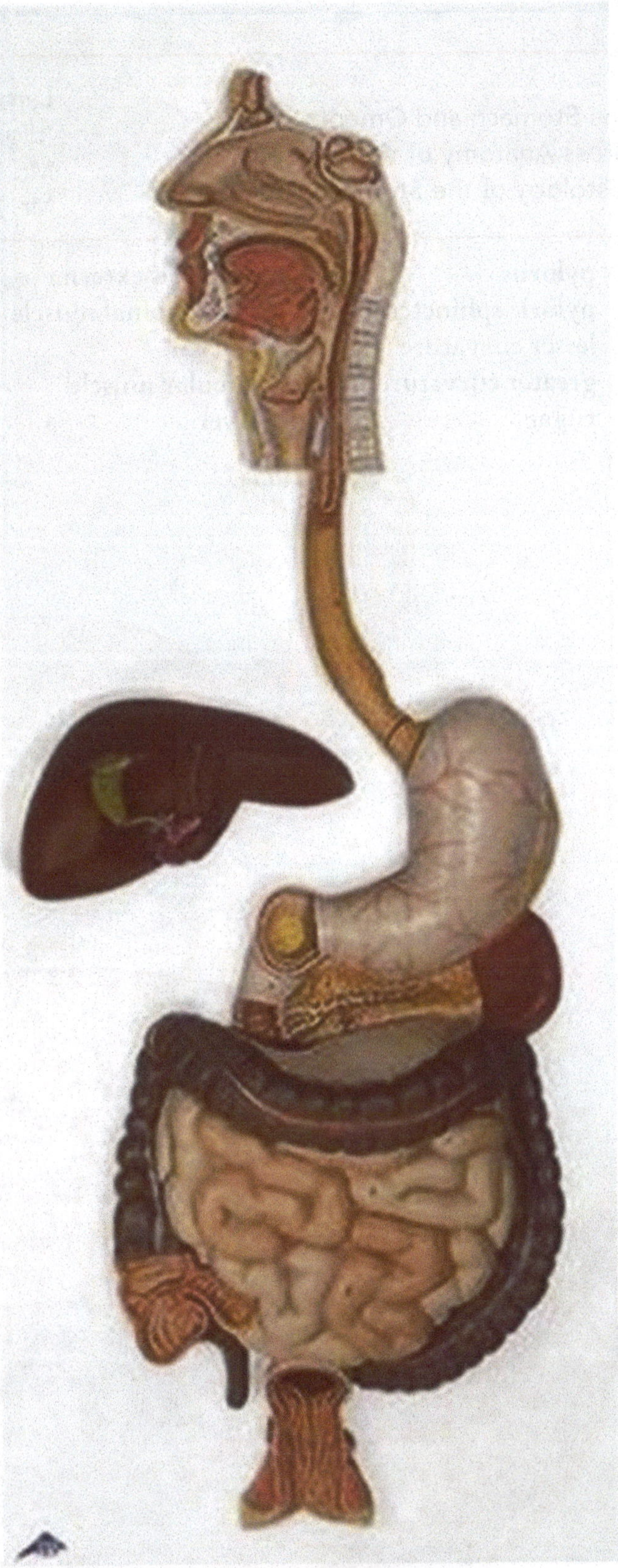

OUTCOME 2: Locate, label, and identify the details of the stomach, the small intestine, and the large intestine.

TOOLBOX

The Stomach and Omenta
Gross Anatomy of the Stomach
Histology of the Stomach Wall

Large plaque of digestive system
Mr. Flip Chart
Mr. Muscle

- esophagus
- cardiac sphincter
- cardia
- fundus
- body
- pylorus
- pyloric sphincter
- lesser curvature
- greater curvature
- rugae
- muscularis externa
  - longitudinal muscle layer
  - circular muscle layer
  - oblique muscle layer
- serosa (visceral peritoneum)

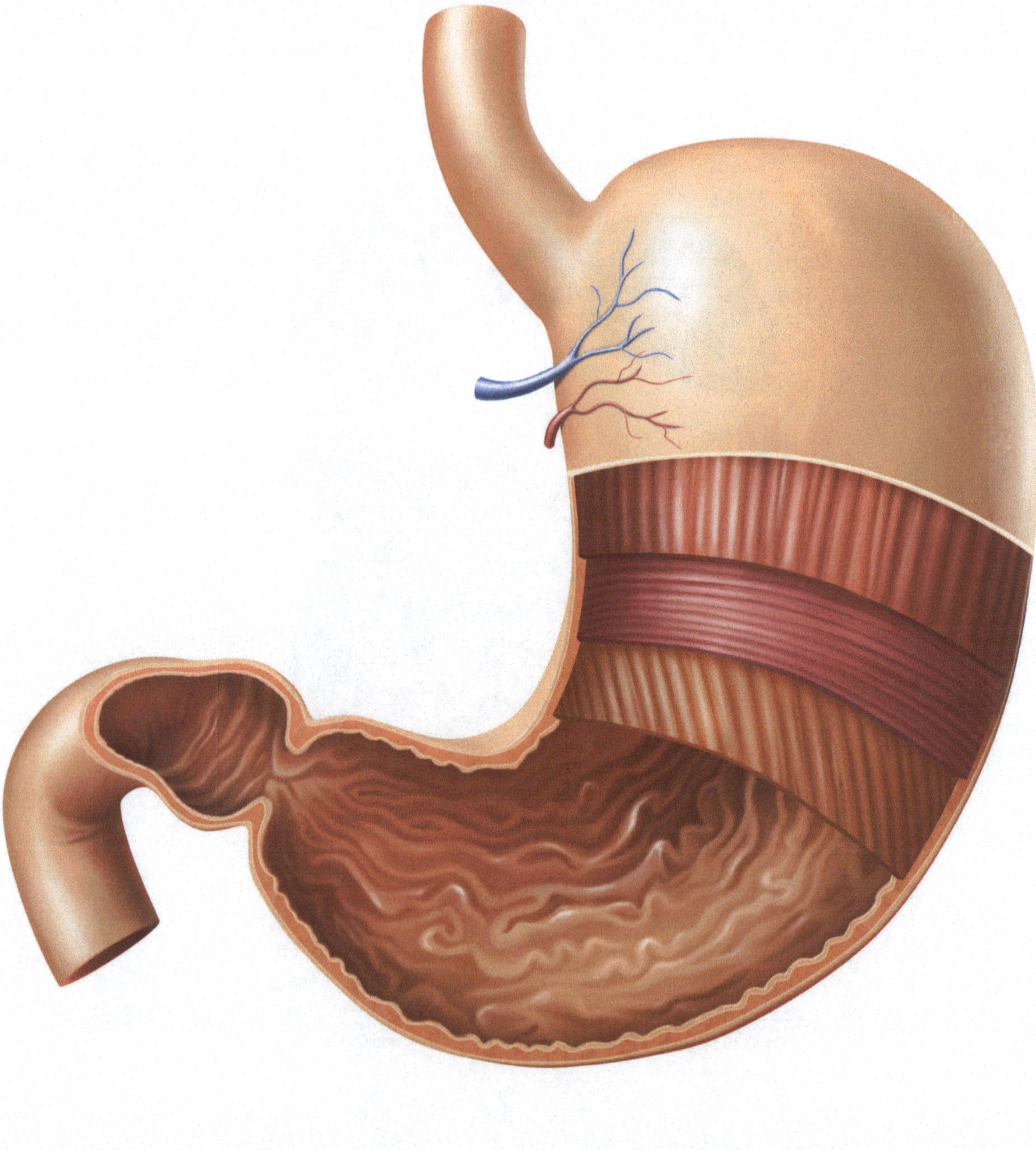

mucosa

simple columnar epithelium

gastric pits

gastric glands

surface mucous cells

mucous neck cells

chief cells

parietal cells

muscularis mucosae

submucosa

muscularis externa

oblique muscle

circular muscle

longitudinal muscle

serosa

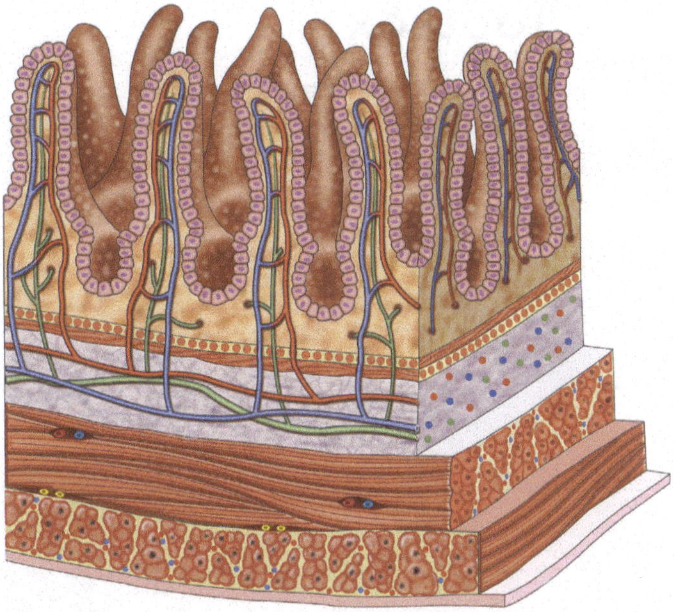

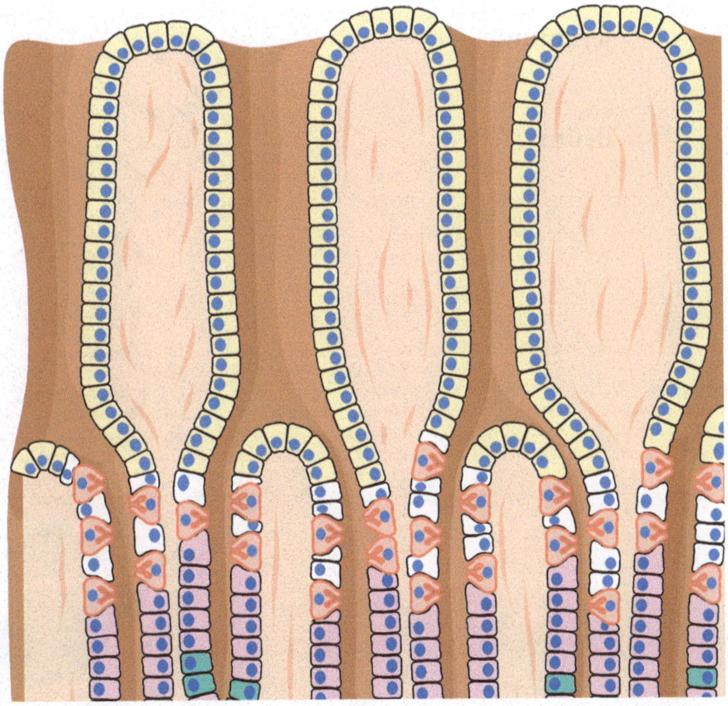

TOOLBOX

Regions of the Small Intestine

Large plaque of digestive system
Mr. Flip Chart
Mr. Muscle

duodenum

jejunum

ileum

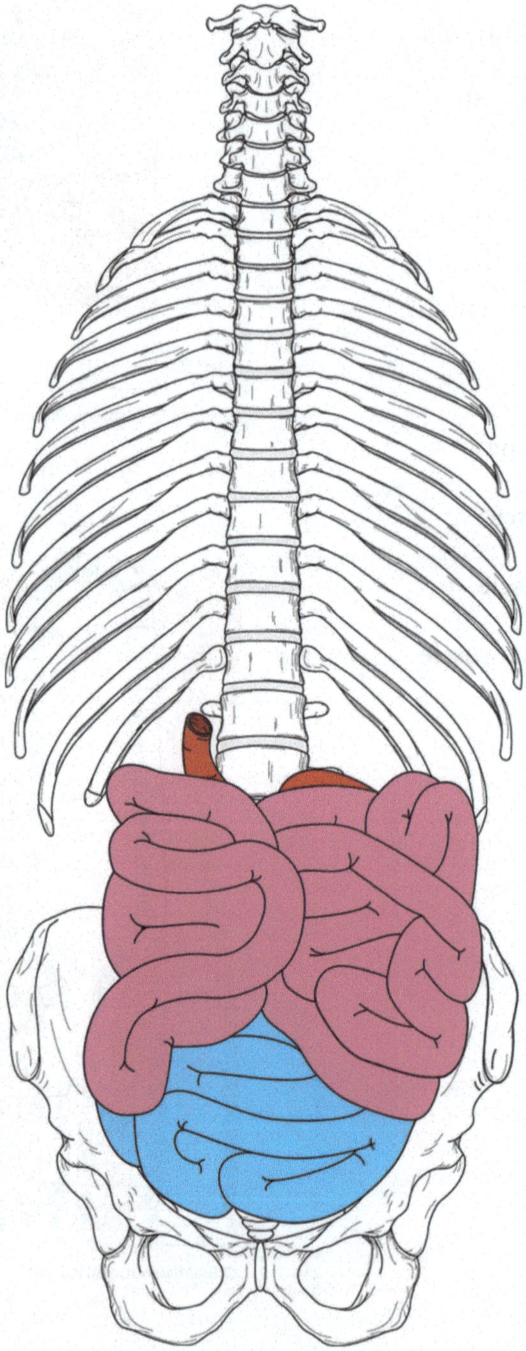

Illustration by Jamey Garbett. © 2003 Mark Nielsen.

TOOLBOX

The Large Intestine

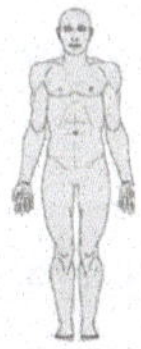

Large plaque of digestive system
Mr. Flip Chart
Mr. Muscle

cecum of the large intestine

vermiform appendix

colon of the large intestine

ascending colon

transverse colon

descending colon

sigmoid colon

haustra

rectum of the large intestine

anal canal

internal anal sphincter (smooth m.)

external anal sphincter (skeletal m.)

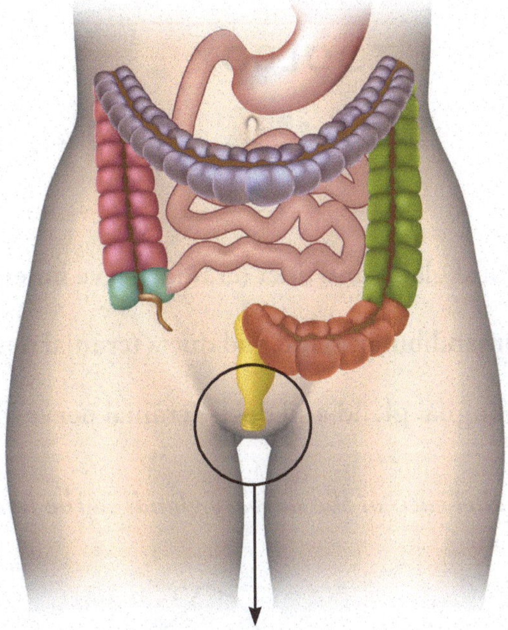

© Alila Medical Media/shutterstock.com

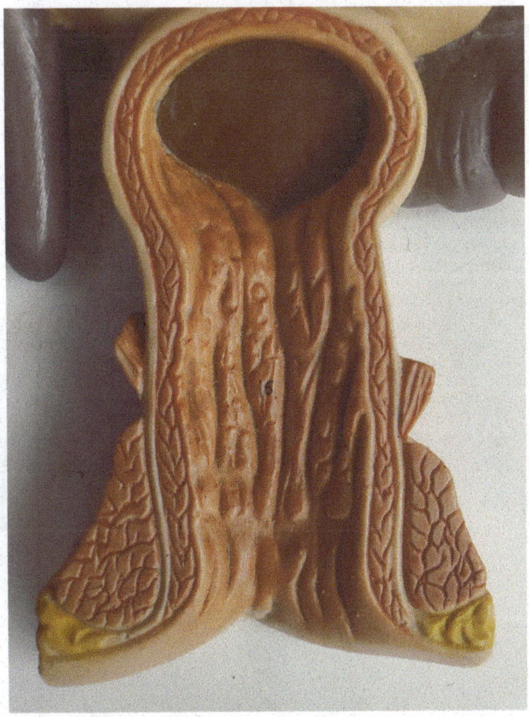

K21 © 3B Scientific GmbH, Germany, 2021. www.3bscientific.com

D. *Accessory Organs of the Digestive System*

OUTCOME 3: Locate, label, and identify the details of the accessory organs of the digestive system.

TOOLBOX

The Salivary Glands

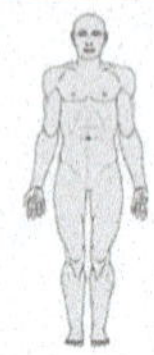

Mr. Flip Chart
Mr. Muscle

parotid gland and duct (cranial nerve innervation?) ______________________________

submandibular gland and ducts (cranial nerve innervation?) ______________________________

sublingual gland and ducts (cranial nerve innervation?) ______________________________

■ ***Label each of the salivary glands in the image.***

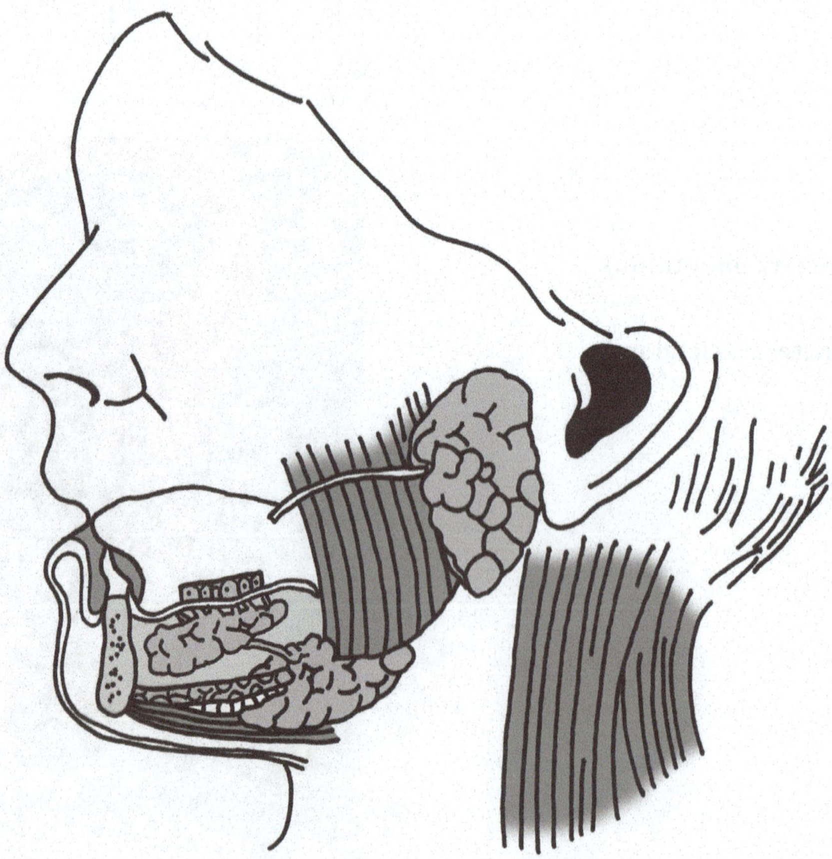

Illustration by Lucia Tranel, ©2015.

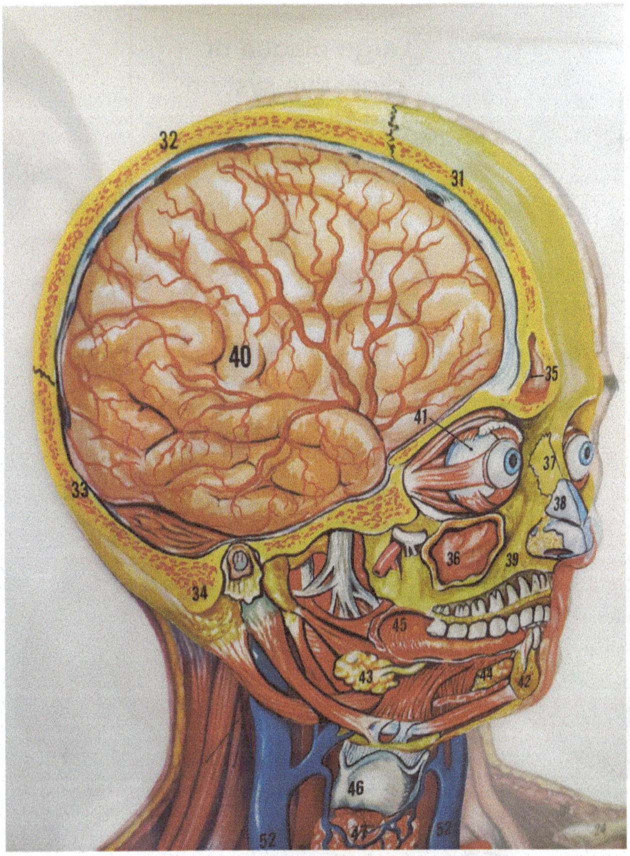
32
31
40
35
41
37
33
38
36
39
34
45
43
44
42
46
52
47
52

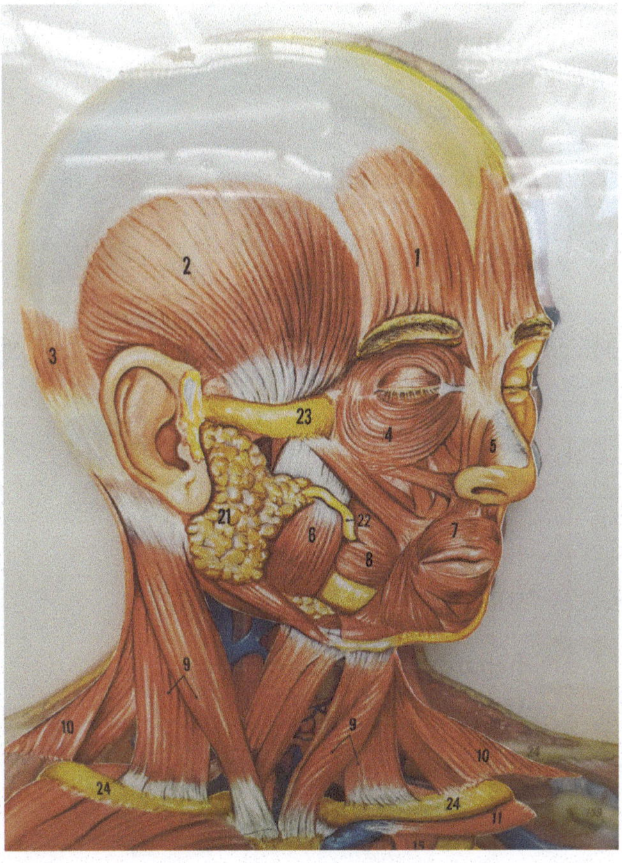
2
1
3
23
4
5
21
22
6
7
8
9
10
9
10
24
24
11

TOOLBOX

The Gallbladder and Associated Bile Ducts

Large plaque of digestive system
Mr. Flip Chart
Mr. Muscle

Gross Anatomical Structures:

- liver
- left hepatic duct
- right hepatic duct
- common hepatic duct
- gall bladder
- cystic duct
- common bile duct
- pancreas
- pancreatic duct
- duodenal papilla
- duodenum

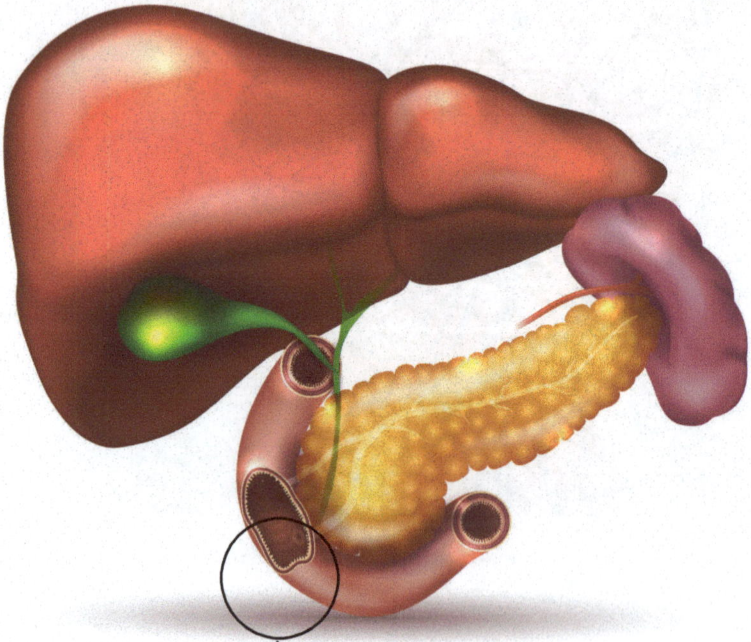

© Tefi/shutterstock.com

Microscopic Anatomical Structures:

- common bile duct
- pancreatic duct
- hepatopancreatic sphincter (sphincter of Oddi)
- hepatopancreatic ampulla (ampulla of Vater)
- duodenal papilla
- duodenal mucosal surface

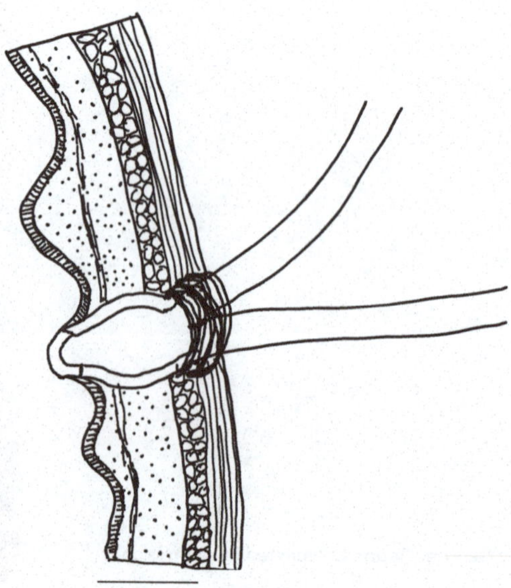

Illustrations by Lucia Tranel, ©2015.

E. *Comparative Anatomy of the Digestive System - Dissection*

OUTCOME 4: Dissect the digestive system in the preserved animal and locate the structures described in the text. Be able to determine where the food was immediately before and after a specific structure.

TOOLBOX

Previous dissection exercises have exposed most of the major digestive organs and accessory structures. Many of the structures to be examined will be a review of the material completed in OUTCOMES 1-3.

Identification of the Major Organs

1. Use one hand to get behind and beneath the pig's left lung and its heart, then gently pull these organs toward the animal's right. Upon doing so, you should be able to see the thick white **thoracic aorta** attached to the dorsal body wall.
   - **esophagus**—the muscular tube that lies alongside the thoracic aorta and immediately dorsal to the trachea. The esophagus begins at the regions of the oropharynx and dorsal laryngopharynx and travels through the thoracic cavity to the cardiac portion of the stomach. It must pass through the diaphragm in order to connect to the stomach's cardia.

Label the **esophagus** and **thoracic aorta** in the following image.

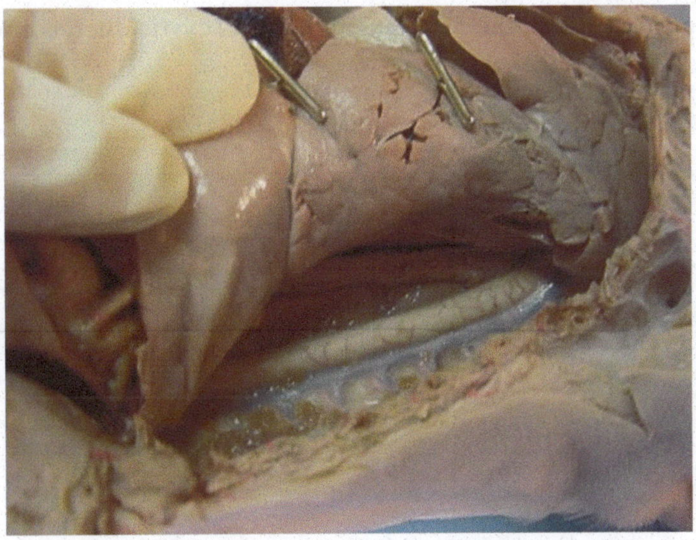

2. In the abdomen, identify the following:
    - peritoneal cavity—the cavity within the abdominal cavity that contains many of the abdominal organs. A few abdominal organs lie **retroperitoneal**; for example, the kidneys.
    - parietal peritoneum—serous membrane adhering to the wall of the peritoneal cavity
    - visceral peritoneum—serous membrane adhering to the surface of the organs within the peritoneal cavity
3. Located immediately posterior diaphragm,
    - **liver**—the large, brown organ that takes up a large portion of the upper right quadrant of the peritoneal cavity. The liver has many functions, but one of its digestive functions is to produce **bile**, which is needed to emulsify lipids. The liver is divided into smaller sections, called **lobes**.
    - **gallbladder**—find a sac-like structure located beneath the right lobe of the liver that functions to store the bile produced by the liver. Often the bile makes the gallbladder look green.
4. Locate and identify the ducts that transport bile from the liver and gallbladder to the duodenum:
    - **cystic duct**—thin tube extending from the distal end of the gallbladder. Bile travels from the gallbladder through the cystic duct to the common bile duct.
    - **common bile duct**—located by looking between where the stomach meets the first part of the small intestine. It is a natural extension from the cystic duct and carries bile to the duodenum.
    - **common hepatic duct**—bile also comes from the liver through this duct. It is very delicate, so gently remove some of the tissue with the forceps to see if you can locate the common hepatic duct.

These three ducts—the cystic duct, the common hepatic duct, the common bile duct—create a create a **Y**.
Label the following structures in the picture: **gallbladder, cystic duct, common hepatic duct,**

common bile duct, esophagus. (The liver and stomach have been labeled).

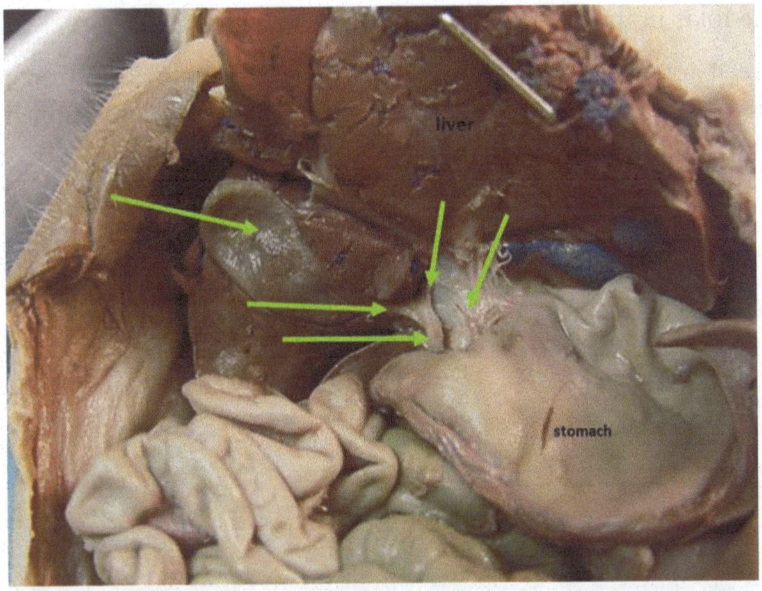

5. Examine the stomach and associated structures.

- **stomach**—balloon-like structure in the upper left abdominal quadrant of the peritoneal cavity. In humans the stomach is **J**-shaped. Regardless of its apparent shape, the regions are analogous a human stomach.
- **lesser curvature**—small, more medial curved margin of the cranial side of the stomach. Normally, you would find a small area of fatty tissue, the **lesser omentum**, adhering to this surface. The lesser omentum is an extension of the visceral peritoneum.
- **greater curvature**—large, outer curved margin of the caudal side of the stomach. Normally, you would find the **greater** omentum, a large apron of fatty tissue, extending from the entire greater curvature and draping over the intestines. This is not generally visible in this preserved animal, except maybe as a thin membrane connecting the stomach and spleen. The greater omentum is also an extension of the visceral peritoneum.
- **spleen**—brown, tongue-shaped structure that hugs the greater curvature of the stomach.
- **cardia**—the region that is connected to the esophagus. The **cardiac sphincter** is an internal smooth muscle at this junction that functions to regulate the entry of chyme into the stomach from the esophagus and also prevents backflow from the stomach into the esophagus.
- **fundus**—the anterior, rounded portion of the stomach
- **body or corpus**—the main portion of the stomach
- **pylorus**—the posterior portion of the stomach that connect to the duodenum. Internally, there is a ring of smooth muscle, the **pyloric sphincter**, which regulates the passage of chyme into the duodenum.

- **internal view of the sphincter muscles.** Cut the stomach open along its greater curvature. Once opened, and the contents carefully cleaned out, identify both the **cardiac sphincter** and the **pyloric sphincter**. Depending upon how full the stomach was, you may also be able to identify the folds in its internal surface, called **rugae.**

In the picture, the green arrows indicate (from left to right), the **pyloric sphincter, cardiac sphincter, diverticulum ventriculi, and the spleen.** The diverticulum ventriculi is mentioned here only so you do not confuse it with the cardiac sphincter. Label the arrows in the image below.

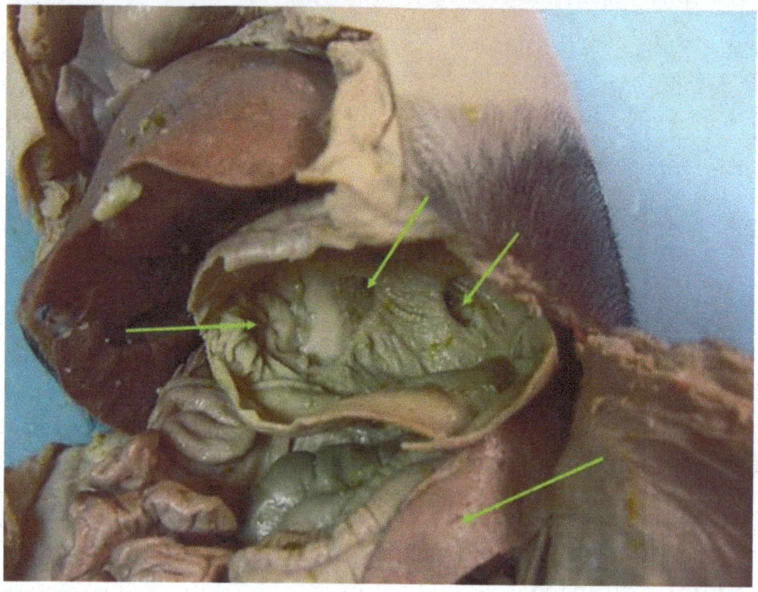

6. Examine the small and large intestines and associated structures.

   - **small intestine**—begins where the pyloric portion of the stomach connects and forms a mass of *loosely coiled* tube of the small intestine. It can be manipulated and spread apart. Do not confuse it with the more tightly coiled tube of the colon.
   - **mesentery**—a thin serous membrane extending from the visceral peritoneum of the small intestine. Gently spread the coils of the small intestine apart to locate the mesentery. Notice the many small blood vessels in this membrane.
   - **cecum**—the beginning of the large intestine. Locate this finger-like structure at the most distal end of the small intestine, the ileum. Two structures, the ileum and the cecum, form a T-shaped connection. Internally, there is another smooth muscle sphincter, the **ileocecal sphincter**.
   - **colon**—very tightly coiled mass
   - **rectum of the large intestine**—straight portion of the tube extending from the colon to the anus.

Locate these structures and then label them in the following image:

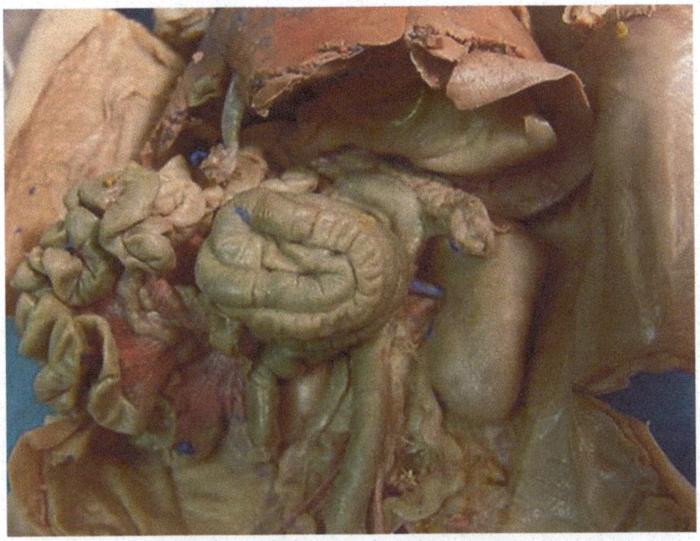

Contributed by Sara Schlereth. © Kendall Hunt Publishing Company.

7. Identification of Accessory Organs

In order to locate and identify the accessory organs, you will need to look in the region of the peritoneal cavity that is posterior to the stomach. The bulk of this region contains intestines: the loosely-coiled **small intestine** and the tightly-coiled **colon** of the **large intestine**.

- **liver**—re-examine. Notice the umbilical vein that you severed when you made the first opening into the cavity.
- **pancreas**—a tail-like structure with a lumpy surface. Gently pull the stomach and liver anteriorly and the small intestines and colon posteriorly. You will likely see a thin membrane between them. Break through this membrane to expose the lumpy tail-like pancreas. Notice that the pancreas extends from its head end (at the duodenum of the small intestine), across the body to its tail, (just beneath the spleen). The tail of the pancreas is usually the part that you will see first.

Label the **spleen, liver, stomach, gallbladder, pancreas, small intestine,** and **colon of the large intestine** in the following image.

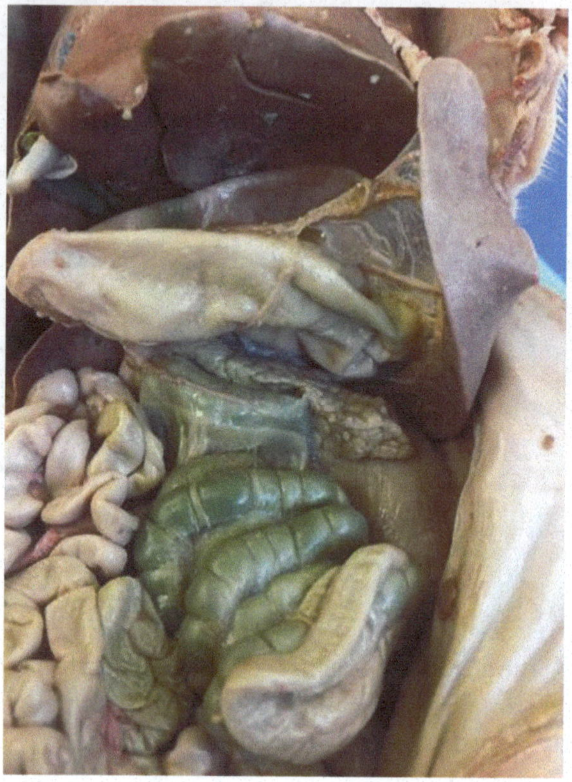

Contributed by Sara Schlereth. © Kendall Hunt Publishing Company.

F. *Histological Organization of Digestive Tract and the Accessory Organs*

TOOLBOX

Histological Organization of the Digestive Tract
Liver Histology
The Pancreas

OUTCOME 5: Review the histological organization of the digestive tract. Name the four layers. Identify the tissues that compose each layer. Describe the function of each layer.

■ ***Label the images.***

mucosa
- mucosal epithelium (simple columnar epithelium)
- lamina propria
- muscularis mucosae (smooth m.)

submucosa (irregular C.T.)
- artery
- vein
- lymph vessel

muscularis externa (smooth m.)
- circular layer
- myenteric plexus
- longitudinal layer

serosa (simple squamous epithelium)

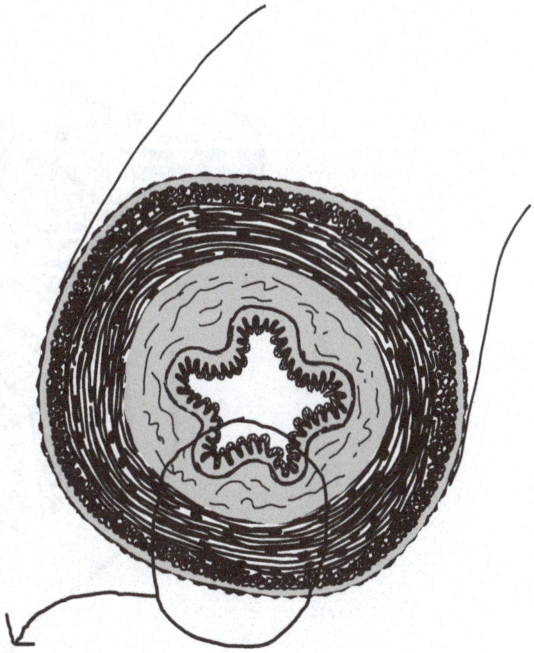

Illustration by Lucia Tranel, 2015.

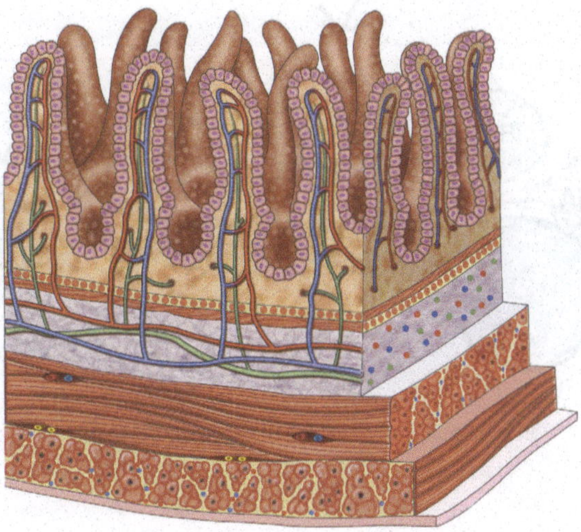

© ilusmedical/shutterstock.com

OUTCOME 6: Review the histological organization of a liver lobule. Determine direction of blood and bile flow.

■ ***Label the parts of a liver lobule.***

liver lobule
- hepatocytes (plates of hepatocytes)
- sinusoid capillary
- central vein

portal triad
- branch of hepatic portal vein
- bile duct
- branch of hepatic artery

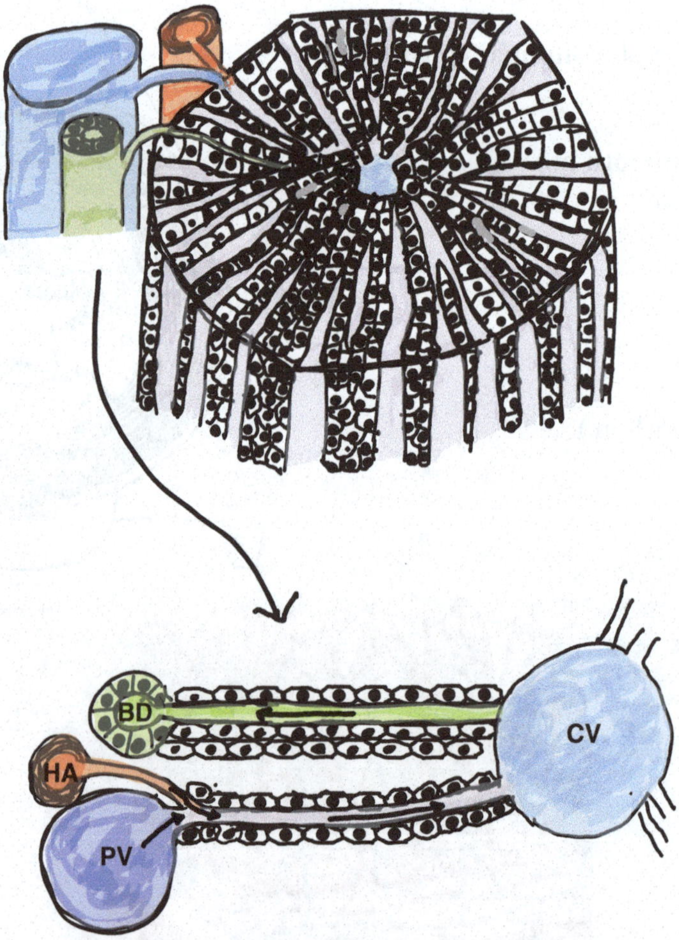

Both Illustrations by Lucia Tranel, © 2015.

OUTCOME 7: Review the histological organization of the pancreatic lobule. Determine which cells are exocrine and which cells are endocrine.

■ ***Label the listed structures on the image and include which is endocrine and which is exocrine.***

pancreatic lobule

clusters of acinar cells

islet of Langerhans

pancreatic duct

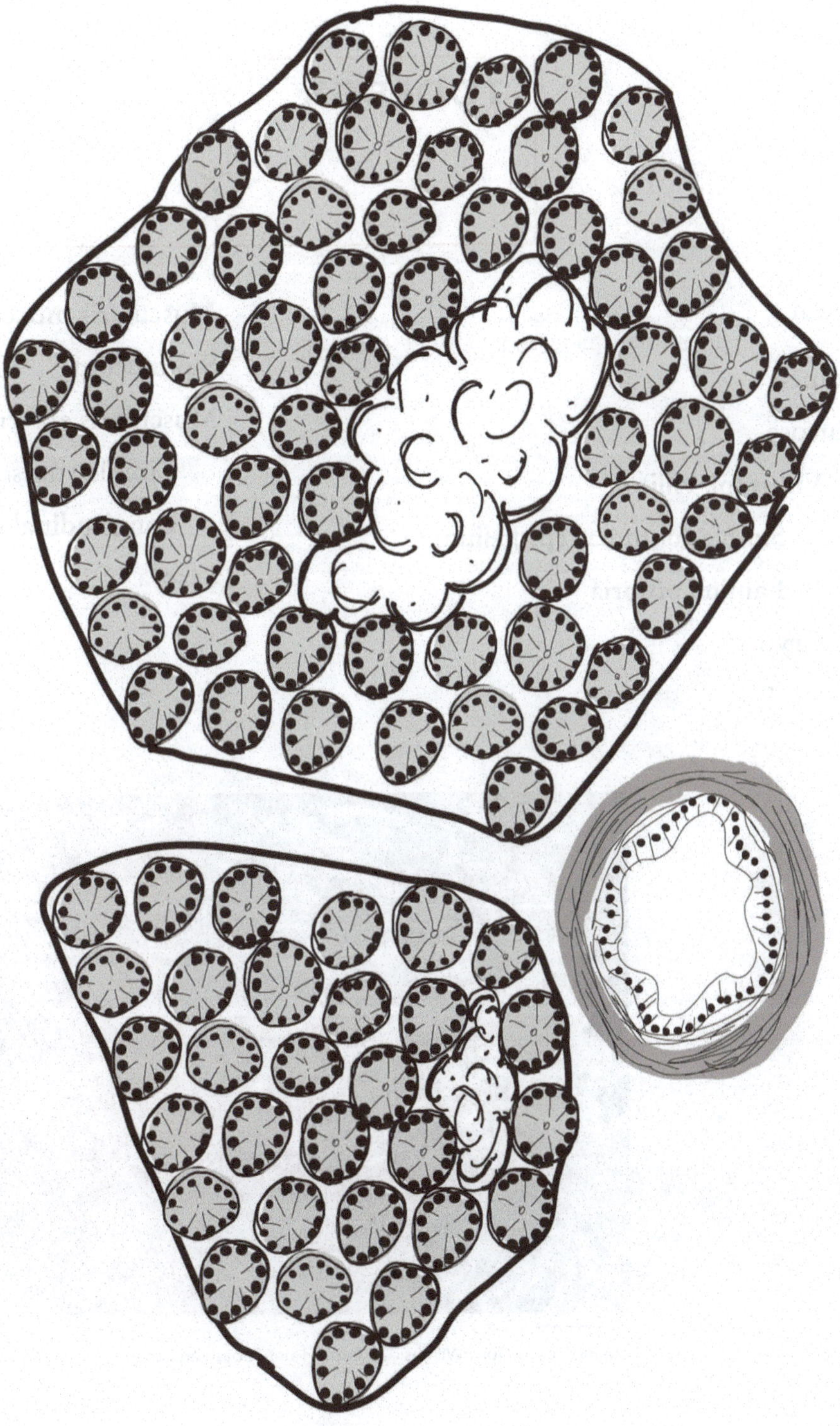

G. *Histological Examination*

OUTCOME 8: Examine a slide of the jejunum under the microscope. Identify the four layers. Identify the tissues that compose each layer. Draw and label what you see.

TOOLBOX

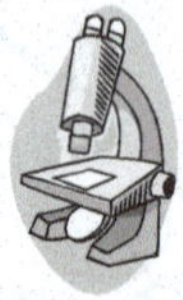

Jejunum
- Lumen
- Mucosa
  - Villus (pl. villi)
    - Simple columnar epithelium
    - Lamina propria
  - Crypts
- Muscularis mucosae
- Submucosa
- Muscularis externa
  - Circular muscle
  - Longitudinal muscle
- Serosa

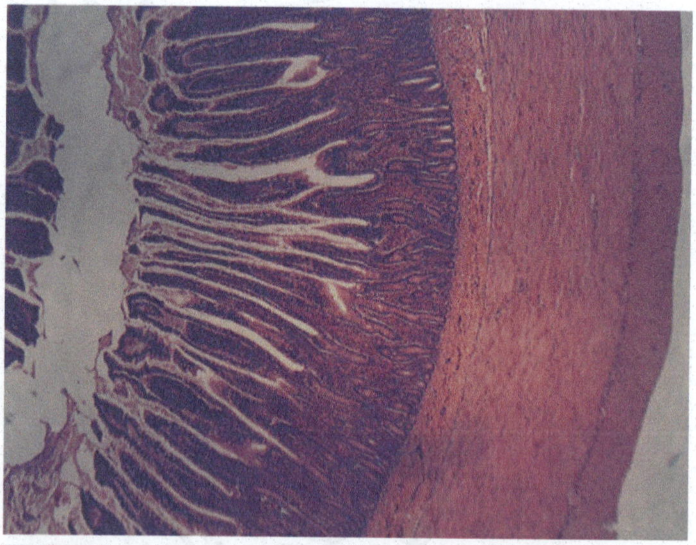

Jejunum

Locate each layer in the wall of the intestine. Note the positional relationship of each layer to the **lumen.** Make sure you can determine the tissue composition and the characteristics of each layer as listed in the table below.

LOCATE LAYER	TISSUE COMPOSITION	NOTES
mucosal epithelium	simple columnar epithelium	goblet cells, villi, microvilli
submucosa	areolar CT	glands
muscularis externa	smooth muscle	circular, longitudinal layers
serosa	simple squamous epithelium	same as visceral peritoneum

OUTCOME 9: Examine two slides of the liver under the microscope. The first slide is stained to enhance the view of a liver lobule. The second slide will allow observation of a portal triad. Identify and draw the details listed.

Pig Liver, trichrome stain

Examine this slide to get a sense of what a **liver lobule** looks like. Blue-stained **interlobular septa** separate the red lobules. Find the **central vein,** located in the center of the lobule. The red cells are the hepatocytes, and the thin, white spaces between the **rows of hepatocytes** are **sinusoid capillaries.** The **portal triads** are in the periphery of the lobule within the blue septa, although their details are hard to see on this slide. Examine the details of the triads on the next slide. Fill in the table.

LOCATE	NOTES	DRAWING
general lobule	Draw a lobule. Label plates of hepatocytes sinusoids central vein portal vein bile duct hepatic artery	

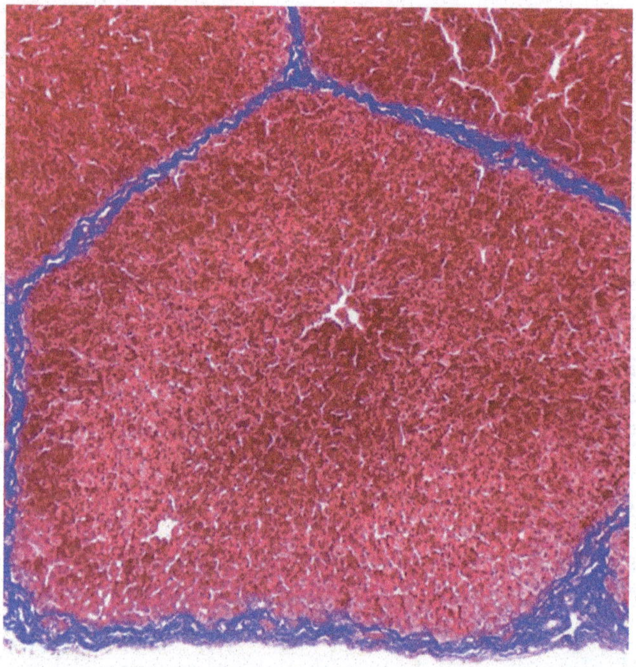

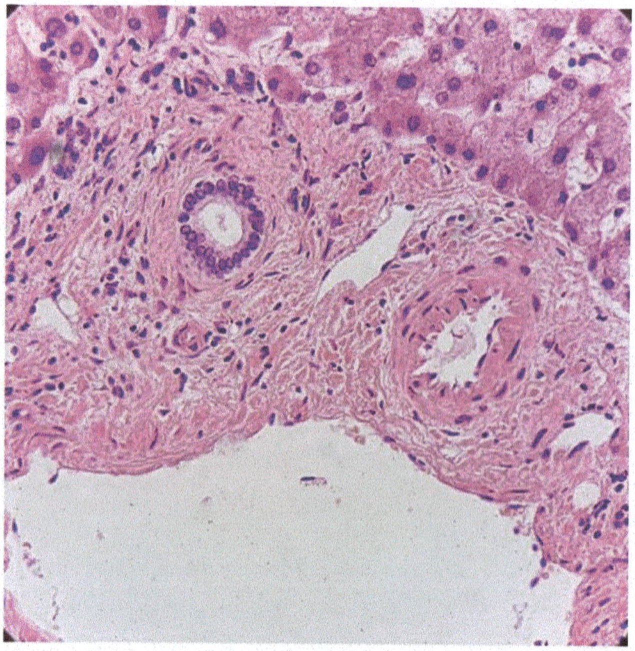

Contributed by Tony Wilmes.

Human Liver, H&E stain

The **portal triad** will appear as a cluster of blood vessels and ducts. The smaller triads are a bit easier to examine, since there are fewer branches. Once you locate a triad, identify the structure with the very large lumen but thin wall. This is a branch of the **hepatic portal vein.** Now find the structure surrounded by **simple cuboidal** or **simple columnar epithelium.** This is a branch of the **bile duct.** The branch of the **hepatic artery** is the vessel with the very small lumen but relatively thick **smooth muscle** layer in the **tunica media.** Locate a **central vein.** You will know when you find one because the vein will appear simple and isolated. You might be able to see the sinusoids that open into the vein on high magnification. Complete the details of your drawing in the above table.

- *Which vessel carries O_2-rich blood into the sinusoids of the liver lobule toward the central vein?*

- *Which vessel carries O_2-poor blood into the sinusoids of the liver lobule toward the central vein?*

- *Which vessel carries O_2-poor blood out of the sinusoids toward the hepatic veins?*

OUTCOME 10: Examine a slide of the pancreas under the microscope. Locate the acinar cells. Locate the islets of Langerhans. Which is exocrine? Which is endocrine? Locate the pancreatic duct. Draw and label what you see.

Pancreas

The pancreas is both **exocrine** and **endocrine** in function. The name of the exocrine cells comes from their "grape-like" cluster formation—**acinar cells.** These cells produce **secretions containing digestive enzymes.** The body releases the enzymes into ducts that empty into the main pancreatic duct that then leads to the duodenum. The endocrine cells appear in small, scattered "islands," known as **Islets of Langerhans** cells. The blood absorbs **hormones (including insulin and glucagon)** from the endocrine cells directly.

LOCATE	NOTES	DRAWING	HISTOLOGY
acinar cells	exocrine		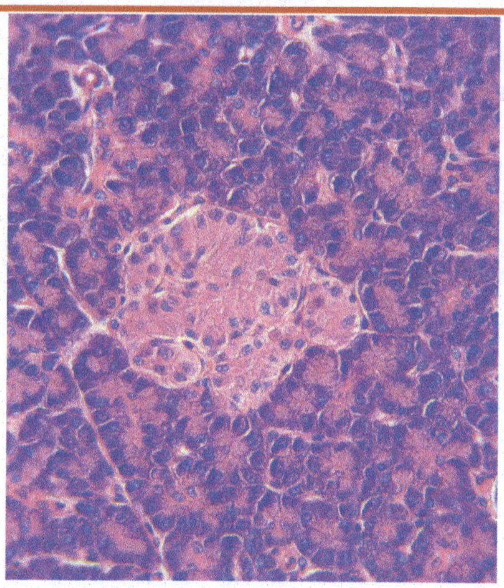 © Jubal Harshaw/shutterstock.com
Islet of Langerhans	endocrine		
pancreatic duct	simple cuboidal → simple columnar epithelium		

CHECKLIST: Summary Checklist Digestive

TERM/STRUCTURE	YOUR NOTES
Mucosa	
Submucosa	
Muscularis Externa	
Adventitia/Serosa	
Esophagus	
Nonkeratinized Stratified Squamous Epithelium	
Muscularis Mucosae	
Smooth Muscle	
Simple Columnar Epithelium	
Goblet Cells	
Loose Irregular Connective Tissue	
Internal Circular Muscle Layer	
External Longitudinal Muscle Layer	
Gastric Pits	
Gastric Glands	
Surface Mucous Cells	
Mucous Neck Cells	
Chief Cells	
Parietal Cells	
Oblique Muscle Layer (Only in Stomach)	

(continued)

TERM/STRUCTURE	YOUR NOTES
Mesothelium (Simple Squamous Epithelium)	
Fibrous Connective Tissue	
Small Intestine	
Ileum	
Plicae Circulares	
Villi	
Microvilli	
Jejunum	
Duodenum	
Peyer's Patches	
Hepatocyte Plate/Sheet	
Sinusoid Capillary of the Liver	
Portal Triad	
Hepatic Portal Vein	
Hepatic Artery	
Bile Duct	
Pancreas	
Liver	
Acinar Cells (Acini)	
Islets of Langerhans	

TERM/STRUCTURE	YOUR NOTES
Pancreatic Ducts	
Parietal Peritoneum	
Visceral Peritoneum	
Common Hepatic Duct	
Gallbladder	
Cystic Duct	
Stomach	
Cardiac Sphincter	
Cardia	
Fundus	
Body (of Stomach)	
Rugae	
Pylorus	
Pyloric Sphincter	
Lesser Curvature	
Greater Curvature	
Large Intestine	
Cecum	
Colon	
Rectum of Large Intestine	
Anus	

(continued)

TERM/STRUCTURE	YOUR NOTES
Common Bile Duct	
Hepatopancreatic Ampulla (Ampulla of Vater)	
Hepatopancreatic Sphincter (Sphincter of Oddi)	
Hepatic Ducts (Left and Right)	
Duodenal Papilla	
Oral Cavity	
Oropharynx	
Laryngopharynx	
Ileocecal Valve/Sphincter	
Ascending Colon of Large Intestine	
Transverse Colon of Large Intestine	
Descending Colon of Large Intestine	
Sigmoid Colon of Large Intestine	
Anal Canal	
Internal Anal Sphincter (Composed of Smooth Muscle)	
External Anal Sphincter (Composed of Skeletal Muscle)	
Parotid Salivary Glands	
Sublingual Salivary Glands	
Submandibular Salivary Glands	

EXERCISE 12

Reproductive System

A. Introduction

In general, the reproductive system consists of three categories of organs: (1) **gonads,** (2) **ductules,** (3) and **accessory organs.** The gonads, the **ovaries** and the **testes,** are the site of gametogenesis. **Diploid stem cells** mature within those structures through a **series of mitotic divisions** and become **diploid primary spermatocytes** and **diploid primary oocytes.** These cells then go through **meiosis,** and the end result is **haploid gametes.** You will identify the cells and tissues involved in the different stages of gametogenesis by examining the gonads under the microscope The **ductules** refer to the series of tubules that conduct the gametes through the system. The **accessory organs** include glands, ligaments, and supportive organs.

B. *Outcomes for Exercise 12*

1. Locate, label, and identify the organs of the female reproductive system. Include the details of each organ and the related structures.

2. Locate, label, and identify the organs of the male reproductive system. Include the details of each organ and the related structures.

3. Review the process of spermatogenesis. Locate, label, and identify the structures related to spermatogenesis. Determine which cells are 2N; which cells are 1N. List the progression of development from the spermatogonium to the mature spermatocyte. Name the cellular division process that occurs in each step of the progression.

4. Review the process of oogenesis. Locate, label, and identify the structures related to oogenesis. Determine which cells are 2N; which cells are 1N. List the progression of development from the oogonium to the mature ovum. Name the cellular division process that occurs in each step of the progression.

5. Examine the slide of the testis under the microscope. Locate the structures listed. Draw and label what you see. Determine which cell division process creates each new cell. Is it 2N or 1N? Describe the function of the interstitial cells and the nurse cells.

6. Examine the slides of the ovary (mature Graafian follicle), the corpus luteum, and the corpus albicans. Locate the structures listed. Draw and label what you see. Determine which cell division process creates each new cell. Is it 2N or 1N? Describe the functions of the corpus luteum.

C. *Anatomy of the Female Reproductive System*

OUTCOME 1: Locate, label, and identify the organs of the female reproductive system. Include the details of each organ and the related structures.

TOOLBOX

The Female Reproductive System

female pelvis

ovary

uterine (fallopian) tube

uterus

- perimetrium
- myometrium
- endometrium
- cervix

vagina

labia minora

labia majora

clitoris

mons pubis

pubic symphysis

urinary bladder

urethra

sigmoid colon

rectum

anus

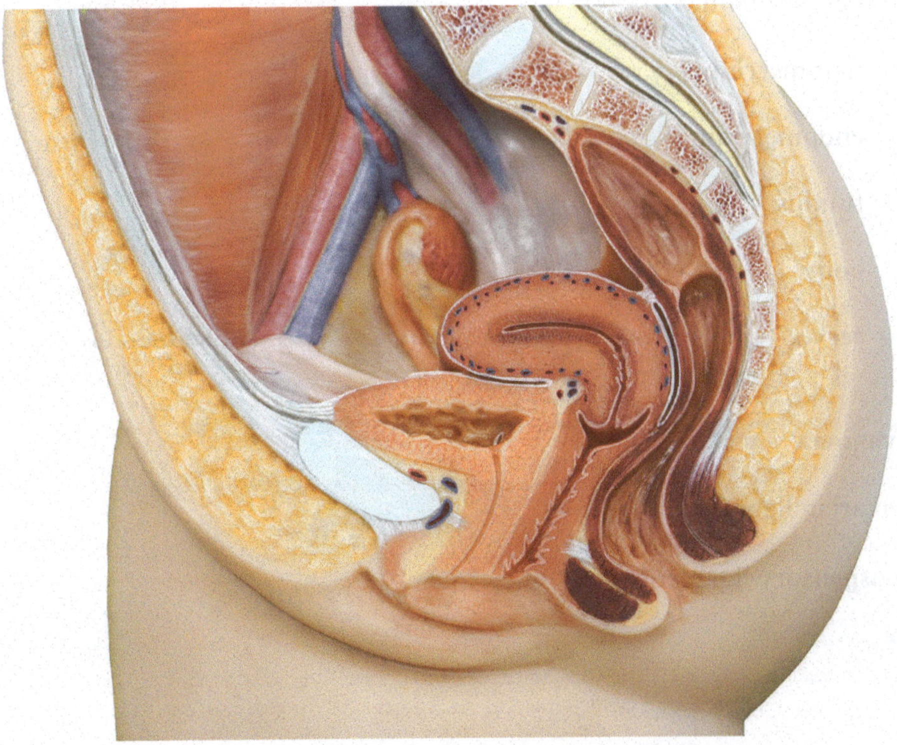

© Choksawatdikorn/shutterstock.com

TOOLBOX

The Uterus

female pelvis

ovary

uterine (fallopian) tube

- fimbriae
- infundibulum
- ampulla
- isthmus

uterus

- uterine cavity
- perimetrium
- myometrium
- endometrium
- fundus
- body
- cervix—internal os, cervical canal, external os

vagina

ovarian ligament

suspensory ligament

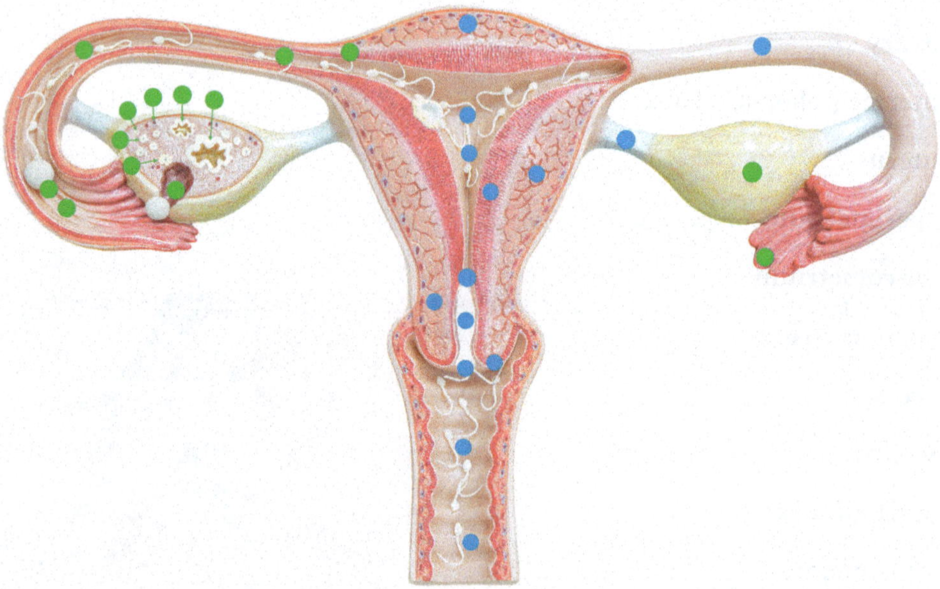

D. *Anatomy of the Male Reproductive System*

OUTCOME 2: Locate, label, and identify the organs of the female reproductive system. Include the details of each organ and the related structures.

TOOLBOX

Male Reproductive System

male pelvis

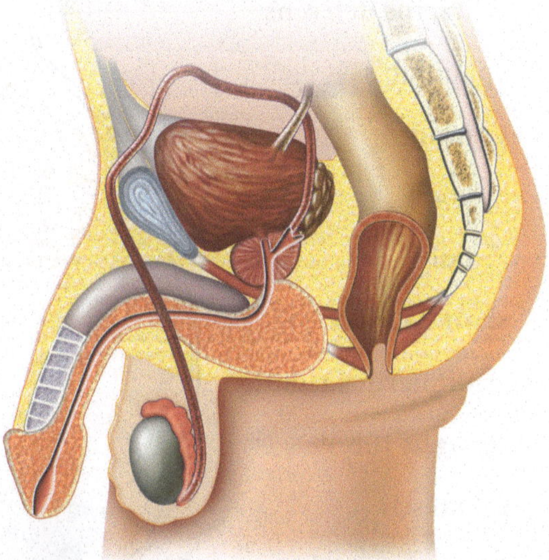

scrotum

testis

epididymis

inguinal canal

ductus deferens

seminal vesicle (gland)

prostate gland (seen whole)

bulbourethral gland

(not visible on diagram; locate in textbook)

prostatic urethra

membranous urethra

spongy (penile) urethra

penis

external urethral orifice

corpus cavernosum

corpus spongiosum

urinary bladder (seen whole)

pubic symphysis

ureter

rectum

anus

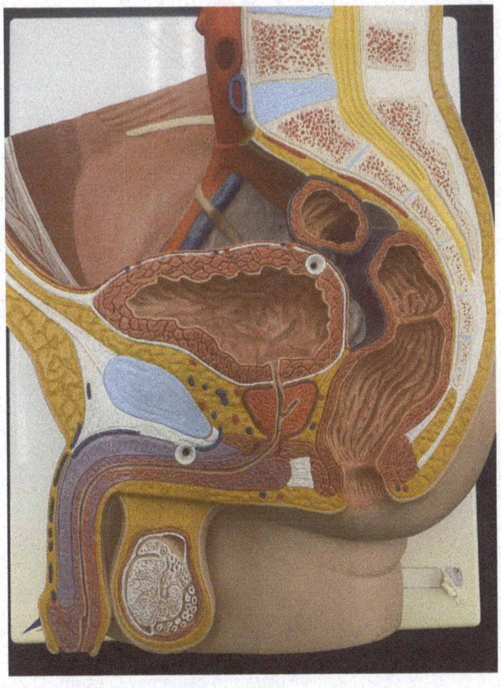

H11 © 3B Scientific GmbH, Germany, 2021. www.3bscientific.com

TOOLBOX

Structure of the Testes
The Epididymis
Ductus Deferens and Accessory Organs

male pelvis
testis model

spermatic cord

spermatic artery

spermatic vein

ductus deferens

epididymis

head

body

tail

scrotal cavity

seminiferous tubule

straight tubule

rete testis

efferent ductules

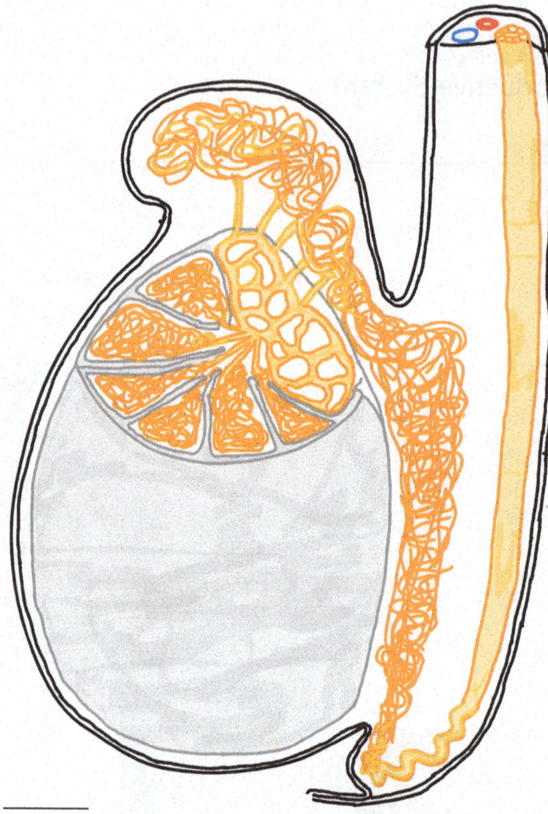

Illustration by Lucia Tranel, © 2015.

ureter

urinary bladder

ductus deferens

seminal glands/vesicles

ampulla of ductus deferens

ejaculatory duct

prostate gland

bulbourethral glands

urogenital diaphragm

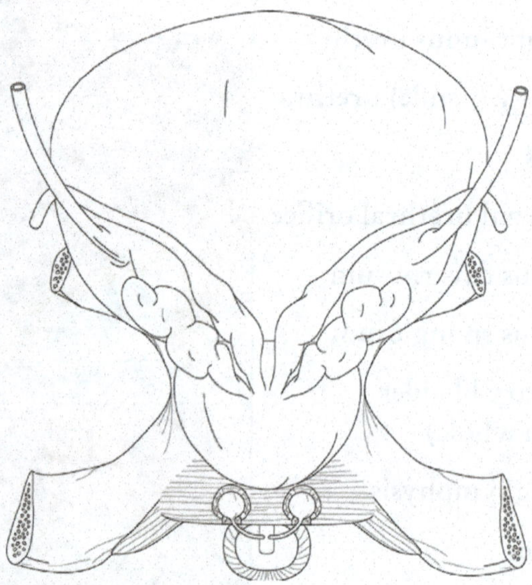

Illustration by Jamey Garbett. © 2003 Mark Nielsen.

TOOLBOX

The Penis

male pelvis

corpora cavernosa, left and right
(corpus cavernosum, sing.)

corpus spongiosum

penile/spongy urethra

central artery

dorsal artery, vein, nerve

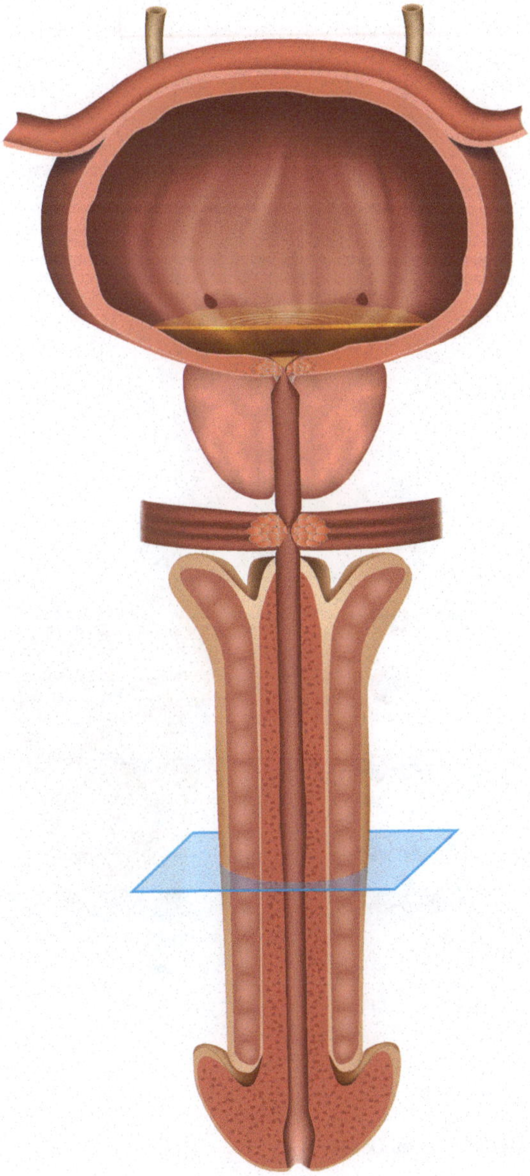

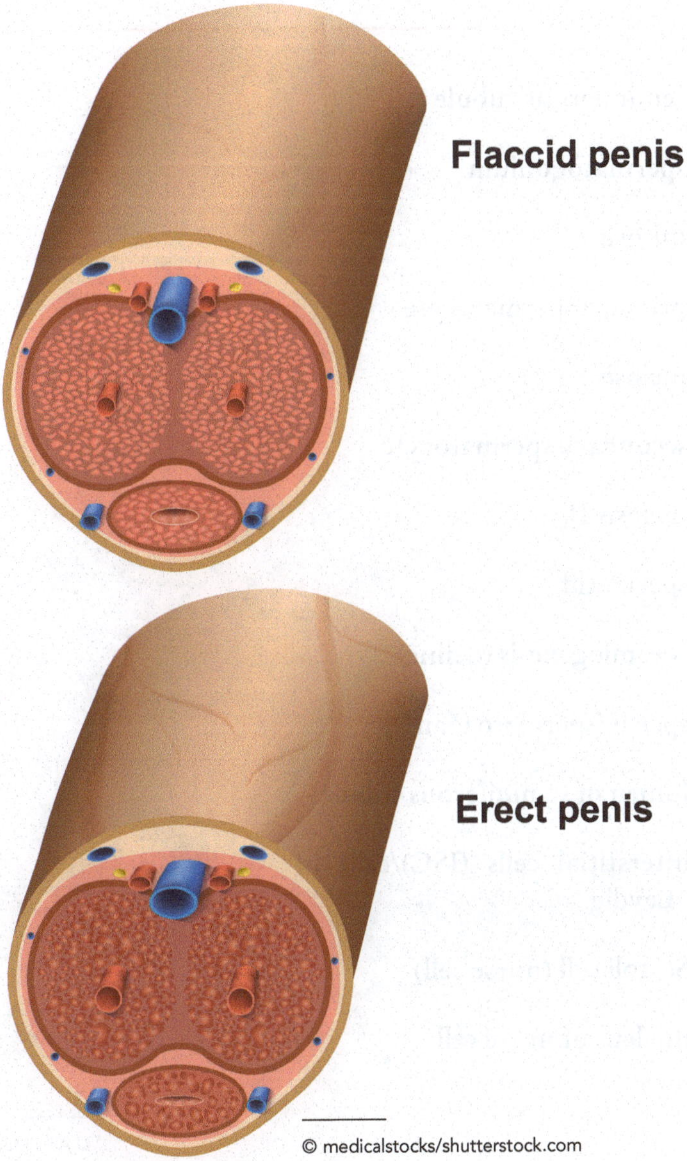

E. *Spermatogenesis and Oogenesis*

OUTCOME 3: Review the process of spermatogenesis. Locate, label, and identify the structures related to spermatogenesis. Determine which cells are 2N; which cells are 1N. List the progression of development from the spermatogonium to the mature spermatocyte. Name the cellular division process that occurs in each step of the progression.

TOOLBOX

The Seminiferous Tubules
Spermatogenesis

testis model

seminiferous tubule

spermatogonium

mitosis

primary spermatocyte

meiosis I

secondary spermatocyte

meiosis II

spermatid

spermiogenesis (define)

sperm (spermatozoan)

lumen of seminiferous tubule

interstitial cells (ISC)/cells of Leydig

Sertoli cell (nurse cell)

nucleus of nurse cell

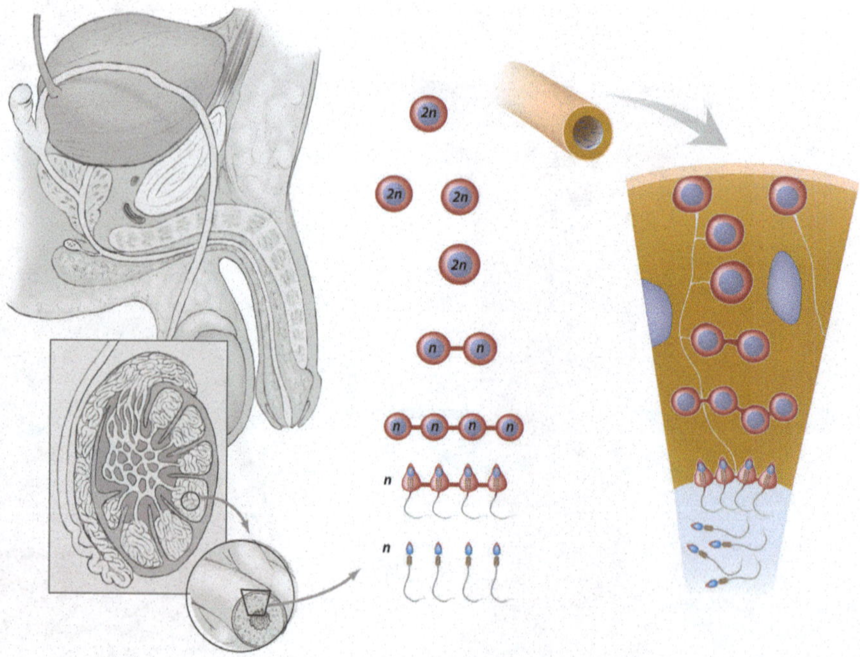

OUTCOME 4: Review the process of oogenesis. Locate, label, and identify the structures related to oogenesis. Determine which cells are 2N; which cells are 1N. List the progression of development from the oogonium to the mature ovum. Name the cellular division process that occurs in each step of the progression.

TOOLBOX

Oogenesis

ovary
(oogonia develop into primary oocytes during fetal development)

germinal epithelium

OVARY PHASE I:
primary oocyte
within primordial follicle

growing follicle

meiosis I

secondary oocyte
within mature follicle

polar body
(may not develop)

ovulation of secondary oocyte

docking of sperm
meiosis II
ovum + sperm
(fertilization)
} may or may not happen

OVARY PHASE II:
empty Graafian follicle
↓
corpus luteum

OVARY PHASE III:
(if fertilization doesn't take place)

corpus luteum atrophies
↓
corpus albicans

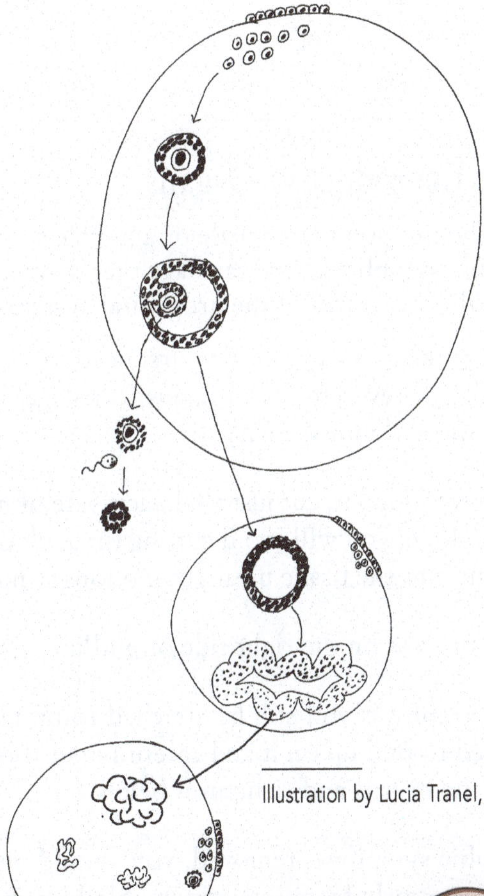

Illustration by Lucia Tranel, 2015.

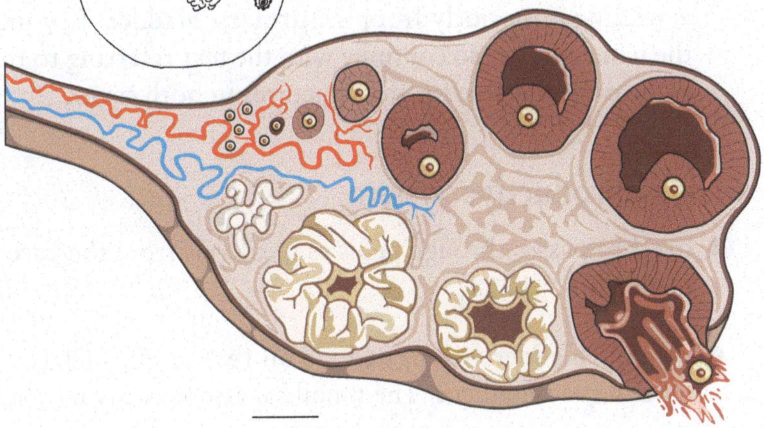

F. Comparative Anatomy Reproductive System—Dissection

OUTCOME 5: Dissect the reproductive system in both male and female preserved animals. Locate and provide the functions of the structures listed in the text. Trace the flow of the pathways—from testis to urethra in the male and from ovary to vagina in the female. Be able to indicate where in the female fertilization and implantation and development of the embryo occurs.

TOOLBOX

Exposing the Reproductive Organs

Regardless of whether you have a male or a female specimen, you will need to make a longitudinal cut through the **pubic symphysis** (the cartilage that connects the right and left pubic bones) to completely expose some of the structures of the urogenital system.

1. Before you make this cut, you will first need to locate (by touch) the **pubic symphysis**—it will feel like a small bone (even though it is just cartilage) in the pelvic region, along the midline between the tail and the abdominal cavity.
2. Once you have found it, cut just to the left side of midline, NOT directly in the middle. By cutting the left of midline, you will avoid cutting through the urethra and other soft structures underneath. Carefully cut into the tissue until you no longer meet with resistance
3. Repeat this process on the right side of midline.
4. The pubic symphysis will still be attached to underlying structures by soft tissue connections, so you will need to pull up on it and carefully cut those connections to remove the pubic symphysis and expose the organs and structures below it.
5. With the pubic symphysis removed, you should be able to see a thick muscular tube, the **urethra,** extending posteriorly from the **urinary bladder**. Beyond this point, the anatomy of the male and the female differ, so continue with the text referring to the appropriate sex. Keep in mind, you will be held responsible for the structures in both sexes.

Male Reproductive System

- **urethra** passes through penis as it extends from the base of the **urinary bladder** to the **urogenital orifice**
- **urinary bladder**—elongated sac on the ventral surface of the flap you made with our initial cuts around the umbilicus. The umbilical arteries are on either side.

- **ductus deferens**—pull the flap with the urinary bladder posteriorly so that you can view the pelvic cavity of the animal. The left and right ductus deferens can be seen passing into the pelvic cavity, arching over the ureters, and disappearing behind the urinary bladder. The opening that they pass through into the cavity is the **inguinal canal**. The ductus carries the sperm from the testis to the ejaculatory duct at the start of the penis.

- **spermatic cord**—before the ductus deferens passes through the inguinal canal, it is part of a collection of structures called the **spermatic cord,** which contains a **ductus deferens, spermatic artery, spermatic vein,** and **spermatic nerve,** all bundled neatly together like wires into a cord. To expose the spermatic cord, continue the dissection on the external genitalia by carefully making an incision through the scrotal sac. Locate the **spermatic cord** within.

- **testis**—at the distal end of the spermatic cord, you will see an enlargement, also surrounded by a membrane. If you carefully remove the membrane from that end, you will expose the bean-shaped **testis** and its surrounding **epididymis.** As was previously mentioned, the bean-shaped testis is the site of spermatogenesis.

- **epididymis**—the C-shaped structure that hugs the testis. Sperm pass from the testis into the epididymis, which is actually a highly coiled tubule. From the epididymis, sperm travel to the ductus deferens→the ejaculatory duct→the prostatic urethra→membranous urethra→penile urethra→urethral orifice. NOTE: even though the ductus deferens is part of the spermatic cord, when you trace the flow of sperm you should include the ductus deferens, since sperm are physically in contact with the vas deferens. Likewise, when tracing sperm out of the body, include the parts of the urethra since this structure conducts the sperm, even though it is within the penis.

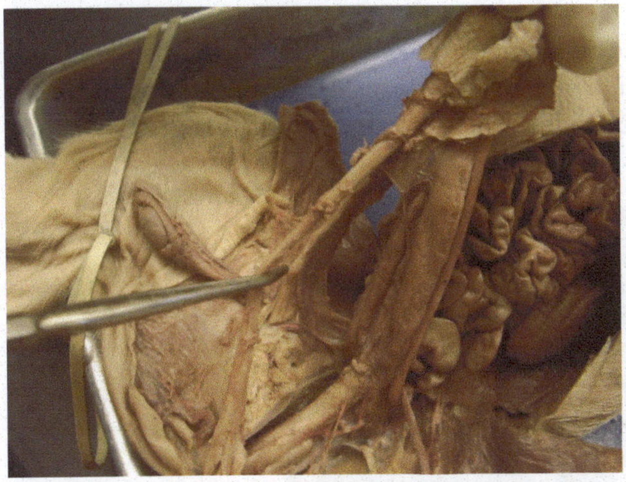

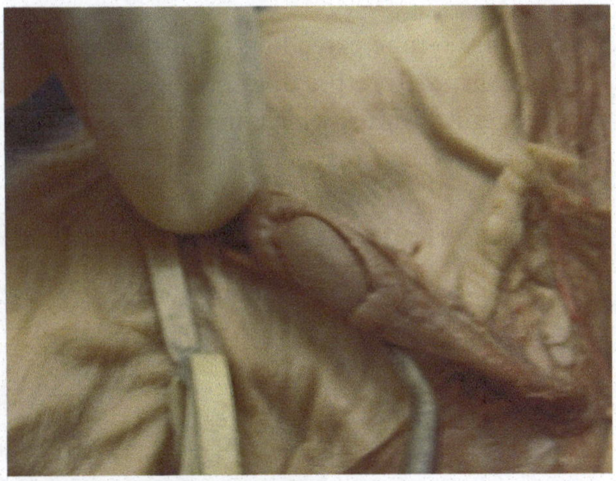

Female Reproductive System

The urogenital systems of female human and the female specimen you are working with are different. Since we are dissecting the pigs, we will talk about those structures first, but since you will more likely be dealing with humans in your health sciences career, we will also discuss the differences.

- **ovaries**—the female gonads. Locate the small, bean-like structures, left and right in the pelvic cavity.
- **uterine tubes**—also known as the **oviducts** or **Fallopian tubes**. A tiny, coiled tubule that hugs each ovary but is NOT attached to the ovary. Transport the egg to the uterine horns.
- **uterine horns**—larger, coiled tubes that are continuous with the uterine tubes. *These structures are part of the uterus, NOT the uterine tube.* In animals that typically have litters or clutches (more than one fetus at a time), these provide extra surface area for the eggs to implant. It is important to remember that they function like the uterus and are not involved with transporting the egg.
- **body of the uterus**—the junction of the two uterine horns.
- **cervix**—the "neck" of the uterus. Examine the distal end of the body of the uterus. You might be able to see a slight bulge in the wall of the uterus. That is the cervix. If you gently rub the area between your fingers, you will notice that there is a slightly perceptible lump.
- **vagina**—the continuation from the uterus and cervix. Unlike humans, the vagina does not open to the outside of the body. Instead, it meets the urethra at the **urogenital sinus**, which, in turn, passes to the outside of the body through the **urogenital orifice**, ventral to the anus.
- **urethra**—tube extending from the urinary bladder and positioned ventral to the vagina. Unlike humans, the urethra does not open to the outside of the body. Instead, it meets the vagina at the **urogenital sinus**, which, in turn, passes to the outside of the body through the **urogenital orifice**, ventral to the anus.

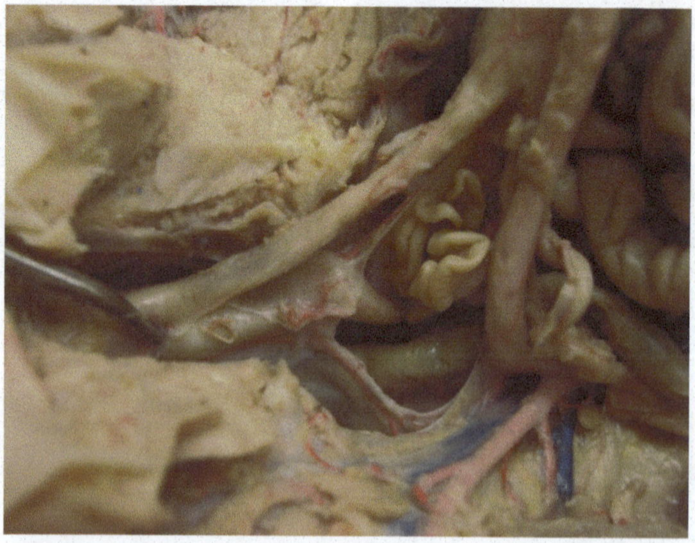

G. *Histology of Testis and Ovary*

OUTCOME 6: Examine the slide of the testis under the microscope. Locate the structures listed. Draw and label what you see. Determine which cell division process creates each new cell. Is it 2N or 1N? Describe the function of the interstitial cells and the nurse cells.

TOOLBOX

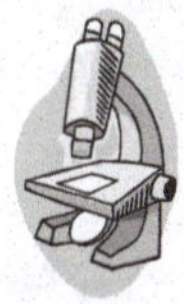

STRUCTURE	YOUR NOTES	LABELED DRAWING
seminiferous tubule		
spermatogonium		
1° and 2° spermatocytes (estimated location)		
spermatids		
sperm		
lumen		
interstitial cells (cells of Leydig)		

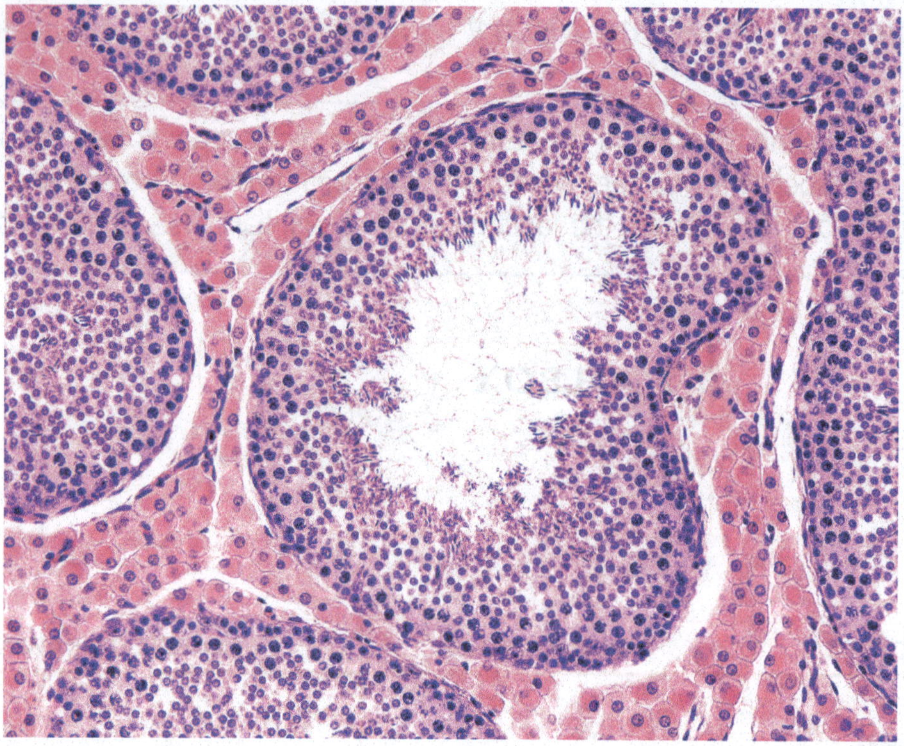

Interstitial or Leydig cells located among seminiferous tubules. They are polyhedral epithelioid cells with a round nucleus and an acidophilic cytoplasm.

OUTCOME 7: Examine the slides of the ovary (mature Graafian follicle), the corpus luteum, and the corpus albicans. Locate the structures listed. Draw and label what you see. Determine which cell division process creates each new cell. Is it 2N or 1N? Describe the function of the corpus luteum.

TOOLBOX

The Ovarian Cycle

Oogenesis occurs in monthly phases, known as the **ovarian cycle.** This is unlike spermatogenesis, which begins at puberty and continues through the entire life of the male. The appearance of the ovary changes during the month, so you will have to examine **three different slides** in order to see what the ovary looks like during an entire ovarian cycle.

a. Ovary, mature Graafian follicle. (Lo Mag)

STRUCTURE	YOUR NOTES	LABELED DRAWING
germinal epithelium		
primordial follicles containing primary oocytes		
growing follicles (primary, secondary follicles)		
mature Graafian follicle (tertiary follicle) containing secondary oocyte		

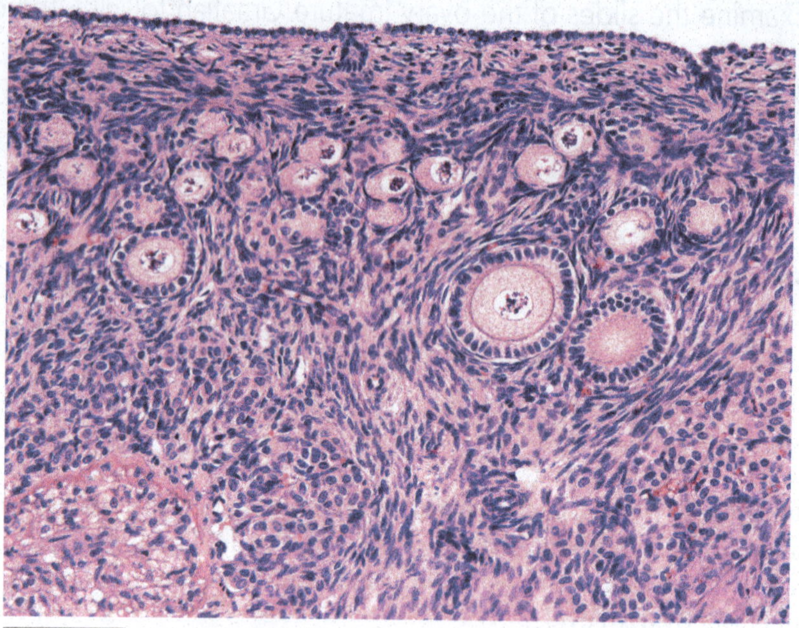

Ovarian cortex showing, under the germinal epithelium, a thin fibrous layer (tunica albuginea), several layers of primordial and primary follicles. In the bottom left corner, an atretic folllicle.

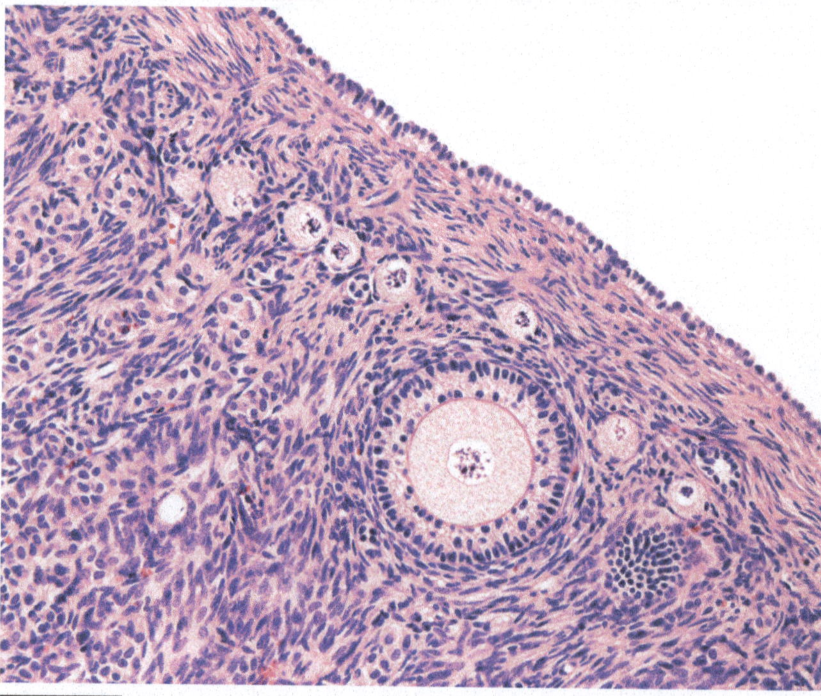

Ovarian cortex showing, from top to bottom, the ovarian germinal epithelium, a thin tunica albuginea, a layer of primordial follicles and a secondary follicle with two layers of granulosa cells.

Ovary, mature Graafian follicle. (Hi Mag) Details of follicle.

STRUCTURE	YOUR NOTES	LABELED DRAWING
antrum		
secondary oocyte		
zona pellucida		
corona radiata		

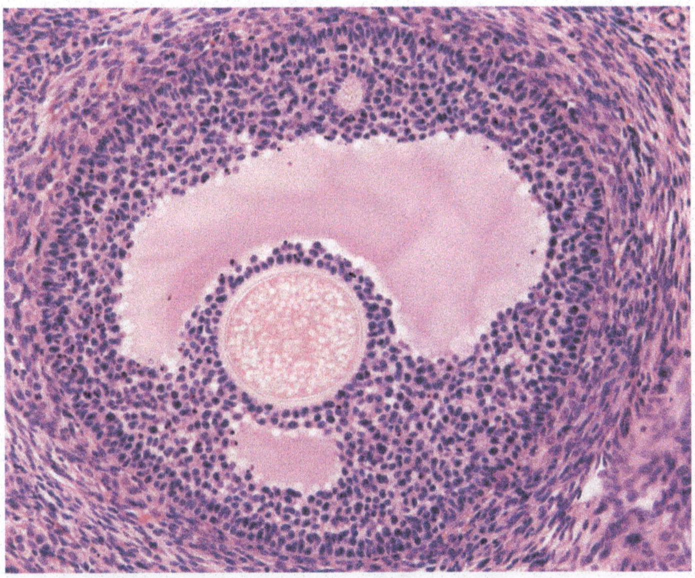

Early tertiary follicle with an antrum well visible. The oocyte, located in a cumulus oophorous is surrounded by zona pellucida and corona radiata. The nucleus of oocyte is in another cut plane.

b. Ovary, corpus luteum.

STRUCTURE	YOUR NOTES	LABELED DRAWING
corpus luteum		

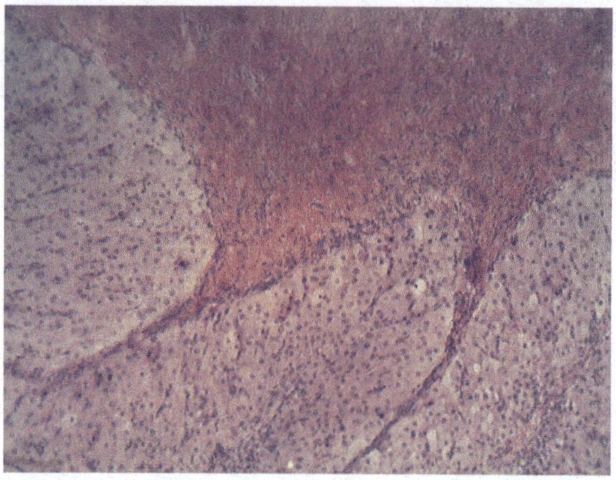

Contributed by Tony Wilmes. © Kendall Hunt Publishing Company.

c. Ovary, corpus albicans

STRUCTURE	YOUR NOTES	LABELED DRAWING
corpus albicans		

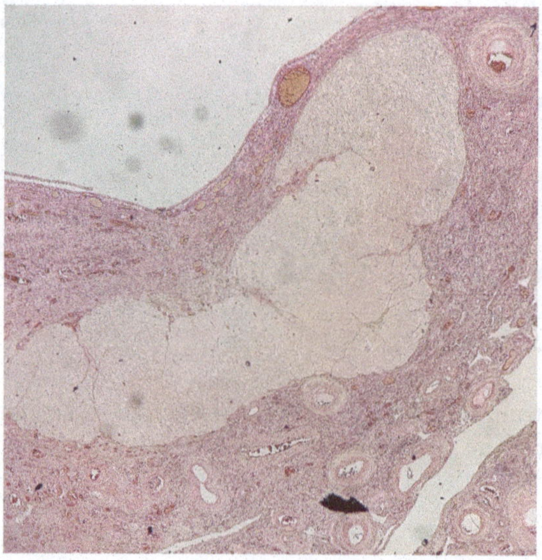

Contributed by Tony Wilmes. © Kendall Hunt Publishing Company.

CHECKLIST: Summary Checklist Reproductive

TERM/STRUCTURE	YOUR NOTES
Haploid (1N)	
Diploid (2N)	
Mitosis	
Meiosis I	
Meiosis II	
Spermiogenesis	
Spertmatogenesis	
Oogenesis	
Spermatozoa (Mature Sperm)	
Spermatogonium	
Primary Spermatocyte	
Secondary Spermatocyte	
Spermatid	
Interstitial Cells (Cells of Leydig)	
Oogonium	
Primary Oocyte	
Secondary Oocyte	
Primordial Follicle	
Primary Follicle	

(*Continued*)

TERM/STRUCTURE	YOUR NOTES
Maturing Follicle	
Graafian/Vesicular Follicle	
Zona Pellucida	
Corona Radiata	
Antrum	
Granulosa Cells of the Graafian Follicle	
Corpus Luteum	
Corpus Albicans	
Seminal Vesicle/Gland	
Bulbourethral/Cowper's Gland	
Vas Deferens/Ductus Deferens	
Inguinal Canal	
Spermatic Cord (Contains Spermatic Artery, Vein, and Nerve)	
Ampulla of the Vas Deferens	
Prostate Gland	
Glans Penis	
Scrotum/Scrotal Sac	
Testis	
Epididymis (Head, Body, and Tail Regions)	

TERM/STRUCTURE	YOUR NOTES
Penis	
Urogenital Orifice	
Ovary	
Ovarian Ligament	
Suspensory Ligament	
Uterine Tube (Fimbriae, Infundibulum, and Isthmus)	
Uterine Horn	
Body of Uterus	
Cervix of Uterus	
Vagina	
Urogenital Sinus	
Seminiferous Tubules	
Straight Tubules/Afferent Ductules	
Rete Testis	
Efferent Ductules	
Ejaculatory Duct	
Urethra (Prostatic, Membranous, and Penile Regions)	
Mesovarium	
Broad Ligament	

TERM/STRUCTURE	YOUR NOTES
Fundus of Uterus	
Uterine Cavity	
Endometrium	
Myometrium	
Perimetrium	
Round Ligament	
Internal Os of Cervix	
External Os of Cervix	
Cervical Canal	
Vaginal Orifice	
Labium Minora	
Labium Majora	
Clitoris	
Mons Pubis	
Pubic Symphysis	
Prepuce/Foreskin	
Corpus Spongiosum	
Corpus Cavernosum	

EXERCISE 13

Urinary System

A. Introduction

The anatomy of the urinary system may seem simplistic at first glance. Its organs include the **kidneys, ureters, urinary bladder,** and **urethra.** Delving deeper however, you will discover that the internal anatomy of the kidney, including the micro-architecture, is quite intricate. Note that the **kidneys attach to the cardiovascular system, thus renal and cardiovascular issues are commonly related.** The tendency is to associate the urinary system with the digestive system, perhaps because the waste products of each system deliver into the same receptacle and flush away together. Have you ever heard someone say, "I can't drink beer; it goes right through me"? You will discover that such a statement is anatomically impossible as you study the urinary system and think about what you already have learned about the digestive system!

B. *Outcomes for Exercise 13*

1. Locate, label, and identify the organs of the urinary system. Identify all structures listed.
2. Locate, label, and identify the details of the kidney: its gross anatomy, blood vessels, and the structures of a renal lobule.
3. Locate, label, and identify the details of a nephron.
4. Trace blood through the vessels and filtrate through the tubules. Be able to determine where blood or filtrate was immediately before and after a specific structure. Be able to locate where the four primary functions of the nephron occur.
5. Examine a dissected pig kidney. Locate the structures listed. Be able to determine where blood or filtrate was immediately before and after a specific structure.
6. Dissect the urinary system in both male and female preserved specimens. Locate and provide the functions of the structures described. Be able to indicate whether the fluid passing through a particular structure is blood, filtrate, or urine. Be able to indicate where a particular fluid was immediately before and after a specific structure.
7. Examine the histology of the kidney. Locate the structures listed. Draw and label what you see. Distinguish between the renal cortex and the renal medulla. Identify the specific tissues that are associated with each structure.

C. *Anatomy of the Urinary System*

OUTCOME 1: Locate, label, and identify the organs of the urinary system. Identify all structures listed.

TOOLBOX

An introduction to the Urinary System

Organs Responsible for Conduction and Storage

♂ Urinary organ model
♂ and ♀ Pelvic models
Mr. Muscle
Mr Flip Chart

kidneys
ureters
urinary bladder
urethra
renal arteries
renal veins
aorta

inferior vena cava
ureters
urinary bladder
ureteral openings
urethra
internal urethral sphincter (note any anatomical differences for location in females and males)

external urethral sphincter (note any anatomical differences for location in females and males)

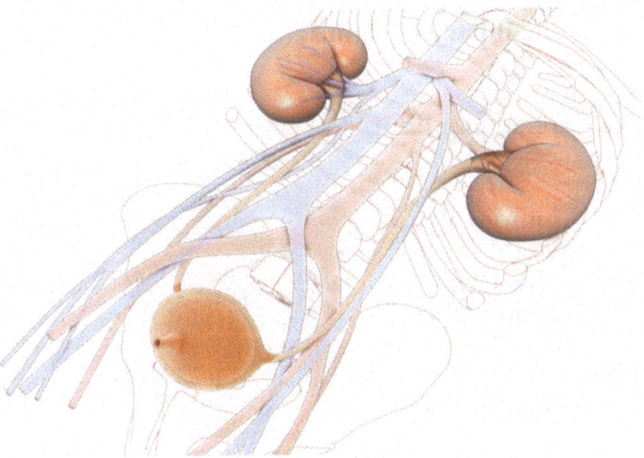

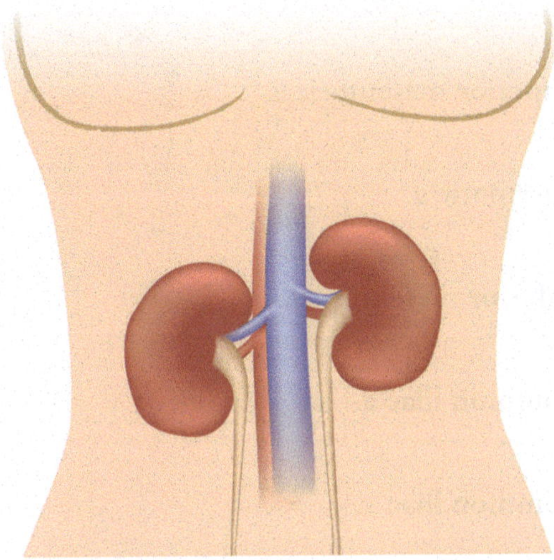

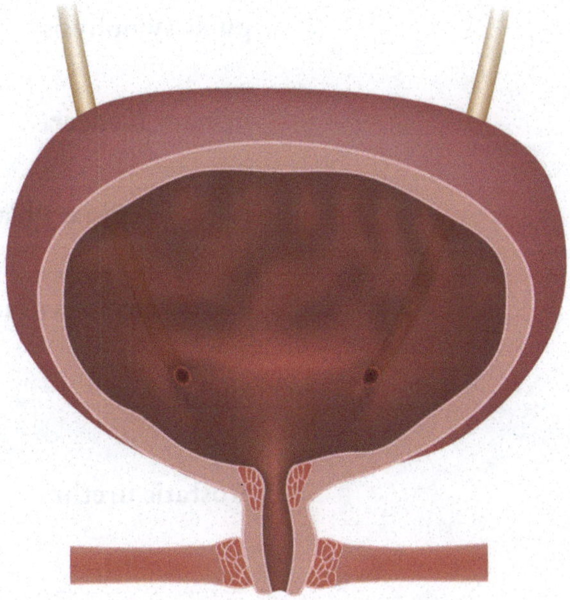

TOOLBOX

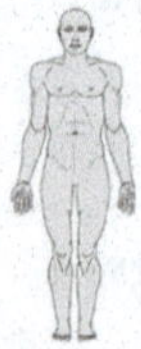

♂ Urinary organ model

■ ***All these structures can be found on the male urinary model. Some are related to the urinary system; some are review. Locate all listed structures.***

celiac a.

superior mesenteric a.

spermatic a.

inferior mesenteric a.

common iliac a.

common iliac v.

spermatic v.

adrenal glands

ductus deferens

seminal glands

prostate gland

renal capsule

renal cortex

renal medulla

medullary pyramid

renal papilla

minor calyx

major calyx

renal pelvis

ureters

pubic bones

pubic symphysis

urinary bladder

rugae of urinary bladder

ureter openings of urinary bladder

trigone

prostatic urethra

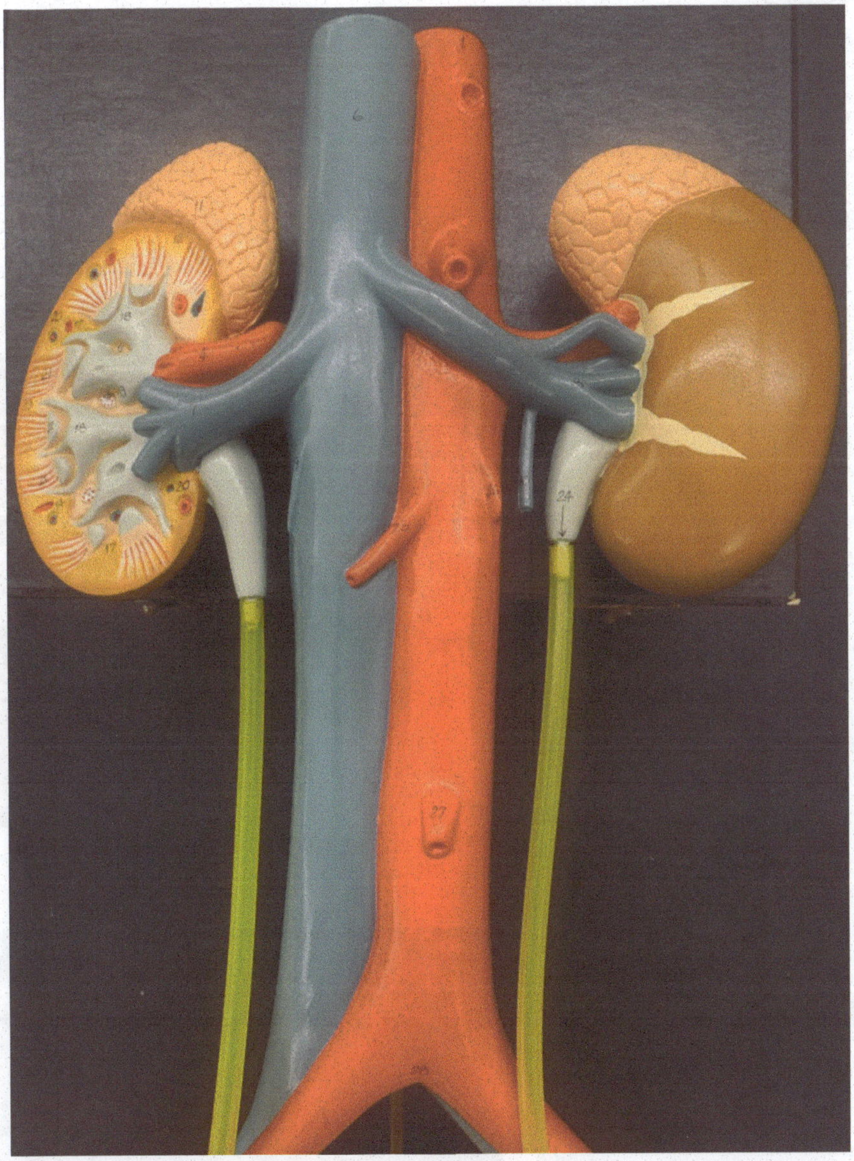
20
24

OUTCOME 2: Locate, label, and identify the details of the kidney: its gross anatomy, its blood vessels, and the structures of a renal lobule.

TOOLBOX

Structure of the Kidney

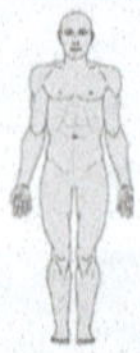

♂ Urinary organ model

adrenal (suprarenal) gland

renal capsule

hilum

renal cortex

renal medulla

medullary pyramids

renal lobe

renal papilla

renal columns

minor calyx

major calyx

renal pelvis

ureter

renal artery

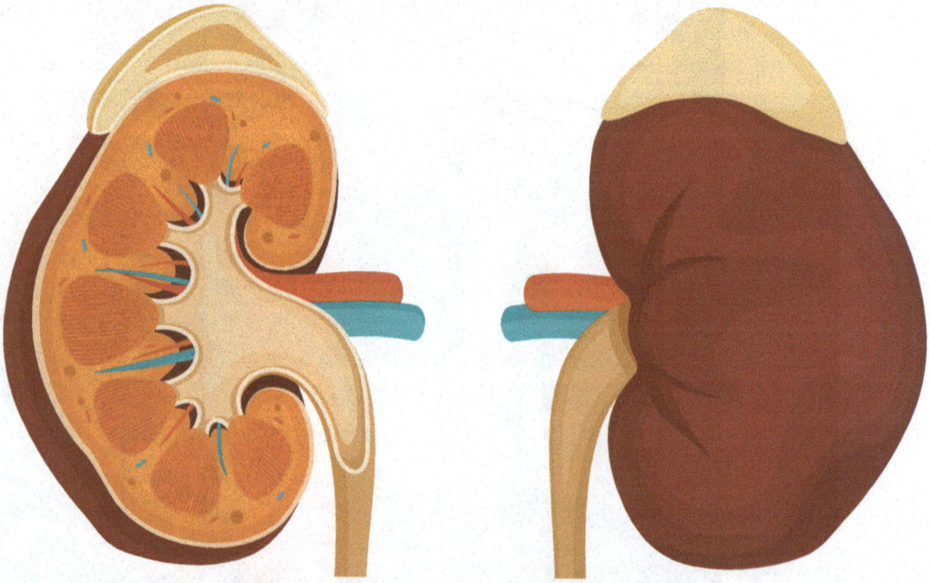

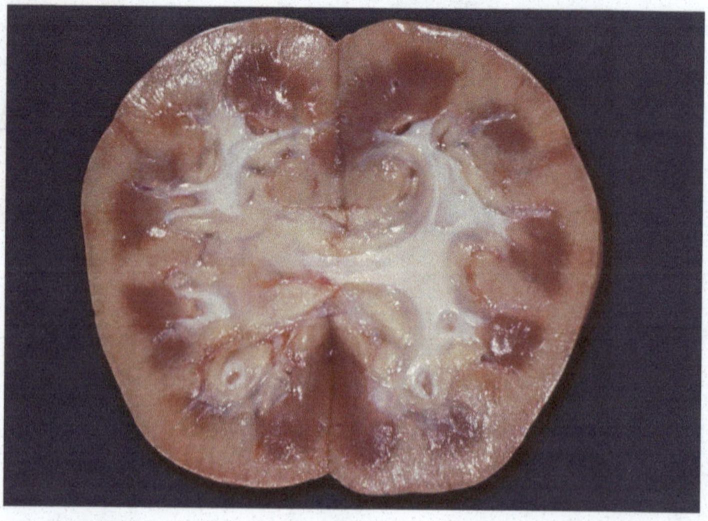

Illustration by Jamey Garbett. © 2003 Mark Nielsen.

© Lightspring/Shutterstock.com

Human kidney cut along its middle plane. In the center is the renal pelvis, divided into calyces that adapt to the tip of the Malpighi's pyramids (renal medulla). On the outside is the renal cortex.

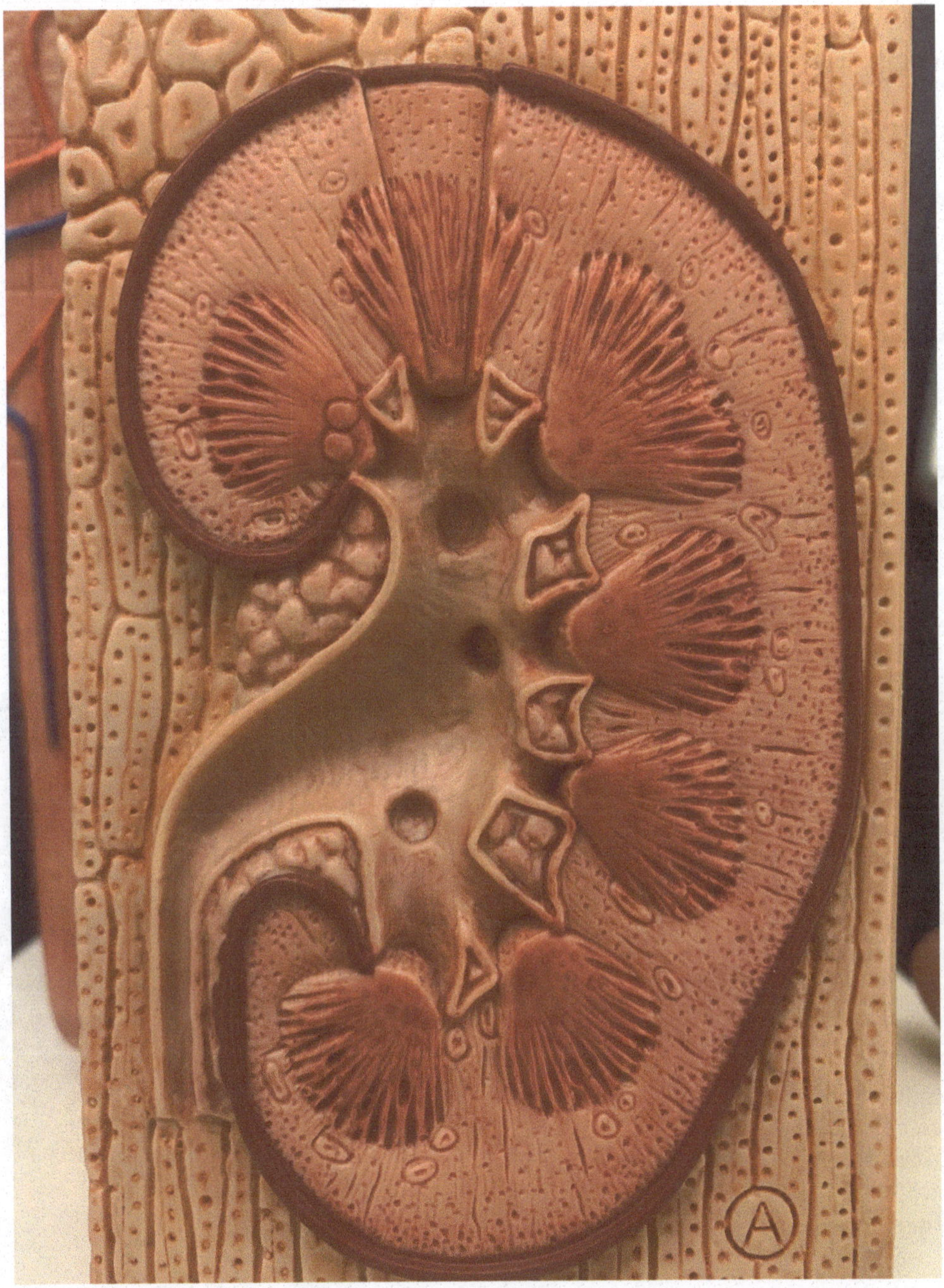

K13 © 3B Scientific GmbH, Germany, 2021. www.3bscientific.com

TOOLBOX

Blood Supply of the Kidney

Urinary plaque

lobar (segmental) artery

interlobar artery (cortical radiate artery)

arcuate artery

interlobular artery

afferent arteriole

glomerulus

efferent arteriole

peritubular capillaries

interlobular vein

arcuate vein

interlobar vein

renal vein

abdominal aorta

inferior vena cava

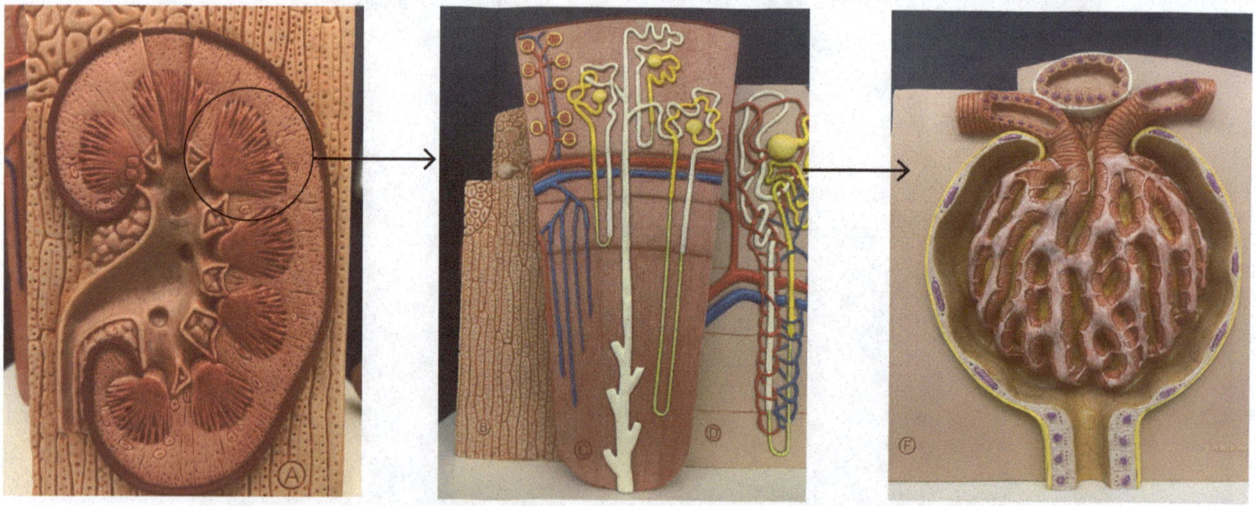

OUTCOME 3: Locate, label, and identify the details of a nephron.

TOOLBOX

Blood Supply of the Kidney
Histology of the Nephron

Urinary plaque

- *Identify all structures*
- *color arteries RED, veins BLUE, and filtrate structure YELLOW*

BLOOD STRUCTURES

arcuate artery

interlobular artery
(cortical radiate artery)

afferent arteriole

glomerulus

efferent arteriole

peritubular capillary

venule

interlobular vein

arcuate vein

FILTRATE STRUCTURES

Bowman's capsule

proximal convoluted tubules

loop of Henle

descending limb

ascending limb

distal convoluted tubule

collecting tubule

collecting duct

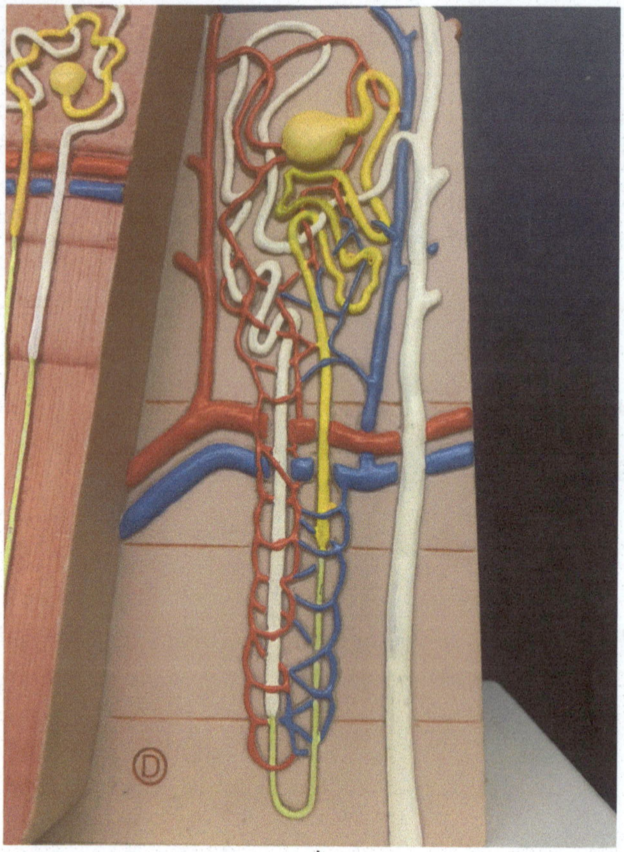
D

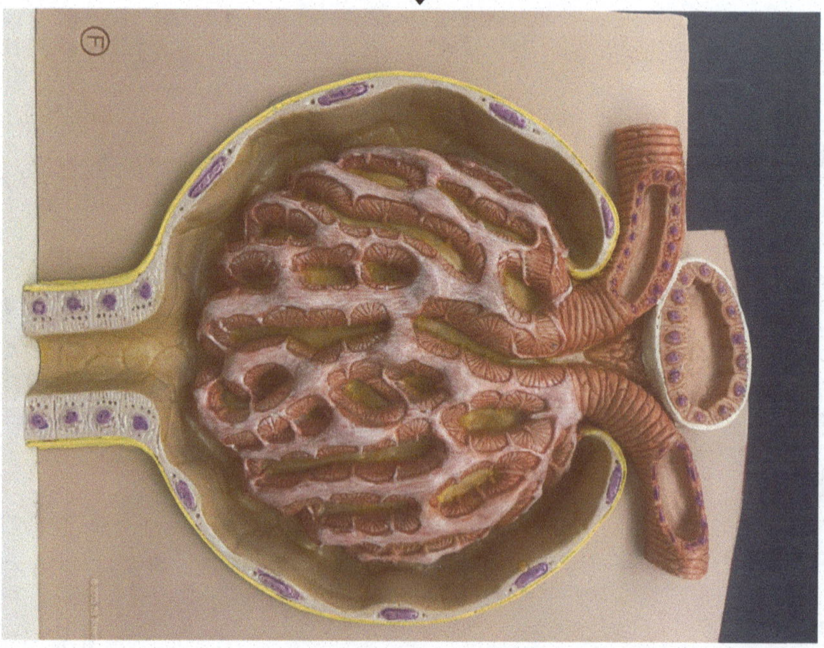
F

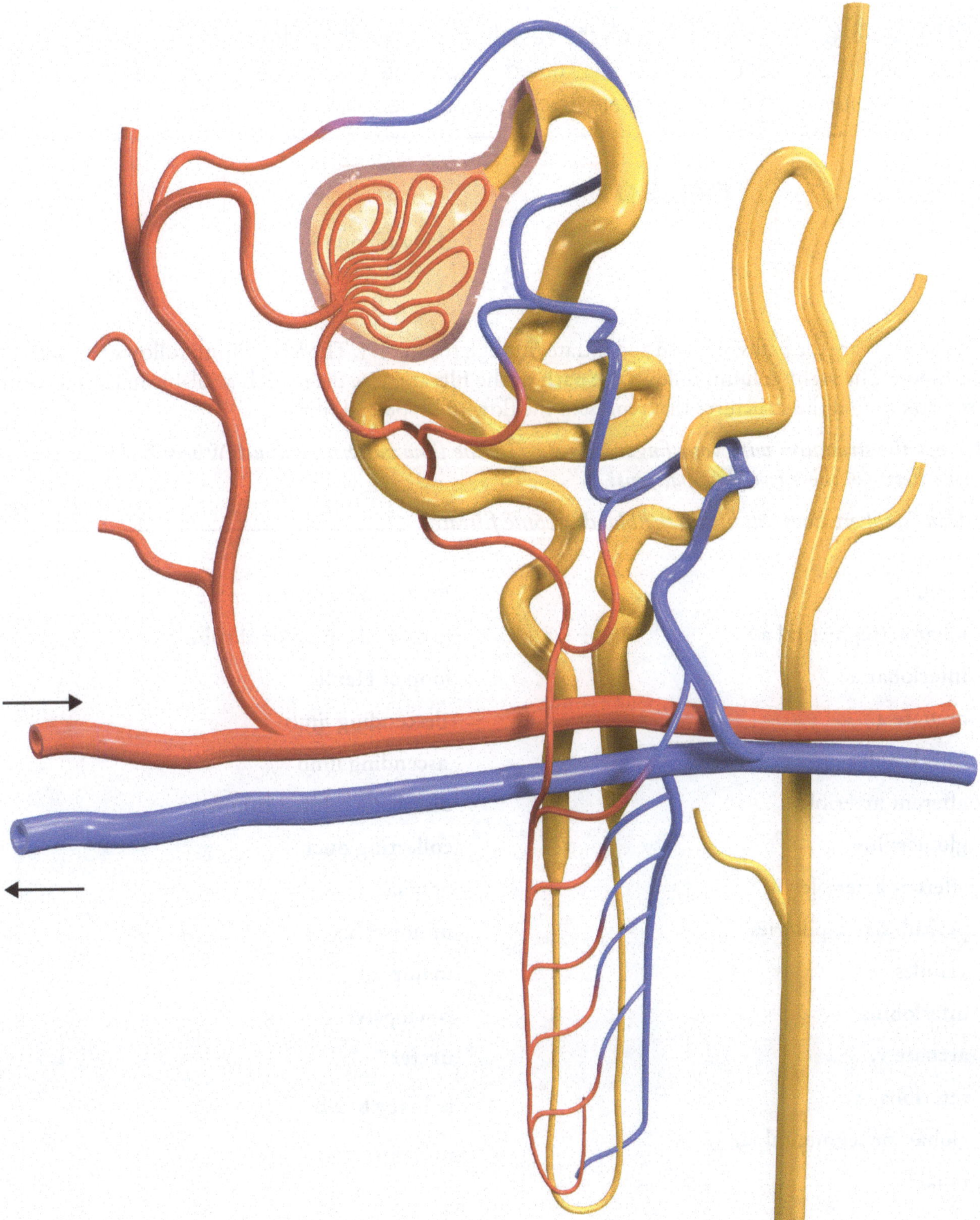

D. *Piecing (Pissing?) It All Together*

OUTCOME 4: Trace blood through the vessels and filtrate through the tubules. Be able to determine where blood or filtrate was immediately before and after a specific structure. Be able to locate where the four primary functions of the nephron occur.

TOOLBOX

Kidney plaque

Listed below are the paths of blood and filtrate through the kidney. **The left column follows the path of the blood. The right column follows the path of the filtrate.** Use the models available in lab and your textbook to find the structures that represent the flow of blood and filtrate.

- ***Trace the structures with your finger as you recite the flow pattern out loud. Where is the connection between these two flow patterns?***

  That is, where does fluid from the blood become filtrate? ______________________

renal a.
lobar a. (segmental a.)
interlobar a.
arcuate a.
interlobular a.
afferent arteriole
glomerulus
efferent arteriole
peritubular capillaries
venules
interlobular v.
arcuate v.
interlobar v.
(lobar or segmental v.)
renal v.

Bowman's capsular space
proximal convoluted tubule
loop of Henle
descending limb
ascending limb
distal convoluted tubule
collecting duct
papilla
minor calyx
major calyx
renal pelvis
ureter
urinary bladder
urethra

E. *Comparative Anatomy—Dissection*

OUTCOME 5: Examine a dissected pig kidney. Locate the structures listed. Be able to determine where blood or filtrate was immediately before and after a specific structure.

TOOLBOX

✂

The pig kidney has been removed from the animal; therefore, it will not have the renal fascia, adipose capsule, and renal capsule in place. Locate the structures listed in the table in the coronally sectioned kidney.

STRUCTURE	YOUR NOTES
renal hilum	
renal artery	
renal vein	
renal cortex	
renal medulla	
renal papilla	
minor calyx	
major calyx	
renal pelvis	
ureter	

F. *Comparative Anatomy Urinary System—Dissection*

OUTCOME 6: Dissect the urinary system in both male and female preserved specimens. Locate and provide the functions of the structures described. Be able to indicate whether the fluid passing through a particular structure is blood, filtrate, or urine. Be able to indicate where a particular fluid was immediately before and after a specific structure.

TOOLBOX

Structures of the Urinary Pathway

Although the anatomies of the male and female urinary systems are almost identical, make sure to examine the system in each so that you are familiar with the varying shapes, sizes, and orientation of the organs.

1. Gently move all the viscerae of the abdominal cavity toward the right, so they do not obstruct your view. Locate:
   - **kidneys**—bean-shaped organs located to the left and right of mid-line, against the dorsal body wall, just posterior to the diaphragm, but anterior enough to be partially protected by the ribcage. Generally, the right kidney is slightly lower and more lateral than the left kidney.
2. The kidneys lie retroperitoneal. If you did not remove the parietal peritoneum and three renal membranes at the time you dissected the renal arteries and veins, you might have to do so now. In order from the most superficial to the surface of the kidney you will find the following:
   - **renal fascia**—a delicate connective tissue sheath that covers each kidney and adrenal gland and anchors the kidneys against the dorsal abdominal wall.
   - **adipose capsule**—layer of adipose tissue. It functions to support the kidneys and provides a cushion protecting them from trauma.
   - **renal capsule**—tough fibrous connective tissue, adhering to the surface of each kidney, providing support and structure to the soft kidney tissue inside.
3. Notice that the medial surface of the kidney is indented.
   - **hilum**—term for the indented region. At the hilum, locate blood vessels and the ureters.
   - **renal arteries**—branch directly from the abdominal aorta bringing blood into the kidneys.
   - **renal veins**—from same location and carry blood directly into the posterior vena cava.

- **ureter**—slightly coiled, flaccid tube that extends from each kidney's hilum posteriorly to the urinary bladder. Each **ureter** transports urine from the kidney to the urinary bladder. If you pull the umbilical flap posteriorly, you will find the connection of the ureters to the urinary bladder.

4. Structures that store and pass urine to the outside of the body.

- **urinary bladder**—elongated, sac-like structure located on the ventral surface of the umbilical flap that was made at the initial dissection. It is flanked laterally by the two umbilical arteries.
- **urethra**—single, medial tube at the distal end of the urinary bladder. Urine must pass through this tube to leave the body. The length and position of the urethra differ between male and female specimen.

Male. The urethra passes through the **three parts** of the penis: **prostatic urethra, membranous urethra, and penile urethra.** The urine passes from the penile urethra to the outside of the body via the opening called the **urogenital orifice.**

Female. Urine travels through the **urethra,** which is located dorsal to the body of the uterus, the cervix, and the vagina. The urethra and vagina meet at the **urogenital sinus** before they leave the body through the **urogenital orifice.**

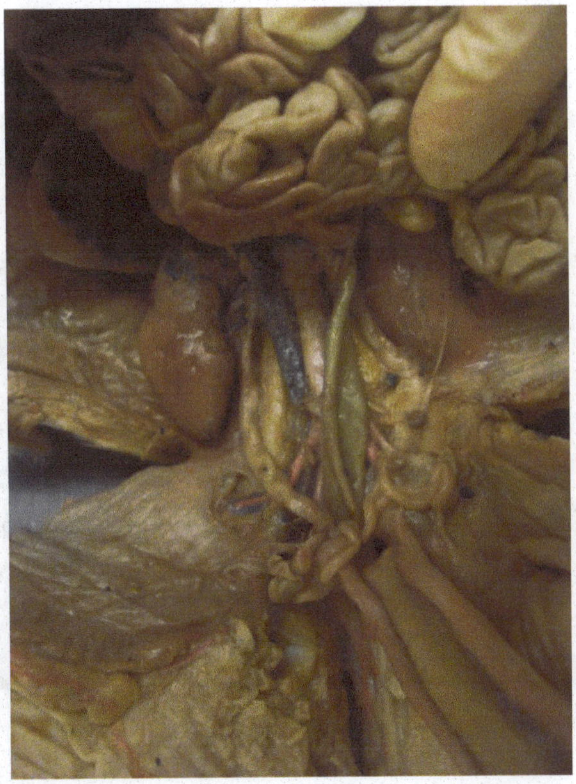

Details of the Kidney

5. Locate the left kidney. Make sure that the renal fascia and adipose capsule have been removed from the ventral surface of the kidney. You will now see the white, fibrous connective tissue of the renal capsule lying directly on the surface of the kidney. Create a **coronal, hemi-section** through the kidney (see image below). Use your scalpel to carefully cut along the lateral margin of the kidney. This mid-coronal section will allow you separate the two kidney halves like a book. Be careful not to cut the kidney completely in half, as you simply want to be able to fold it open. This will expose the three main regions of the kidney. These regions are visually different due to the microscopic arrangement of the tissue.

   - **renal cortex**—*lighter colored outermost region of the cut surface.* As blood flows into the kidney, the vessels branch to the site of filtration located in this region.
   - **renal medulla**—*the darker, reddish-brown region on the cut surface.* Consists of small triangular structures known as **renal pyramids.** The pyramids end in small nipple-like structures, renal papillae. This region collects the filtrate, and it drips from the papillae to the renal pelvis.
   - **renal pelvis**—*the most medial region on the cut surface.* It appears as a thick, white tissue in the dissection. The renal pelvis will funnel the filtrate, now urine, into the ureter. Carefully, insert the tip of your probe the renal pelvis into the ureter to illustrate the connection.

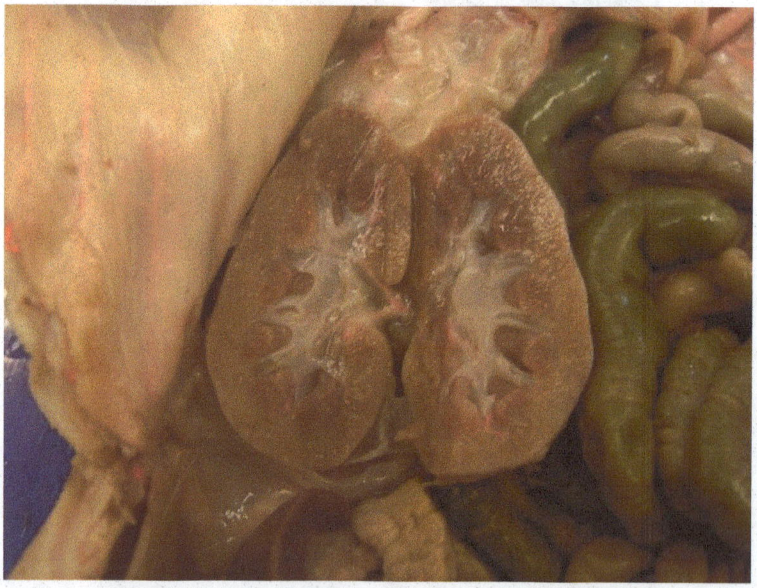

G. *Histology of the Kidney*

OUTCOME 7: Examine the histology of the kidney. Locate the structures listed. Draw and label what you see. Distinguish between the renal cortex and the renal medulla. Identify the specific tissues that are associated with each structure.

TOOLBOX

A Typical Nephron
Histology of the Nephron

Examine the kidney slide on low magnification to ascertain the major regions of the kidney: the renal cortex and the renal medulla.

Renal Cortex

Examine the renal cortex under high magnification. You know that you are viewing the cortex if you see renal corpuscles. Locate a corpuscle and identify the structures listed in the table. Use the diagram of the renal corpuscle as an aid. Record the specific tissue that composes each of the structures.

STRUCTURE	YOUR NOTES
renal corpuscle	
glomerulus	
Bowman's capsular space	
parietal membrane of Bowman's capsule	
proximal convoluted tubule	
distal convoluted tubule	
renal capsule	

Renal Medulla

Examine the renal medulla under high magnification. If you see renal tubules running parallel to one another, you know that you are viewing the medulla. Identify the structures listed in the table. Record the specific tissue that makes up each of the structures listed.

STRUCTURE	YOUR NOTES
collecting tubule	
thick branch of the loop of Henle	
thin branch of the loop of Henle	

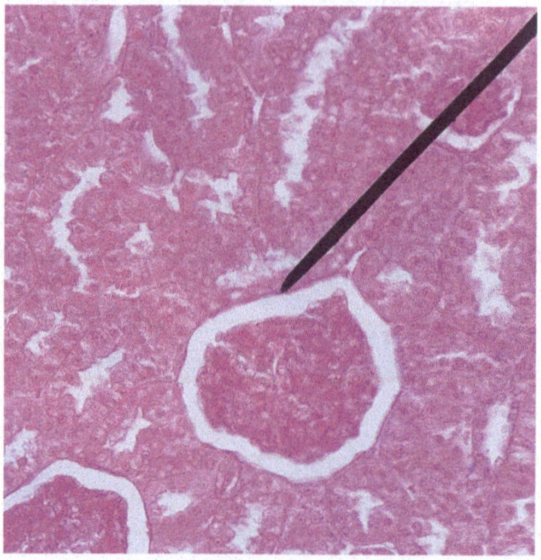

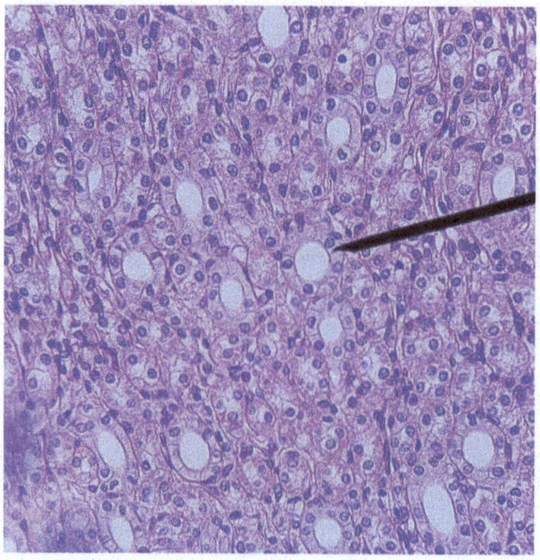

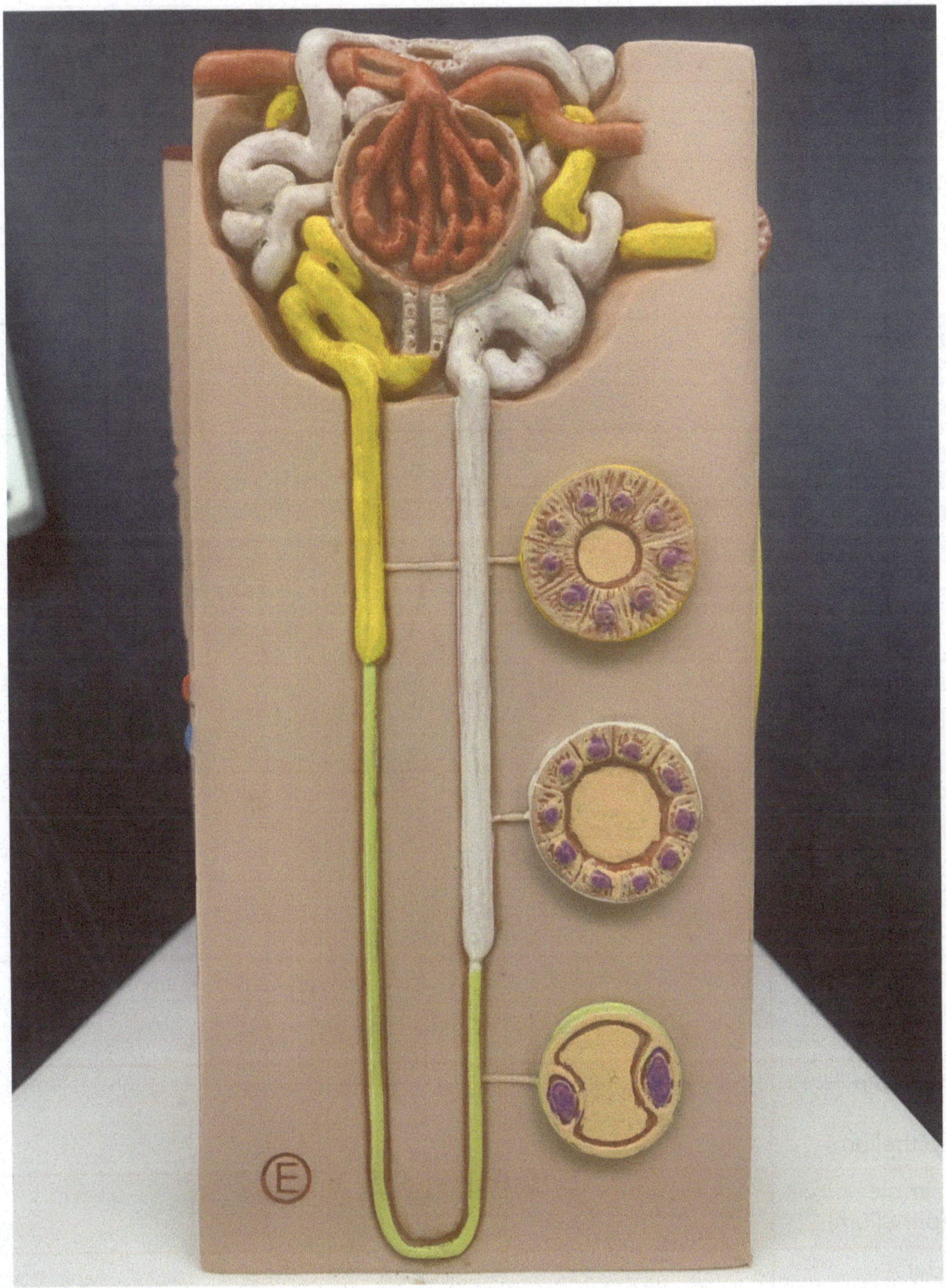

---

image is from a 3B model (#K13).

CHECKLIST: Summary Checklist Urinary

TERM/STRUCTURE	YOUR NOTES
Kidney	
Renal Cortex	
Renal Medulla	
Renal Corpuscle	
Glomerulus	
Visceral Membrane of Bowman's Capsule (Podocytes)	
Parietal Membrane of Bowman's Capsule (Simple Squamous Epithelium)	
Bowman's Capsular Space	
Proximal Convoluted Tubule (Simple Cuboidal Epithelium)	
Distal Convoluted Tubule (Simple Cuboidal Epithelium	
Collecting Duct (Simple Cuboidal Epithelium)	
Loop of Henle: Thick Portion (Simple Cuboidal Epithelium)	
Loop of Henle: Thin Portion (Simple Squamous Epithelium)	
Vasa Recta (Simple Squamous Epithelium)	
Urinary Bladder	
Detrusor Muscle of Urinary Bladder (Smooth Muscle Tissue)	

TERM/STRUCTURE	YOUR NOTES
Mucosa of Urinary Bladder (Transitional Epithelium)	
Lamina Propria of Urinary Bladder	
Adventitia of Urinary Bladder	
Renal Capsule	
Medullary Pyramid	
Renal Column	
Renal Papilla	
Minor Calyx	
Major Calyx	
Renal Pelvis	
Renal Artery	
Renal Vein	
Ureter	
Urethra	
Urogenital Orifice	
Adrenal Gland	
Hilum	
Bowman's Capsule	
Loop of Henle: Ascending and Descending Branches	
Collecting Tubule	

(continued)

TERM/STRUCTURE	YOUR NOTES
Rugae of Urinary Bladder	
Openings of Ureter	
Trigone Region	
Internal Urethral Sphincter	
External Urethral Sphincter	
Urethral Orifice	
Lobar/Segmental Artery	
Interlobar Artery	
Arcuate Artery	
Interlobular Artery	
Afferent Arteriole	
Efferent Arteriole	
Peritubular Capillaries	
Venules	
Interlobular Vein	
Arcuate Vein	
Interlobar Vein	
Lobar Vein	
Abdominal Aorta	
Inferior Vena Cava	